Understanding Society and N

Michael J. Manfredo • Jerry J. Vaske
Andreas Rechkemmer • Esther A. Duke
Editors

Understanding Society and Natural Resources

Forging New Strands of Integration Across the Social Sciences

Editors
Michael J. Manfredo
Jerry J. Vaske
Esther A. Duke
Department of Human Dimensions
  of Natural Resources
Colorado State University
Fort Collins, CO, USA

Andreas Rechkemmer
Graduate School of Social Work
University of Denver
Denver, CO, USA

This book is a publication of the International Association for Society and Natural Resources

ISBN 978-94-017-8967-7   ISBN 978-94-017-8959-2 (eBook)
DOI 10.1007/978-94-017-8959-2
Springer Dordrecht Heidelberg New York London

Library of Congress Control Number: 2014938625

Printed on acid-free paper

Springer is part of Springer Science+Business Media (www.springer.com)

Foreword

This book covers a wide range of subjects which have enormous relevance to the interface between human society and the use and conservation of natural resources. This is a theme on which perhaps much more could have been done by researchers and academics, but possibly the integration of various disciplines, particularly through those dealing with the physical sciences and researchers involved in the social sciences, does not take place with adequate facility in most parts of the world. This volume is clearly an important contribution to the literature with a proper blending of different disciplines that would help us understand the interface between human society and natural resources in an integral manner.

The very first pages beginning with the introduction set out the case for trans-disciplinarity. This theme is then dealt with elaborately in subsequent chapters in a manner that would appeal to all the disciplines represented in the chapters of the book. I would hope that this effort can also be replicated through integration of disciplines dealing with the subject of climate change. As was logical, the initial work of scientists dealing with climate change focused largely on the biophysical and geophysical aspects of this problem. This, of course, was essential because it was important for society to understand what really was happening with changes in the physical system given that emissions of greenhouse gases have been increasing, and as a result the concentration of these gases going up significantly since industrialization. It was also essential to understand the physical nature of impacts of climate change, such as those involving the entire water cycle and how it would be affected as a result, as well as to assess the physical impacts of climate change in the form of extreme events and disasters. The Intergovernmental Panel on Climate Change (IPCC) brought out a special report in 2011 on Managing the Risks of Extreme Events and Disasters to Advance Climate Change Adaptation. This provided an in-depth assessment not only of extreme events and how their frequency and intensity would change as a result of climate change but also various human dimensions of the problem. One of the observations that was brought out in the report stated that between 1970 and 2008, 95 % of the fatalities that took place around the world as a result of all kinds of disasters occurred in the developing countries. There was also an elaboration of several other implications for human society from increase in extreme events and disasters, which clearly brought out far better integration of the physical sciences with the social sciences than was perhaps possible some years ago. However, much more research and a greater extent of published material would help our understanding of the human aspects of climate change, if such work were to be carried out through the combining of various disciplines and by blending the physical sciences with the social sciences.

There have, of course, been some outstanding examples of brilliant researchers in one set of these disciplines or the other making a foray into another set of disciplines. A prominent example of this was the seminal work of Garrett Hardin, who explained the basis of what he termed as the tragedy of the commons. Hardin was a biologist but he mapped out a reality which entered right into the territory of the social sciences. And that piece of work, published in 1968, was a remarkable but simple way of explaining the nexus between human actions and the state of the global commons.

I believe this present volume is really a trail blazer because quite apart from the substance that the following pages contain, in my view, the inspiration that this provides for persons from diverse disciplines focusing on common problem areas is in itself a major contribution. I am sure the readers of this book would find it of enormous value in providing a comprehensive understanding of a complex subject which cannot be produced by any single discipline.

Chairman, Intergovernmental Panel on Climate   Rajendra Kumar Pachauri
Change (IPCC) and Director General, The Energy
and Resources Institute (TERI), New Delhi, India

Preface: AND not OR

Since the release of the Brundtland Report, issued by The World Commission on Environment and Development in 1987, increased attention has been placed on the role and place of sustainability in development plans. Much of this literature draws its definition of sustainable development from that report (1987:41):

> Sustainable development is development that meets the needs of the present without compromising the ability of future generations to meet their own needs.

Generally, the challenges of attaining sustainability are posed under a competing set of *OR* conditions that reflect extreme positions. For example, in many political contexts (i.e., at a macro level), the choice between sustaining current styles of living and quality of life are contrasted with having fewer of the conveniences currently enjoyed – the latter occurring as a result of a modification of lifestyles, desired or not. Similarly, since the debates over prevailing strategies towards production stimulated by the Club of Rome's sponsoring of *The Limits to Growth* (Meadows et al. 1972, 2004; Cole et al. 1973), choices between maximizing yields at the exponential cost of exhausting nonrenewable natural resources (including oil, gas, coal, and nuclear energy) *or* identifying and introducing acceptable limits to industrial and agricultural-technological processes have been discussed. Work by Gever et al. (1986), informed through the use of a kcal conversion factor (which provides an understandable constant for analysis) in a systems model framework, extended these arguments to renewable resources (wood, wind, and hydroelectricity). Both systems approach this issue from a declining resource-efficiency framework – essentially, many modern production and economic practices are shown to be nonsustainable. As Daly (1988:13) suggested, if the economy was conceptualized as a subsystem of an ecosystem that was finite and non-growing, the economy must at some point become non-growing or it will eventually overrun the "…the regeneration and absorption capacities of the environment."

The expanding natural resource and environmental literature is also marked by its use of choices between opposing perspectives. Much of the literature refers to a schism between advocates of a utilitarian perspective (use the resource) and those championing a preservation viewpoint (leave the resource alone). More recent debates among those wanting to preserve the natural environment because it has a right to exist in of itself (it is part of the natural order) and those who wish to preserve it for our use (if we take it all now we will contribute to our own destruction later) echo the utilitarian-preservation framework.

Dunlap and Catton (1979) framed the latter concern in what commonly became known as the HEP/NEP debate. HEP (the Human Exemptionalism Paradigm) referred to those who believed in the centrality of human systems. This was expressed in their domain assumption which viewed the physical environment as being largely irrelevant for use in understanding social behavior (1979:250). Holders of this perspective were anthropocentric and espoused a view that placed human society at the center of the natural world. Adherents of this perspective also believed current and foreseeable problems would be addressed by technological improvements and inventions. From this perspective, we could offset the depletion of stocks of natural capital by humanely created capital. Julian Simon (1996:588) explained this line of reasoning well. He wrote:

> Increased population and a higher standard of living cause actual and expected shortages, and hence price rises. A higher price represents an opportunity that attracts profit-minded entrepreneurs and socially minded inventors to seek new ways to satisfy the shortages. Some fail, at cost to themselves. A few succeed, and the final result is that we end up *better off* than if the original shortage problems had never arisen.

In other words, sustainability need not even be an issue because human ingenuity will guarantee that any problems associated with resource depletion will be addressed through market mechanisms.

NEP, the New Environmental Paradigm and alternative framework (Dunlap and Van Liere 1978; Dunlap et al. 2000), placed human society into a larger gestalt that viewed it as a part of the natural order. A NEP perspective suggested: (1) there were real and finite limits to what technology could do; (2) there was a need to accept the limits on human affairs imposed by the biophysical environment through physical and biological constraints; and (3) human survival was dependent upon the health of the environment.

Much of the current sustainable development literature directs attention to the unresolved tensions between environmental protection and economic development regularly treating these issues as separate policy concerns. Analyses of the environmental protection movement emphasized changes in social values that occurred more or less explicitly in response to this tension (cf., Humphrey and Buttel 1982; Schnaiberg 1980; Buttel 1992). Hays (1991), for example, in his discussion of the post-World War II history of forest planning and management, noted a shift in public attitudes from one that viewed forestlands primarily as a source of useful products (such as wood and wood products) towards one that viewed the forest as a setting for home, work, and play. More generally, Field and Burch (1988) noted a shift from control and exploitation of nature as dominant themes to the emergence of a view that partners nature and society.

The selection of any option based on an OR scenario is problematic since **no** choice is made with impunity. All choices have costs. Moreover, whereas we might agree that sustainable development is not only an honorable but a necessary goal, its implementation will not be easy. Current thinking has been dominated by those who offer extreme positions, often posed in terms of black and white – the tyranny of *OR* condition. What are needed are efforts that seek to strike a balance between extremes. Such work will lead to the possibility of an *AND* scenario. As a result, we may find greater acceptance of new efforts to implement sustainable strategies that draw on the best from each of a series of alternatives.

The key question, of course, is how do we effectively move from the commonplace *OR* situation towards an *AND* scenario. In our view, this shift requires two interrelated tasks: (1) the creation of a true dialogue based on a fusion of perspectives within and across disciplinary boundaries; and (2) the development of a model of scholarship that provides both academic and civic benefits while creating real partnerships between experts and citizens. Both of these tasks deserve increased critical attention from natural resource researchers. The authors in this edited volume represent leading thinkers on many topics impacted by the issues associated with moving from an *OR* to an *AND* framework. To varying degrees, each chapter provides insights into this process. We thank the authors for their efforts and encourage researchers to take the lessons learned in this edited volume seriously by incorporating them in their current and future scholarship.

University Park, PA, USA	A.E. Luloff
University Park, PA, USA	Jeffrey C. Bridger
Huntsville, TX, USA	Gene L. Theodori

References

Brundtland, G. (1987). Chairman, our common future (The Brundtland report). In *World Commission on environment and development*. Oxford: Oxford University Press.

Buttel, F. (1992). Environmentalization: Origins, processes, and implications for social change. *Rural Sociology, 57*, 1–27.

Cole, H. S. D., Freeman, C., Jahoda, M., & Pavitt, K. L. R. (1973). *Thinking about the future: A critique of the limits to growth*. London: Chatto and Windus for Sussex University Press.

Daly, H. E. (1988). *Beyond growth*. Boston: Beacon Press.

Dunlap, R. E., & Catton, W. R., Jr. (1979). Environmental sociology. *Annual Review of Sociology, 5*, 243–273.

Dunlap, R., & Van Liere, K. D. (1978). The 'new environmental paradigm': A proposed measuring instrument and preliminary results. *The Journal of Environmental Education, 9*(4), 10–19.

Dunlap, R., Van Liere, K. D., Mertig, A. G., & Jones, R. E. (2000). Measuring endorsement of the new ecological paradigm: A revised NEP scale. *Journal of Social Issues, 56*(3), 425–442.

Field, D. R., & Burch, W. R., Jr. (1988). *Rural sociology and the environment*. Middleton: Social Ecology Press.

Gever, J., Kaufmann, R., Skole, D., & Vorosmarty, C. (1986). *Beyond oil: The threat to food and fuel in the coming decades*. Cambridge: Ballinger.

Hays, S. P. (1991). Human choice in the Great Lakes wildlands. In R. G. Lee, D. R. Field, & W. R. Burch (Eds.), *Community and forestry: Continuities in the sociology of natural resources* (pp. 41–51). Boulder: Westview.

Humphrey, C. R., & Buttel, F. R. (1982). *Environment, energy, and society*. Belmont: Wadsworth.

Meadows, D. H., Meadows, D. L., Randers, J., & Behrens, W. W., III. (1972). *The limits to growth*. New York: Universe Books.

Meadows, D. H., Randers, J., & Meadows, D. H. (2004). *Limits to growth: The 30-year update*. White River Junction: Chelsea Green Publishing Company.

Schnaiberg, A. (1980). *The environment: From surplus to scarcity*. New York: Oxford University.

Simon, J. (1996). *The ultimate resource 2*. Princeton: Princeton University Press.

Acknowledgments

The editors and contributing authors thank Colorado State University Warner College of Natural Resources alumnus Ying Lee and the International Association for Society and Natural Resources for their generous financial support of this book. Thanks also to Dr. R.K. Pachauri, Chairman of the Intergovernmental Panel on Climate Change (IPCC) for his support and encouragement.

Contents

Contributors

Leslie Acton Nicholas School of the Environment, Duke Marine Lab, Duke University, Beaufort, NC, USA

Abigail Bennett Nicholas School of the Environment, Duke Marine Lab, Duke University, Beaufort, NC, USA

Randall B. Boone Natural Resource Ecology Laboratory, Colorado State University, Fort Collins, CO, USA

Jeffrey C. Bridger Department of Agricultural Economics, Sociology, and Education, Penn State University, University Park, PA, USA

Jeffrey Broadbent Department of Sociology, Institute for Global Studies, University of Minnesota, Minneapolis, MN, USA

Tom R. Burns Department of Sociology, University of Uppsala, Uppsala, Sweden

Kevin Collins Department of Engineering and Innovation, Open University, Milton Keynes, UK

Robert Costanza Crawford School of Public Policy, The Australian National University, Canberra, ACT, Australia

Joan M. Diamond Nautilus Institute for Security and Sustainability, Berkeley, CA, USA

Esther A. Duke Department of Human Dimensions of Natural Resources, Colorado State University, Fort Collins, CO, USA

Paul Ehrlich Department of Biological Sciences, Stanford University, Stanford, CA, USA

Graham Epstein School of Public Affairs and Department of Political Science, The Vincent and Elinor Ostrom Workshop in Political Theory and Policy Analysis, Indiana University, Bloomington, IN, USA

Melissa L. Finucane Behavioral and Policy Sciences, RAND Corporation, Pittsburgh, PA, USA

Jefferson Fox The East-West Center, Honolulu, HI, USA

Jörg Friedrichs Oxford Department of International Development, Queen Elizabeth House, Oxford, UK

David Fulton Department of Fisheries, Wildlife and Conservation Biology, University of Minnesota, St Paul, MN, USA

Kathleen A. Galvin Department of Anthropology, Colorado State University, Fort Collins, CO, USA

Michael C. Gavin Department of Human Dimensions of Natural Resources, Colorado State University, Fort Collins, CO, USA

Rebecca Gruby Department of Human Dimensions of Natural Resources, Colorado State University, Fort Collins, CO, USA

Ilan Kelman Institute for Risk & Disaster Reduction (IRDR) and Institute for Global Health (IGH), University College London (UCL), London, UK

Donald Kennedy The Center for Environmental Science and Policy (CESP), Stanford University, Stanford, CA, USA

Kristin Ludwig Natural Hazards Mission Area, U.S. Geological Survey, Reston, VA, USA

A.E. Luloff Department of Agricultural Economics, Sociology, and Education, Penn State University, University Park, PA, USA

Nora Machado Center for Research and Studies in Sociology, Lisbon University Institute, Lisbon, Portugal

Gary E. Machlis School of Agricultural, Forest, and Environmental Sciences, Clemson University, Clemson, SC, USA

Michael J. Manfredo Department of Human Dimensions of Natural Resources, Colorado State University, Fort Collins, CO, USA

Mateja Nenadovic Nicholas School of the Environment, Duke Marine Lab, Duke University, Beaufort, NC, USA

Lennart Olsson Lund University Centre for Sustainability Studies (LUCSUS), Lund University, Lund, Sweden

Rajendra Kumar Pachauri Chairman, Intergovernmental Panel on Climate Change (IPCC) and Director General, The Energy and Resources Institute (TERI), New Delhi, India

Andreas Rechkemmer Graduate School of Social Work, University of Denver, Denver, CO, USA

Eugene A. Rosa (Deceased) Moscow, ID, USA

Sumeet Saksena The East-West Center, Honolulu, HI, USA

James H. Spencer Department of Planning, Development and Preservation, Clemson University, Clemson, SC, USA

Tara L. Teel Department of Human Dimensions of Natural Resources, Colorado State University, Fort Collins, CO, USA

Gene L. Theodori Department of Sociology, Sam Houston State University, Huntsville, TX, USA

Jerry J. Vaske Department of Human Dimensions of Natural Resources, Colorado State University, Fort Collins, CO, USA

Philip Vaughter College of Education, University of Saskatchewan, Saskatoon, SK, Canada

Peter H. Verburg Institute for Environmental Studies, VU University Amsterdam, Amsterdam, The Netherlands

Introduction

Multidisciplinarity is a goal of the sciences (McMichael et al. 2003; Poteete et al. 2010). Problems such as climate change, biodiversity loss, land degradation, food security, and water availability are complex and threaten human sustainability. "It is ridiculous to think that the way to understand complexity is to dig deeper and narrower at one spatial and temporal scale in a single field of science alone" (Giraudoux et al., 2007:294). To address complex problems the social sciences must address critical questions in modeling social-ecological systems. For example, how much progress has been made in developing integrated, multi-scale representation of social phenomenon that can be interwoven with the biological and physical sciences? What are the best approaches for pursing social science integration? Is true consilience possible, where one discipline builds upon another? Is a theory of "everything social" to be developed from game theory as Varoufakis (2008) advocates? Or should we continue on a path of disciplinary strength/integrity with the philosophy that "the gains from disciplinary and methodological cross-fertilization are greatest when scholars with a solid command of their own disciplines and methods interact with each other" (Poteete et al. 2010:271)? Perhaps social science integration is simply impractical as suggested by Elster (2010)? As we consider answers to these questions, it will be important to consider if there are ways to alter our approaches to present a more holistic social science.

In this edited volume, leading scholars from different disciplinary backgrounds wrestle with the answers to these questions. This is a critical time to set a vision for the future of integrative science (Costanza 2009). This book explores the growing concern of how best to achieve effective integration of the social science disciplines in addressing natural resource issues.

An Enduring Concern

The quest for integration among the social sciences is not new. Rorty (2001) contends that unity of science was the undertone of logical positivism, which was ubiquitous some 60 years ago. For example, Talcott Parson's interdisciplinary volume

Toward a General Theory of Action (Parson and Shils 1951/2001) was born from an optimism for a "unified theory of science". The 2001 re-publication of this book, concluded that "in the past half century, American behavioral and social scientists have come to shun such efforts at such general social theorizing, and have chosen more modest, though not necessarily more successful, ways to advance their science" (Smelser in Parson and Shils 2001: xix).

E.O. Wilson raised similar concerns about the lack of integration efforts in the social sciences in his 1998 book *Consilience*. Wilson (1998) describes consilience, as the "jumping together" of knowledge by the linking of "facts and fact-based theory to create a common groundwork of explanation" (p. 8). Consilience, he contends, is a critical step in the advancement of science and is the greatest of all intellectual challenges as we enter an age of synthesis among biology, the social sciences and the humanities. The social sciences have lagged in their advancements and contributions to society due to their lack of consilience. He argues that, "…it is obvious to even casual inspection that the efforts of the social sciences are snarled by disunity and a failure of vision… Split into independent cadres, they stress precision in words within their specialty but seldom speak the same technical language from one specialty to the next" (Wilson 1998:198). This proposal has not been without criticism, particularly the recommendation for a recommitment to positivistic, reductionistic science (Berry 2000; Ceccarelli 2001; Gould 2003). While there is disagreement on Wilson's fundamentalist approach, few disagree with the importance of pursuing the (re)unification of social science knowledge or with the slow progress so far being made on this topic.

The National Science and Technology Council (NSTC 2009) recognized the need for collaboration among social sciences. Understanding humans from individual behavior to societal systems is a difficult and wide-ranging quest (p. 3). The questions posed by the social sciences are best answered using methods from disciplines that cut across traditional academic boundaries. Advances in explaining human diversity is constrained by traditional disciplinary approaches that focus on one level of analysis (Norenzayan 2011). Collaborative teams that cross disciplinary boundaries will open up new horizons in the behavioral sciences.

The 2010 UNESCO's World Social Science Report identified social science integration as a major issue consideration. The social sciences are at a critical juncture. The direction might be toward a new integration with the hard sciences, or towards local, context-dependent problem-solving, integrated into 'epistemic communities' with actors originating from different social activities outside science" (p. 189).

Debates about integration are not new among social scientists working in natural resources. Belsky (2002), for example, addressed integration in the context of whether environmental and natural resources sociology are separate sub-disciplines. Integration merits debate for a number of reasons. First, natural resource issues are complex and are affected by multiple proximate driving social factors. Single disciplinary studies focused at one level are unlikely to provide explanations that represent this complexity and are limited in their ability to inform policy recommendations. Complex problems are best explored across disciplines that examine social-ecological phenomenon from different scales.

Second, multi-disciplinary initiatives such as those with physical and biological scientists are necessary to understand the scope of the social sciences. Too frequently there is a belief that one social scientist on an multi-disciplinary team is adequate social science representation. Third, more complete models of human behavior will be achieved through a synthesis of diverse social science perspectives.

Overview of Book

This book summarizes, compares, and contrasts important social science integration movements, conversations, and experiments as they relate to environmental problem solving. The focus is on recent developments, examples of successful integration efforts, and methodological advances for facilitating social science integration. Diverse viewpoints are brought into the conversation through chapters from leading scholars from a variety of backgrounds. Thirty-eight authors at the forefront of integration have contributed to the 11 chapters. The book is structured as follows.

Part I evaluates the status of *integration*. Costanza opens this section by presenting a vision of a desired future where the study of humans is reintegrated with the study of the rest of nature and the barriers between traditional disciplines dissolve to allow for "consilience" across natural and social sciences as well as the humanities. This will require reestablishment of a balance between synthesis and analysis. He points to nascent efforts to encourage an increase in focus on synthesis within the sciences in research and education and a shift from the logical positivist view to a pragmatic view. Costanza also takes on issues of scale/aggregation, and discusses how hierarchy theory and a complex systems approach coupled with the development of a theory of biological and co-evolution will lead to increased understanding of humans' place in nature. Costanza suggests that humanity might be prepared to develop a shared vision of a desirable and sustainable future and implement adaptive management systems to get us there.

In Chap. 2 Kelman et al. provide an overview of one arena for dialogue and collaboration amongst scientists, humanists, and non-scientists in the context of public policy engagement and outreach – the Millennium Alliance for Humanity and the Biosphere (MAHB). MAHB seeks to provide a large-scale synthesis that fuses knowledge about physical and social systems into blueprints for acceptable sustainable action that Costanza referred to in Chap. 1. The authors define and describe MAHB, including research, application, and a research agenda. They emphasize that the key is not to await full knowledge before acting on sustainability challenges. Instead, it is about using multiple disciplines to monitor and evaluate ongoing process, to ensure that actions do not exacerbate the existing problem or cause new problems.

Part II of the book presents *topics in integration*. In Chap. 3, Machlis et al. explore the challenge of collaboration and interdisciplinary teams as well as the importance of coupled human-natural systems. They present distinctive characteristics of science during environmental crises through two case studies – the Deepwater Horizon oil spill and Hurricane Sandy. Machlis et al. suggest that a research agenda which includes integration efforts needs to be developed for understanding and improving science during crisis.

Friedrichs (Chap. 4) makes the case for modified Malthusian theories to ground the study of resource management through science integration. He contends that the main impediment to integration both between various social scientific disciplines and between the social and the physical sciences is a refusal of social scientists to appreciate how deeply the societal sphere is embedded in wider biophysical and social-ecological systems. The chapter begins with a classical Malthusian framework and gradually adds complexity to it, showing how its logical structure is reproduced by simple neo-Malthusian theories that have been developed to account for contemporary global challenges. He demonstrates the potential of more sophisticated neo-Malthusian models and modified Malthusian theories contributing to better science integration.

Finucane et al. (Chap. 5) present a conceptual framework for analyzing social-ecological models of emerging infectious diseases. Specifically they examine whether risks, and perceptions of risk, associated with highly pathogenic avian influenza (HPAI) caused by the H5N1 virus can be associated with anthropogenic environmental changes produced by urbanization, agricultural change, and natural habitat alterations in the context of Vietnam. To address multi-scale issues within the framework, they draw upon multiple social science theories and methods. Finucane et al. conclude that no single theory or method is sufficient to explain complex phenomena such as emerging infectious disease and the relationships between factors influencing disease outbreaks. Thus, they argue that integrated approaches are the best way to provide an in-depth description and analysis of a complex problem.

Esptein et al. (Chap. 6) use the social-ecological systems (SES) framework to study power. They explore the long-standing divide among social scientists regarding power and its effects on the sustainability of social-ecological systems. They argue that there has been little constructive interaction between power-centered and institution-centered approaches. The authors use the SES framework as a tool to confront interdisciplinary puzzles that bridges the gap between social and ecological research. The chapter outlines a systematic approach for integrating diverse conceptualizations of power with the SES framework and then applies this approach to study the relationship between power and social-ecological outcomes. The analysis suggests that the SES framework is a promising tool for social science integration, but also that important questions remain concerning the validity of classifications, measurement, and statistical tests.

Manfredo et al. (Chap. 7) conclude Part II by making the case that increased integration of the human individual into dynamic, multi-level models is essential to understanding agency, innovation, and adaptation in social-ecological systems. They use the social-ecological systems framework introduced in the previous chapter as a starting point to examine how conservation science with a focus on the human individual – particularly the tradition of social science research known as "human dimensions of natural resources" – might fit within a systems approach. They suggest the implications for how ecosystem sciences can integrate the human individual into dynamic, multi-level models, and how human dimensions research can envision the individual and direct new research initiatives in a broader social-ecological context. They argue that the complexity of social systems is in need of more attention in SES models, and that these models will remain poorly specified until there is a representation of the multi-level context of human individuals. Ecosystem science sees the system as hierarchies nested within broader hierarchies, each operating at different speeds and cycles of change, Manfredo et al. propose how to use this same approach for examining individuals in their social-ecological context.

Part III focuses on *Methodological Advances for Facilitating Social Science Integration* For example, Verburg (Chap. 8) reviews how human-environment interactions are conceptualized in land change modeling at different scales and discusses the prospect for using land change models as a platform for integrating social science knowledge.

Boone et al. (Chap. 9) explore simulation as an approach to social-ecological integration; with an emphasis on agent-based modeling. They argue that questions regarding sustainability are broad in scope and that understanding the linkages between ecological and social systems has become paramount to society. They focus on computer simulations that use process-based or rule-based approaches to simulate events or behaviors through time. A case-study from Samburu District, Kenya, is presented as an example of agent-based modeling that illustrates network structures and provides an analysis of wet- versus dry-season livestock dispersal. They conclude that the inclusion of complexity calls for mixed methods research that is no longer tied to mainstream disciplinary methods.

Broadbent et al. (Chap. 10) present social network analysis (SNA) to examine the interactive effects of the social and natural sciences as well as the humanities to enable to the study of societal patterns and dynamics of unified systems. They explore traditional applications of SNA as an Integrative Structurational Analysis (ISA), a method that incorporates advances in discourse network analysis (DNA). They draw upon the international research project Comparing Climate Change Policy Networks (Compon) to illustrate the application of this ISA method and approach to the mitigation policy-formation processes of a set of nation-states and one region.

Collins (Chap. 11) examines the pressure that researchers and policy-makers are under to integrate natural and social sciences with policy. This pressure arises because of the complexity of environmental situations characterized by uncertainties, interdependencies and multiple stakeholders. Collins emphasizes the importance of framing natural resource management and explores links between ideas of integration and systems thinking. He introduces social learning systems as a conceptual and methodological innovation to enable integration. Water management research is used as an example to explore practical issues and findings. The chapter concludes with a short commentary on the constraints and opportunities for designing social learning systems.

Integration is not a straightforward and linear process. This book offers a structured overview of the integration opportunities and challenges. The chapters provide an overview of the history, vision, advances, examples, and methods that could lead to natural resource social science integration. While more work is necessary, this book provides an insight into the current state of social science integration.

Fort Collins, CO, USA

Michael J. Manfredo
Esther A. Duke

References

Belsky, J. M. (2002). Beyond the natural resource and environmental sociology divide: Insights from a transdisciplinary perspective. *Society and Natural Resources, 15*, 69–280.

Berry, W. (2000). *Life is a miracle: An essay against modern superstition.* Counterpoint: Washington D.C.

Ceccarelli, L. (2001). Rhetorical criticism and the rhetoric of science. *Western Journal of Communication, 65*(3), 314.

Costanza, R. (2009). Science and ecological economics: Integrating of the study of humans and the rest of nature. *Bulletin of Science Technology & Society, 29*, 358.

Elster, J. (2010). One social science or many? In United Nations Educational, Scientific and Cultural Organization (UNESCO) (Ed.), *World social science report: Knowledge divides* (pp. 199–203). Paris: UNESCO Publishing. Retrieved from www.unesco.org/publishing

Giraudoux, P., Pleydell, D., Raoul, F., Vaniscotte, A., Ito, A., & Craig, P. S. (2007). Echinoccus Multilocularis: Why are multidisciplinary and multiscale approaches essential in infectious disease ecology? *Tropical Medicine and Health, 35*(4), 293–299.

Gould, S. J. (2003). *The hedgehog, the fox, and the Magister's pox: Mending the gap between science and the humanities.* New York: Harmony Books.

McMichael, A. J., Butler, C. D., & Folke, C. (2003). New visions for addressing sustainability. *Science, 12*, 1919–1920.

National Science and Technology Council. (2009). *Social, behavioral and economic research in the federal context* (Report by the Subcommittee on social, behavioral and economic sciences). Washington, DC: Executive Office of the President, National Science and Technology Council.

Norenzayan, A. (2011). Explaining human behavioral diversity. *Science, 27*, 1041–1042.

Parsons, T., & Shils, E. A. (1951/2001). *Toward a general theory of action: Theoretical foundations for the social sciences.* New Brunswick: Transaction Publishers.

Poteete, A. R., Janssen, M. A., & Ostrom, E. (2010). *Working together: Collective action, the commons, and multiple methods in practice*. Princeton, New Jersey: Princeton University Press.

Rorty, R. (2001). Studied ambiguity. *Science, 293*, 2399–2400.

Varoufakis, Y. (2008). Capitalism according to evolutionary game theory: The impossibility of an evolutionary model of historical change. *Science and Society, 72*(1), 63–94.

Wilson, E. O. (1998). *Consilience: The unity of knowledge*. New York: Alfred A. Knopf.

Part I
The Status of Integration

Chapter 1
A Vision of the Future of Science: Reintegrating of the Study of Humans and the Rest of Nature

Robert Costanza

1.1 The Role of Envisioning in Creating the Future

Envisioning is a primary tool in futures studies (Garrett 1993; Slaughter 1993; Kouzes and Posner 1996; Razak 1996; Adesida and Oteh 1998). There has also been significant practical success in using envisioning and "future searches" in organizations and communities around the world (Weisbord 1992; Weisbord and Janoff 1995). This experience has shown that it is quite possible for disparate (even adversarial) groups to collaborate on envisioning a desirable future, given the right forum.

Meadows (1996) discusses why the processes of envisioning and goal setting are so important (at all levels of problem solving); why envisioning and goal setting are so underdeveloped in our society; and how we can begin to train people in the skill of envisioning, and begin to construct shared visions of a sustainable and desirable society. She tells the personal story of her own discovery of that skill and her attempts to use the process of shared envisioning in problem solving. From this experience, several general principles emerged, including:

1. In order to effectively envision, it is necessary to focus on what one really wants, not what one will settle for. For example, the lists below show the kinds of things people really want, compared to the kinds of things they often settle for.

This chapter is a revised version of a paper that first appeared in 2003 in *Futures* (Vol. 35, pp. 651–671).

R. Costanza (✉)
Crawford School of Public Policy, The Australian National University,
J.G. Crawford Building, #132 Lennox Crossing, Canberra, ACT 0200, Australia

M.J. Manfredo et al. (eds.), *Understanding Society and Natural Resources*,
DOI 10.1007/978-94-017-8959-2_1, © The Author(s) 2014

Really want	Settle for
Self-esteem	Fancy car
Serenity	Drugs
Health	Medicine
Human happiness	GNP
Permanent prosperity	Unsustainable growth

2. A vision should be judged by the clarity of its values, not the clarity of its implementation path. Holding to the vision and being flexible about the path is often the only way to find the path.
3. Responsible vision must acknowledge, but not be crushed by, the physical constraints of the real world.
4. It is critical for visions to be shared because only shared visions can be responsible.
5. Vision must be flexible and evolving.

This chapter represents a step in the ongoing process of creating a shared vision of the future of science. It lays out a personal vision of the kind of science I would really want to see in the future and why this new vision of science would be an improvement over what we now have. The paper itself is an attempt to share that vision, without getting bogged down in speculation about how the vision might be achieved or impediments to it's achievement. Hopefully, the ideas presented here will generate a dialogue culminating in a shared vision of the future of science that can motivate movement in the direction of the vision.

1.2 Consilience Among *All* the Sciences

"Consilience" according to Webster, is "a leaping together". Biologist E. O. Wilson's book by that title (Wilson 1998) attempted a grand synthesis, or "leaping together" of our current state of knowledge by "linking facts and fact-based theory across disciplines to create a common groundwork for explanation" and a prediction of where we are headed. Wilson believes that "the Enlightenment thinkers of the seventeenth and eighteenth centuries got it mostly right the first time. The assumptions they made of a lawful material world, the intrinsic unity of knowledge, and the potential of indefinite human progress are the ones we still take most readily into our hearts, suffer without, and find maximally rewarding through intellectual advance. The greatest enterprise of the mind has always been and always will be the attempted linkage of the sciences and humanities. The ongoing fragmentation of knowledge and resulting chaos in philosophy are not reflections of the real world but artifacts of scholarship. The propositions of the original Enlightenment are increasingly favored by objective evidence, especially from the natural sciences" (p. 8). Wilson takes an unabashedly logical positivist and reductionist approach to science and to consilience, arguing that: "The central idea of the consilience world view is that all tangible phenomena, from the birth of stars to the workings of social institutions, are based on material processes that are ultimately reducible, however long and tortuous the sequences, to the laws of physics" (p. 266). Deconstructionists and post-modernists, in this view, are merely gadflys who are nonetheless useful in order to keep the "real" scientists honest.

While there is probably broad agreement that integrating the currently fragmented sciences and humanities is a good idea, many will disagree with Wilson's neo-Enlightenment, reductionist prescription. The problem is that the type of consilience envisioned by Wilson would not be a real "leaping together" of the natural sciences, the social sciences, and the humanities. Rather, it would be a total takeover by the natural sciences and the reductionist approach in general. There are, however, several well-known problems with the strict reductionist approach to science (Williams 1997), and several of its contradictions show up in Wilson's view of consilience.

Wilson recognizes that the real issue in achieving consilience is one of scaling – how do we transfer understanding across the multitude of spatial and temporal scales from quarks to the universe and everything in between. But he seems to fall back on the overly simplistic reductionist approach to doing this – that if we understand phenomena at their most detailed scale we can simply "add up" in linear fashion from there to get the behavior at larger scales. While stating that "The greatest challenge today, not just in cell biology and ecology but in all of science, is the accurate and complete description of complex systems" (p. 85), he puts aside some of the main findings from the study of complex systems – that scaling in adaptive, living systems is neither linear nor easy, and that "emergent properties," which are unpredictable from the smaller scale alone, are important. While acknowledging on the one hand that analysis and synthesis, reductionism and wholism, are as inseparable as breathing out and breathing in, Wilson glosses over the difficulty of actually doing the synthesis in complex adaptive systems and the necessity of studying and understanding phenomena at multiple scales simultaneously, rather than reducing them to the laws of physics.

The consilience we are really searching for, I believe, is a more balanced and pluralistic kind of "leaping together," one in which the natural and social sciences and the humanities all contribute equitably. A science that is truly transdisciplinary and multiscale, rather than either reductionistic or wholistic, is, in fact, evolving, but I think it will be much more sophisticated and multifaceted in its view of the complex world in which we live, the nature of "truth" and the potential for human "progress" than the Enlightenment thinkers of the seventeenth and eighteenth centuries could ever have imagined. The remainder of this paper attempts to flesh out what this new transdisciplinary future for the reintegrated natural and social sciences might look like.

1.3 Reestablishing the Balance Between Synthesis and Analysis

Science, as an activity, requires a balance between two quite dissimilar activities. One is analysis – the ability to break down a problem into its component parts and understand how they function. The second is synthesis – the ability to put the pieces back together in a creative way in order to solve problems. In most of our current university research and education, these capabilities are not developed in a balanced, integrated way. For example, both natural and social science research and education focuses almost exclusively on analysis, while the arts and engineering focus on synthesis. But, as mentioned above, analysis and synthesis, reductionism and wholism, are as inseparable as breathing out and breathing in. It is no wonder that our current approach to science is so dysfunctional. We have been holding our breath for a long time!

In the future, the need for a healthy balance between analysis and synthesis will be recognized at all levels of science education and research. One can already see the beginnings of this development. For example, the National Center for Ecological Analysis and Synthesis (NCEAS – http://www.nceas.ucsb.edu/) was established in response to the recognition in the ecological community that the activity of synthesis was both essential and vastly under-supported. Ecologists recognized that they could only obtain funding and professional recognition for collecting new data. They never had the time, resources, or professional incentives to figure out what their data *meant*, or how it could be effectively used to build a broader understanding of ecosystems or to manage human interactions with them more effectively. The response to NCEAS so far has been overwhelmingly positive, and I expect that synthesis, as a necessary component of the scientific process, will eventually receive its fair share of resources and rewards. Funding for synthesis activities will become available from the major government science funding agencies on an equal footing with analysis activities. For example, NSF has recently established the National Socio-Environmental SYNthesis Center (SESYNC – http://www.sesync.org/) aimed at broadening synthesis activity to better encompass the social sciences and humanities.

In the universities, the curriculum will be restructured to achieve a better balance between synthesis and analysis. More courses will be "problem-based," workshops aimed at collaboratively addressing real problems via creative synthesis. Research has conclusively shown that "problem-based" curricula are very effective not only at supporting synthesis, but also at developing better analytical skills, since students are much more motivated to learn analytical tools if they have a specific problem to solve (Grigg 1995; Scott and Oulton 1999; Wheeler and Lewis 1997). There are already a few entire universities structured around the model of problem-based learning, including Maastricht University in the Netherlands and the University of Aalborg in Denmark. In addition, the capabilities of current and developing electronic communication technology will be more effectively employed in university education in the future. The market will soon be flooded with courses delivered over the Internet, but with little coordination among them and little recognition of the importance of integrating synthesis and communication into the educational process. The university of the future will take full advantage of the Internet, but it will also take much better advantage of the local face-to-face interactions on campus. Analysis courses are most amenable to delivery over the web. They could therefore afford to use the best faculty from around the world to produce them and could be continuously updated and improved. Grading would be internalized in the course, but testing would be proctored by the local host universities. This use of the Internet to provide most basic "tools" courses would free faculty to participate in synthesis courses, rather than repeating the same basic tools courses over and over at all campuses. Synthesis courses would be face-to-face "problem-based" studio or workshop courses focused on interactively solving real, current problems in the field (using the tools from the analysis courses or developing new tools in the process). These courses would be offered at local campuses or at the location of the problem itself, with quality control via the requirement for peer review of the results. Grading would be part of the peer review process and therefore would be performed external to the courses themselves.

This restructuring of research funding and the universities will also break down the strict disciplinary divisions that now exist. In the future, disciplinary boundaries will be as porous as many state and national boundaries are today. Likewise, one's disciplinary background will be noted much as one's place of birth is noted today – an interesting fact about one's path through life, but not a central defining characteristic. By focusing on problems and synthesis (rather than tools) universities will reclaim their role in society as the font of knowledge and wisdom (rather than merely technical expertise).

1.4 A Pragmatic Modeling Philosophy

Practical problem solving requires the integration of three elements: (1) creation of a shared vision of both how the world works and how we would like the world to be; (2) systematic analysis appropriate to and consistent with the vision; and (3) implementation appropriate to the vision. Scientists generally focus on only the second of these steps, but integrating all three is essential to both good science and effective management. "Subjective" values enter in the "vision" element, both in terms of the formation of broad social goals and in the creation of a "pre-analytic vision" which necessarily precedes any form of scientific analysis. Because of this need for vision, completely "objective" scientific analysis is impossible. In the words of Joseph Schumpeter (1954, p. 41):

"In practice we all start our own research from the work of our predecessors, that is, we hardly ever start from scratch. But suppose we did start from scratch, what are the steps we should have to take? Obviously, in order to be able to posit to ourselves any problems at all, we should first have to visualize a distinct set of coherent phenomena as a worthwhile object of our analytic effort. In other words, analytic effort is of necessity preceded by a preanalytic cognitive act that supplies the raw material for the analytic effort. In this book, this preanalytic cognitive act will be called Vision. It is interesting to note that vision of this kind not only must precede historically the emergence of analytic effort in any field, but also may reenter the history of every established science each time somebody teaches us to see things in a light of which the source is not to be found in the facts, methods, and results of the preexisting state of the science."

Nevertheless, it is possible to separate the process into the more subjective (or normative) envisioning component, and the more systematic, less subjective analysis component (which is based on the vision). "Good science" can do no better than to be clear about its underlying pre-analytic vision, and to do analysis that is consistent with that vision.

The task would be simpler if the vision of science were static and unchanging. But as the quote from Schumpeter above makes clear, this vision is itself changing and evolving as we learn more. This does not invalidate science as some deconstructionists would have it. Quite the contrary, by being explicit about its underlying pre-analytic vision, science can enhance its honesty and thereby its credibility. This credibility is a result of honest exposure and discussion of the underlying process and its inherent subjective elements, and a constant pragmatic testing of the results against real world problems, rather than by appeal to a non-existent objectivity.

The pre-analytic vision of science is changing from the "logical positivist" view (which holds that science can discover ultimate "truth" by falsification of hypothesis) to a more pragmatic view that recognizes that we do not have access to any ultimate, universal truths, but only to useful abstract representations (models) of small parts of the world. Science, in both the logical positivist and in this new "pragmatic modeling" vision, works by building models and testing them. But the new vision recognizes that the tests are rarely, if ever, conclusive (especially in the life sciences and the social sciences), the models can only apply to a limited part of the real world, and the ultimate goal is therefore not "truth" but quality and utility. In the words of William Deming "All models are wrong, but some models are useful" (McCoy 1994).

The goal of science is then the creation of useful models whose utility and quality can be tested against real world applications. The criteria by which one judges the utility and quality of models are themselves social constructs that evolve over time. There is, however, fairly broad and consistent consensus in the peer community of scientists about what these criteria are. They include: (1) testablity; (2) repeatability; (3) predictability; and (4) simplicity (i.e. Occam's razor – the model should be as simple as possible – but no simpler!). But, because of the nature of real world problems, there are many applications for which some of these criteria are difficult or impossible to apply. These applications may nevertheless still be judged as "good science". For example, some purely theoretical models are not directly "testable" – but they may provide a fertile ground for thought and debate and lead to more explicit models which *are* testable. Likewise, field studies of watersheds are not, strictly speaking, repeatable because no two watersheds are identical. But there is much we can learn from field studies that can be applied to other watersheds and tested against the other criteria of predictability and simplicity. How simple a model can be depends on the questions being asked. If we ask a more complex or more detailed question, the model will probably have to be more complex and detailed. Complex problems require "complex hypotheses" in the form of models. These complex models are always "false" in the sense that they can never match reality exactly. As science progresses and the range of applications expands, the criteria by which utility and quality are judged must also change and adapt to the changing applications.

1.5   A Multiscale Approach to Science

In understanding and modeling ecological and economic systems exhibiting considerable biocomplexity, the issues of scale and hierarchy are central (Ehleringer and Field 1993; O'Neill et al. 1989). The term "scale" in this context refers to both the resolution (spatial grain size, time step, or degree of complexity of the model) and extent (in time, space, and number of components modeled) of the analysis. The process of "scaling" refers to the application of information or models developed at one scale to problems at other scales. The scale dependence of predictions is increasingly recognized in a broad range of ecological studies, including: landscape ecology (Meentemeyer and Box 1987), physiological ecology (Jarvis and McNaughton 1986), population interactions (Addicott et al. 1987), paleoecology (Delcourt et al. 1983), freshwater ecology (Carpenter and Kitchell 1993), estuarine ecology (Livingston 1987), meteorology and climatology (Steyn et al. 1981) and global change (Rosswall et al. 1988). However, "scaling rules" applicable to biocomplex systems have not yet been adequately developed, and limits to extrapolation have been difficult to identify (Turner et al. 1989). In many of these disciplines primary information and measurements are generally collected at relatively small scales (i.e. small plots in ecology, individuals or single firms in economics) and that information is then often used to build models and make inferences at radically different scales (i.e. regional, national, or global). The process of scaling is directly tied to the problem of aggregation, which in complex, non-linear, discontinuous systems (like ecological and economic systems) is far from a trivial problem.

1.5.1   Aggregation

Aggregation error is inevitable as attempts are made to represent n-dimensional systems with less than n state variables, much like the statistical difficulties associated with sampling a variable population (Bartel et al. 1988, Gardner et al. 1982; Ijiri 1971). Cale et al. (1983) argued that in the absence of linearity and constant proportionality between variables – both of which are rare in ecological systems – aggregation error is inevitable. Rastetter et al. (1992) give a detailed example of scaling a relationship for individual leaf photosynthesis as a function of radiation and leaf efficiency to estimate the productivity of the entire forest canopy. Because of non-linear variability in the way individual leaves process light energy, one cannot simply use the fine scale relationship between photosynthesis and radiation and efficiency along with the mean values for the entire forest to represent total forest productivity without introducing significant aggregation error. Therefore, strategies to minimize aggregation error are necessary.

Jarvis and McNaughton (1986) explain the source of aggregation error shown by Rastetter by highlighting the discrepancy in transpiration control theory between meteorologists and plant physiologists. The meteorologists believe that weather patterns determine transpiration and have developed a series of equations that successfully calculate regional transpiration rates. The plant physiologists believe in stomatal control of transpiration and have demonstrated this with leaf chamber experiments in the field and laboratory. Therefore, it seems that different processes control transpiration at different scales, and aggregation from a single leaf to regional vegetation is impossible without accounting for this scale-dependent variability in transpiration control. One must somehow understand and embed this variability into the coarse scale.

Turner et al. (1989) list four steps for predicting across scales:

1. identify the spatial and temporal scale of the process to be studied;
2. understand the way in which controlling factors (constraints) vary with scale;
3. develop the appropriate methods to translate predictions from one scale to another; and
4. empirically test methods and predictions across multiple scales.

Rastetter et al. (1992) describe and compare four basic methods for scaling that are applicable to complex systems:

1. **partial transformations** of the fine scale relationships to coarse scale using a statistical expectations operator;
2. **moment expansions** as an approximation to 1;
3. **partitioning** or subdividing the system into smaller, more homogeneous parts (see the resolution discussion further on); and
4. **calibration** of the fine scale relationships to coarse scale data.

They go on to suggest a combination of these four methods as the most effective overall method of scaling in complex systems. (Rastetter et al. 1992).

1.5.2 Hierarchy Theory

Hierarchy theory provides an essential conceptual base for building coherent models of complex systems (Allen and Starr 1982; O'Neill et al. 1986; Salthe 1985; Gibson et al. 2000). Hierarchy is an organizational principle that yields models of nature that are partitioned into nested levels that share similar time and space scales. In a constitutive hierarchy, an entity at any level is part of an entity at a higher level and contains entities at a lower level. In an exclusive hierarchy, there is no containment relation between entities, and levels are distinguished by other criteria, e.g. trophic levels. Entities are to a certain extent insulated from entities at other levels in the sense that, as a rule, they do not directly interact; rather they provide mutual constraints. For example, individual organisms see the ecosystem they inhabit as a slowly changing set of external (environmental) constraints and the complex dynamics of component cells as a set of internal (behavioral) constraints.

From the scaling perspective, hierarchy theory is a tool for partitioning complex systems in order to minimize aggregation error (Thiel 1967; Hirata and Ulanowicz

1985). The most important aspect of hierarchy theory is that ecological systems' behavior is limited by both the potential behavior of its components (biotic potential) and environmental constraints imposed by higher levels (O'Neill et al. 1989). The flock of birds that can fly only as fast as its slowest member, or a forested landscape that cannot fix atmospheric nitrogen if specific bacteria are not present are examples of biotic potential limitation. Animal populations limited by available food supply and plant communities limited by nutrient remineralization are examples of limits imposed by environmental constraints. O'Neill et al. (1989) use hierarchy theory to define a 'constraint envelope' based upon the physical, chemical and biological conditions within which a system must operate. They argue that hierarchy theory and the resulting 'constraint envelope' enhance predictive power. Although they may not be able to predict exactly what place the system occupies within the constraint envelope, they can state with confidence that a system will be operating within its constraint envelope.

Viewing biocomplexity through the lens of hierarchy theory should serve to illuminate the general principles of life systems that occur at each level of the hierarchy. While every level will necessarily have unique characteristics, it is possible to define forms and processes that are isomorphic across levels (as are many laws of nature). Troncale (1985) has explored some of these isomorphisms in the context of general system theory. In the context of scaling theory we can seek isomorphisms which assist in the vertical integration of scales. These questions feed into the larger question of scaling, and how to further develop the four basic methods of scaling mentioned above for application to complex systems.

1.5.3 Fractals and Chaos

One well-known isomorphism is the "self-similarity" between scales exhibited by fractal structures (Mandelbrot 1977) which may provide another approach to the problem of scaling. This self-similarity implies a regular and predictable relationship between the scale of measurement (here meaning the resolution of measurement) and the measured phenomenon. For example, the regular relationship between the measured length of a coastline and the resolution at which it is measured is a fundamental, empirically observable one. It can be summarized in the following equation:

$$L = k \cdot s^{(1-D)} \qquad (1.1)$$

where:

L = the length of the coastline or other "fractal" boundary
s = the size of the fundamental unit of measure or the resolution of the measurement
k = a scaling constant
D = the fractal dimension

Primary questions concern the range of applicability of fractals and chaotic systems dynamics to the practical problems of modeling ecological economic systems. The influence of scale, resolution, and hierarchy on the mix of behaviors one observes in systems has not been fully investigated, and this remains a key question for developing coherent models of complex ecological economic systems.

1.5.4   Resolution and Predictability

The significant effects of nonlinearities raise some interesting questions about the influence of resolution (including spatial, temporal, and component) on the performance of models, and in particular their predictability. Costanza and Maxwell (1994) analyzed the relationship between resolution and predictability and found that while increasing resolution provides more descriptive information about the patterns in data, it also increases the difficulty of accurately modeling those patterns. There may be limits to the predictability of natural phenomenon at particular resolutions, and "fractal like" rules that determine how both "data" and "model" predictability change with resolution.

Some limited testing of these ideas was done by resampling land use map data sets at several different spatial resolutions and measuring predictability at each. Colwell (1974) used categorical data to define predictability as the reduction in uncertainty (scaled on a 0–1 range) about one variable given knowledge of others. One can define spatial auto-predictability (P_a) as the reduction in uncertainty about the state of a pixel in a scene, given knowledge of the state of adjacent pixels in that scene, and spatial cross-predictability (P_c) as the reduction in uncertainty about the state of a pixel in a scene, given knowledge of the state of corresponding pixels in other scenes. P_a is a measure of the internal pattern in the data, while P_c is a measure of the ability of some other model to represent that pattern.

A strong linear relationship was found between the log of P_a and the log of resolution (measured as the number of pixels per square kilometer). This fractal-like characteristic of "self-similarity" with decreasing resolution implies that predictability, like the length of a coastline, may be best described using a unitless dimension that summarizes how it changes with resolution. One can define a "fractal predictability dimension" (DP) in a manner analogous to the normal fractal dimension (Mandelbrot 1977, 1983). The resulting DP allows convenient scaling of predictability measurements taken at one resolution to others.

Cross-predictability (P_c) can be used for pattern matching and testing the fit between scenes. In this sense it relates to the predictability of models versus the internal predictability in the data revealed by P_a. While P_a generally increases with increasing resolution (because more information is being included), P_c generally falls or remains stable (because it is easier to model aggregate results than fine grain ones). Thus we can define an optimal resolution for a particular modeling problem that balances the benefit in terms of increasing data predictability (P_a) as one

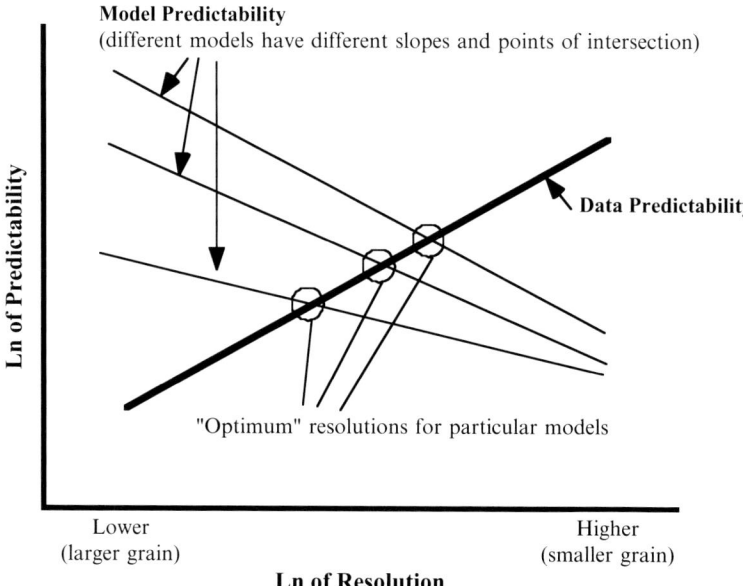

Model Predictability
(different models have different slopes and points of intersection)

Ln of Predictability

Data Predictability

"Optimum" resolutions for particular models

Lower
(larger grain)

Higher
(smaller grain)

Ln of Resolution

Fig. 1.1 Relationship between resolution and predictability for data and models (From Costanza and Maxwell 1994)

increases resolution, with the cost of decreasing model predictability (P_c). Figure 1.1 shows this relationship in generalized form.

1.6   Cultural and Biological Co-evolution

In modeling the dynamics of complex systems it is impossible to ignore the discontinuities and surprises that often characterize these systems, and the fact that they operate far from equilibrium in a state of constant adaptation to changing conditions (Rosser 1991, 1992; Holland and Miller 1991; Lines 1990; Kay 1991). The paradigm of evolution has been broadly applied to both ecological and economic systems (Boulding 1981; Arthur 1988; Lindgren 1991; Maxwell and Costanza 1993) as a way of formalizing understanding of adaptation and learning behaviors in non-equilibrium dynamic systems. The general evolutionary paradigm posits a mechanism for adaptation and learning in complex systems at any scale using three basic interacting processes: (1) information storage and transmission; (2) generation of new alternatives; and (3) selection of superior alternatives according to some performance criteria.

The evolutionary paradigm is different from the conventional optimization paradigm popular in economics in at least four important respects (Arthur 1988): (1) evolution is path dependent, meaning that the detailed history and dynamics of the system are important; (2) evolution can achieve multiple equilibria; (3) there is no guarantee that optimal efficiency or any other optimal performance will be achieved, due in part to path dependence and sensitivity to perturbations; and (4) "lock-in" (survival of the first rather than survival of the fittest) is possible under conditions of increasing returns. While, as Arthur (1988) notes "conventional economic theory is built largely on the assumption of diminishing returns on the margin (local negative feedbacks)" life itself can be characterized as a positive feedback, self-reinforcing, autocatalytic process (Kay 1991; Günther and Folke 1993) and we should expect increasing returns, lock-in, path dependence, multiple equilibria and sub-optimal efficiency to be the rule rather than the exception in economic and ecological systems.

1.6.1 Cultural vs. Genetic Evolution

In biological evolution, the information storage medium is the genes, the generation of new alternatives is by sexual recombination or genetic mutation, and selection is performed by nature according to a criteria of "fitness" based on reproductive success. The same *process* of change occurs in ecological, economic, and cultural systems, but the elements on which the process works are different. For example, in cultural evolution the storage medium is the culture (the oral tradition, books, film or other storage medium for passing on behavioral norms), the generation of new alternatives is through innovation by individual members or groups in the culture, and selection is again based on the reproductive success of the alternatives generated, but reproduction is carried out by the spread and copying of the behavior through the culture rather than biological reproduction. One may also talk of "economic" evolution, a subset of cultural evolution dealing with the generation, storage, and selection of alternative ways of producing things and allocating that which is produced. The field of "evolutionary economics" has grown up in the last decade or so based on these ideas (cf. Day and Groves 1975; Day 1989). Evolutionary theories in economics have already been successfully applied to problems of technical change, to the development of new institutions, and to the evolution of means of payment.

For large, slow-growing animals like humans, genetic evolution has a built-in bias towards the long-run. Changing the genetic structure of a species requires that characteristics (phenotypes) be selected and accumulated by differential reproductive success. Behaviors learned or acquired during the lifetime of an individual cannot be passed on genetically. Genetic evolution is therefore usually a relatively slow process requiring many generations to significantly alter a species' physical and biological characteristics.

Cultural evolution is potentially much faster. Technical change is perhaps the most important and fastest evolving cultural process. Learned behaviors that are successful, at least in the short term, can be almost immediately spread to other members of the culture and passed on in the oral, written, or video record. The increased speed of adaptation that this process allows has been largely responsible for *homo sapiens'* amazing success at appropriating the resources of the planet. Vitousek et al.

(1986) estimate that humans now directly control from 25 to 40 % of the total primary production of the planet's biosphere, and this is beginning to have significant effects on the biosphere, including changes in global climate and in the planet's protective ozone shield.

Both the benefits and the costs of this rapid cultural evolution are potentially significant. Like a car that has increased speed, humans are in more danger of running off the road or over a cliff. Cultural evolution lacks the built-in long-run bias of genetic evolution and is susceptible to being led by its hyper-efficient short-run adaptability over a cliff into the abyss.

Another major difference between cultural and genetic evolution may serve as a countervailing bias, however. As Arrow (1962) has pointed out, cultural and economic evolution, unlike genetic evolution, can at least to some extent employ foresight. If society can see the cliff, perhaps it can be avoided.

While market forces drive adaptive mechanisms (Kaitala and Pohjola 1988), the systems that evolve are not necessarily optimal, so the question remains: What external influences are needed and when should they be applied in order to achieve an optimum economic system via evolutionary adaptation? The challenge faced by ecological economic systems modelers is to first apply the models to gain foresight, and to respond to and manage the system feedbacks in a way that helps avoid any foreseen cliffs (Berkes and Folke 1994). Devising policy instruments and identifying incentives that can translate this foresight into effective modifications of the short-run evolutionary dynamics is the challenge (Costanza 1987).

What is really needed is a coherent and consistent theory of genetic and cultural co-evolution. These two types of evolution interact with each other in complex and subtle ways, each determining and changing the landscape for the other.

1.6.2 Evolutionary Criteria

A critical problem in applying the evolutionary paradigm in dynamic models is defining the selection criteria *a priori*. In its basic form, the theory of evolution is circular and descriptive (Holling 1987). Those species or cultural institutions or economic activities survive which are the most successful at reproducing themselves. But we only know which ones were more successful *after the fact*. To use the evolutionary paradigm in modeling, we require a quantitative measure of fitness (or more generally *performance*) in order to drive the selection process.

Several candidates have been proposed for this function in various systems, ranging from expected economic utility to thermodynamic potential. Thermodynamic potential is interesting as a performance criteria in complex systems because even very simple chemical systems can be seen to evolve complex non-equilibrium structures using this criteria (Prigogine 1972; Nicolis and Prigogine 1977, 1989), and all systems are (at minimum) thermodynamic systems (in addition to their other characteristics) so that thermodynamic constraints and principles are applicable across both ecological and economic systems (Eriksson 1991).

This application of the evolutionary paradigm to thermodynamic systems has led to the development of far-from-equilibrium thermodynamics and the concept of dissipative structures (Prigogine 1972). An important research question is to determine the range of applicability of these principles and their appropriate use in modeling ecological economic systems.

Many dissipative structures follow complicated transient motions. Schneider and Kay (1994) propose a way to analyze these chaotic behaviors and note that, "Away from equilibrium, highly ordered stable complex systems can emerge, develop and grow at the expense of more disorder at higher levels in the system's hierarchy." It has been suggested that the integrity of far-from-equilibrium systems has to do with the ability of the system to attain and maintain its (set of) optimum operating point(s) (Kay 1991). The optimum operating point(s) reflect a state where self-organizing thermodynamic forces and disorganizing forces of environmental change are balanced. This idea has been elaborated and described as "evolution at the edge of chaos" by Kauffman and Johnson (1991).

The concept that a system may evolve through a sequence of stable and unstable stages leading to the formation of new structures seems well suited to ecological economic systems. For example, Gallopin (1989) stresses that to understand the processes of economic impoverishment "...The focus must necessarily shift from the static concept of poverty to the dynamic processes of impoverishment and sustainable development within a context of permanent change. The dimensions of poverty cannot any longer be reduced to only the economic or material conditions of living; the capacity to respond to changes, and the vulnerability of the social groups and ecological systems to change become central." In a similar fashion Robinson (1991) argues that sustainability calls for maintenance of the dynamic capacity to respond adaptively, which implies that we should focus more on basic natural and social processes, than on the particular forms these processes take at any time. Berkes and Folke (1994) have discussed the capacity to respond to changes in ecological economic systems, in terms of institution building, collective actions, cooperation, and social learning. These might be some of the ways to enhance the capacity for resilience (increase the capacity to recover from disturbance) in interconnected ecological economic systems.

As discussed earlier, cultural evolution also has the added element of human foresight. To a certain extent, we can *design* the future that we want by appropriately setting goals and envisioning desired outcomes.

1.7   Creating a Shared Vision of a Desirable and Sustainable Future

Probably the most challenging task facing humanity today is the creation of a shared vision of a sustainable and desirable society, one that can provide permanent prosperity within the biophysical constraints of the real world in a way that is fair and equitable to all of humanity, to other species, and to future generations. This vision does not now exist, although the seeds are there. We all have our own private visions of the world we really want and we need to overcome our fears and skepticism and begin to share these visions and build on them – until we have built a vision of the world we want.

We need to fill in the details of our desired future in order to make it tangible enough to motivate people across the spectrum to work toward achieving it. Nagpal and Foltz (1995) have begun this task by commissioning a range of individual visions of a sustainable world from around the globe. They laid out the following challenge for each of their "envisionaries" :

> Individuals were asked not to try to predict what lies ahead, but rather to imagine a *positive* future for their respective region, defined in any way they chose – village, group of villages, nation, group of nations, or continent. We asked only that people remain within the bounds of plausibility, and set no other restrictive guidelines.

The results were quite revealing. While these independent visions were difficult to generalize, they did seem to share at least one important point. The "default" western vision of continued material growth was not what people envisioned as part of their "positive future." They envisioned a future with "enough" material consumption, but where the focus has shifted to maintaining high quality communities and environments, education, culturally rewarding full employment, and peace.

These results are consistent with surveys about the degree of desirability that people expressed for four hypothetical visions of the future in the year 2100 (Costanza 2000). The four visions derive from two basic world views, whose characteristics are laid out in Fig. 1.2. These world views have been described in many ways (Bossel 1996), but an important distinction has to do with one's degree of faith in technological progress (Costanza 1989). The "technological optimist" world view is one in which technological progress is assumed to be able to solve all current and future social problems. It is a vision of continued expansion of humans and their dominion over nature. This is the "default" vision in our current western society, one that represents continuation of current trends into the indefinite future. It is the "taker" culture as described so eloquently by Daniel Quinn in "Ishmael" (1992).

There are two versions of this vision, however. One that corresponds to the underlying assumptions on which it is based actually being true in the real world, and one that corresponds to those assumptions being false, as shown in Fig. 1.2. The positive version of the "technological optimist" vision was called "Star Trek," after the popular TV series which is its most articulate and vividly fleshed-out manifestation. The negative version of the "technological optimist" vision was called "Mad Max" after the popular movie of several years ago that embodies many aspects of this vision gone bad.

The "technological skeptic" vision is one that depends much less on technological change and more on social and community development. It is not in any sense "anti-technology." But it does not assume that technological change can solve all

Four Visions of the Future

	Real State of the World	
	Optimists Are Right (Resources are unlimited)	**Skeptics Are Right** (Resources are limited)
Technological Optimism Resources are unlimited Technical Progress can deal with any challenge Compitition promotes progress; markets are the guiding principle	**Star Trek** Fusion energy becomes practical, solving many economic and environmental problems. Humans journey to the inner solar system, where population continues to expand (mean rank 2.3)	**Mad Max** Oil production declines and no affordable alternative emerges. Financial markets collapse and governments weaken, too broke to maintain order and control over desperate, impoverished populations. The world is run by transnational corporations. (mean rank -7.7)
Technological Skepticsm Resources are limited Progress depends less on technology and more on social and community development Cooperation promotes progress; markets are the servants of larger goals	**Big Government** Governments sanction companies that fail to pursue the public interest. Fusion energy is slow to develop due to strict saftey standards. Family-planning programs stabilize population growth. Incomes become more equal. (mean rank 0.8)	**EcoTopia** Tax reforms favor ecologically beneficent industries and punish polluters and resource depleters. Habitation patterns reduce need for transportation and energy. A shift away from consumerism increases quality of life and reduces waste. (mean rank 5.1)

World View & Policy (left margin label)

Fig. 1.2 Payoff matrix for technological optimism vs. skepticism

problems. In fact, it assumes that some technologies may create as many problems as they solve and that the key is to view technology as the servant of larger social goals rather than the driving force. The version of this vision that corresponds to the skeptics being right about the nature of the world was called "Ecotopia" after the semi-popular book of the late 1970s (Callenbach 1975). If the optimists turn out to be right about the real state of the world, the "big government" vision comes to pass – Ronald Reagan's worst nightmare of overly protective government policies getting in the way of the free market.

Each of these future visions was described as a narrative from the perspective of the year 2100 (Costanza 2000). A total of 418[1] respondents were read each of the four visions. They were asked: "For each vision, I'd like you to first state, on a scale of −10 to +10, using the scale provided, *how comfortable you would be living in the*

---

[1] The Americans consisted of 17 participants in an Ecological Economics class at the University of Maryland, 260 attendees at a convocation speech at Wartburg College in Waverly, IA, January 27, 1998, and 39 via the world wide web. The Swedes consisted of 71 attendees at a "Keynotes in Natural Resources" Lecture at the Swedish University of Agricultural Science, Uppsala, April 20, 1999 and 31 attendees at a presentation at Stockholm University, April 22, 1999.

Table 1.1 Results of a survey of desirability of each of the four visions on a scale of −10 (least desirable to +10 (most desirable)) for self-selected groups of Americans and Swedes

	Americans (n=316)	Swedes (n=102)	Pooled (n=418)
Star Trek	+2.38 (±5.03)	+2.48 (±5.45)	+2.38 (±5.13)
Mad Max	−7.78 (±3.41)	−9.12 (±2.30)	−8.12 (±3.23)
Big Government	+0.54 (±4.44)	+2.32 (±3.48)	+0.97 (±4.29)
Ecotopia	+5.32 (±4.10)	+7.33 (±3.11)	+5.81 (±3.97)

Standard deviations are given in parentheses after the means

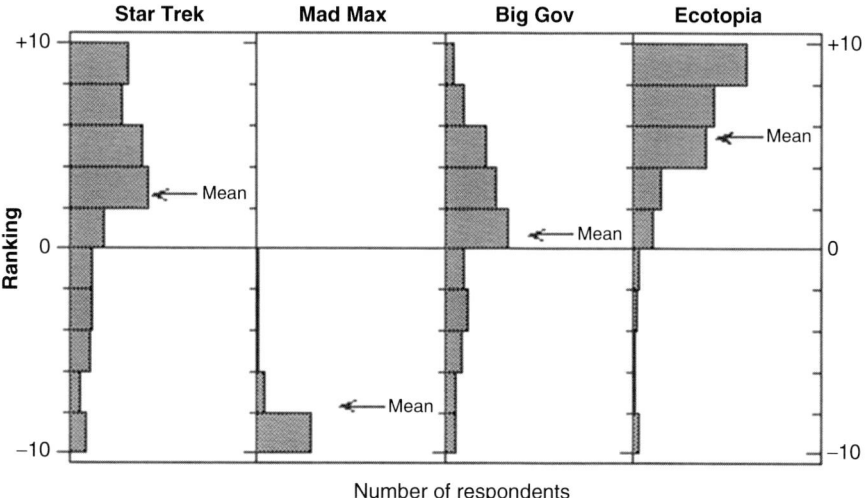

Fig. 1.3 Frequency distributions of the responses to the visions survey

world described. How desirable do you find such a world? I'm not asking you to vote for one vision over the others. *Consider each vision independently*, and just state how desirable (or undesirable) you would find it if you happened to find yourself there." They were also asked to give their age, gender, and household income range on the survey form. The surveys were conducted with groups from both the US and Sweden. The results (mean ± standard deviation) are shown in Table 1.1 for each of these groups and pooled.

Frequency distributions of the results are plotted in Fig. 1.3. The majority of those surveyed found the Star Trek vision positive (mean of +2.48 on a scale from −10 to +10). Given that it represents a logical extension of the currently dominant world view and culture, it is interesting that this vision was rated so low. I had expected this vision to be rated much higher, and this result may indicate the deep ambivalence many people have about the direction society seems to be headed. The frequency plot (and the high standard deviation) also shows this ambivalence toward Star Trek. The responses span the range from +10 to −10, with only a weak preponderance toward the positive side of the scale. This result applied for both the American and Swedish subgroups.

Those surveyed found the Mad Max vision very negative at -8.12 (only about 3 % of participants rated this vision positive). This was as expected. The Americans seemed a bit less averse to Mad Max (-7.78) than the Swedes (-9.12), and with a larger standard deviation.

The Big Government vision was rated on average just positive at 0.97. Many found it appealing, but some found it abhorrent (probably because of the limits on individual freedom implied). Here there were significant differences between the Americans and Swedes, with the Swedes ($+2.32 \pm 3.48$) being much more favorably disposed to Big Government and with a smaller standard deviation than the Americans ($+0.54 \pm 4.44$). This also was as expected, given the cultural differences in attitudes toward government in America and Sweden. Swedes rated Big Government almost as highly as Star Trek.

Finally, most of those surveyed found the Ecotopia vision "very positive" (at 5.81) some wildly so, some only mildly so, but very few (only about 7 % of those surveyed) expressed a negative reaction to such a world. Swedes rated Ecotopia significantly higher than Americans, also as might be expected given cultural differences.

Some other interesting patterns emerged from the survey. All of the visions had large standard deviations, but (especially if one looks at the frequency distributions) the Mad Max vision was consistently very negative and the Ecotopia vision was consistently very positive. Age and gender seemed to play a minor, but interesting role in how individuals rated the visions. Males rated Star Trek higher than females (mean $= 3.66$ vs. 1.90, $p = .0039$). Males also rated Mad Max higher that females (-7.11 vs. -8.20, $p = .0112$). The means were not significantly different by gender for either of the other two visions. Age was not significantly correlated with ranking for any of the visions, but the variance in ranking seemed to decrease somewhat with age, with younger participants showing a higher range of ratings than older participants.

Much more work is necessary to implement living democracy, and within that to create a truly shared vision of a desirable and sustainable future. This ongoing work needs to engage all members of society in a substantative dialogue about the future they desire and the policies and instruments necessary to bring it about. Scientists are a critical stakeholder group to include in this dialogue.

The future, at least to some extent, is amenable to design. As when building a house, a good plan or vision of what the house is intended to look like and how it will function is essential to building a coherent and useful structure. This design process needs to be informed by the reality of the situation – the nature of the complex, adaptive systems within which we are working – but it also needs to express our shared desires. In the future our knowledge about living systems will dramatically improve and we can achieve a true consilence among all the aspects of that knowledge. This will help us understand the constraints within which the design process must work. But we also need to involve our imagination, creativity, and ability to envision in order to design as useful and beautiful a world as we can within those constraints.

1.8   Conclusions

In this vision of the future of science:

- One's discipline will be noted much as one's place of birth is noted today – where one started on life's journey, but not what totally defines one's life.
- Science research and education will balance analysis and synthesis to produce not just data, but knowledge and even wisdom. This will enable vastly improved links with social decision-making.
- The limits of predictability of complex, adaptive, living systems will be recognized, and a "pragmatic modeling" philosophy of science will be adopted. This will allow new, adaptive approaches to environmental management and better links with social decision-making.
- A multiscale approach to understanding, modeling, and managing complex, adaptive, living systems will be the norm, and methods for transferring knowledge across scales will be vastly improved.
- A consistent theory of biological and cultural co-evolution will evolve and increase understanding of humans' place in nature and the possibilities of designing a sustainable and desirable human presence in the biosphere.
- Envisioning and goal setting will be recognized as critical parts of both science and social decision-making. We will create a shared vision of a desirable and sustainable future, and implement adaptive management systems at multiple scales in order to get us there.

References

Addicott, J. F., Aho, J. M., Antolin, M. F., Padilla, D. K., Richardson, J. S., & Soluk, D. K. (1987). Ecological neighborhoods: Scaling environmental patterns. *Oikos, 49*, 340–346.

Adesida, O., & Oteh, A. (1998). Envisioning the future of Nigeria. *Futures, 30*, 569–572.

Allen, T. F. H., & Starr, T. B. (1982). *Hierarchy*. Chicago: University of Chicago Press.

Arrow, K. (1962). The economic implications of learning by doing. *Review of Economic Studies., 29*, 155–173.

Arthur, W. B. (1988). Self-reinforcing mechanisms in economics. In P. W. Anderson, K. J. Arrow, & D. Pines (Eds.), *The economy as an evolving complex system* (pp. 9–31). Redwood City: Addison-Wesley.

Bartel, S. M., Cale, W. G., O'Neill, R. V., & Gardner, R. H. (1988). Aggregation error: Research objectives and relevant model structure. *Ecological Modelling, 41*, 157–168.

Berkes, F., & Folke, C. (1994). Investing in cultural capital for sustainable use of natural capital. In A. M. Jansson, M. Hammer, C. Folke, & R. Costanza (Eds.), *Investing in natural capital: The ecological economics approach to sustainability* (pp. 128–149). Washington, DC: Island press, 504 pp.

Bossel, H. (1996). *20/20 vision: Explorations of sustainable futures*. Kassel: Center for Environmental Systems Research, University of Kassel.

Boulding, K. E. (1981). *Evolutionary economics*. Beverly Hills: Sage.

Cale, W. G., O'Neill, R. V., & Gardner, R. H. (1983). Aggregation error in nonlinear ecological models. *Journal of Theoretical Biology, 100*, 539–550.

Callenbach, E. (1975). *Ecotopia*. New York: Bantam.

Carpenter, S. R., & Kitchell, J. F. (Eds.). (1993). *The trophic cascade in lakes*. Cambridge/New York: Cambridge University Press. 385 pp.

Colwell, R. K. (1974). Predictability, constancy, and contingency of periodic phenomena. *Ecology, 55*, 1148–1153.

Costanza, R. (1987). Social traps and environmental policy. *BioScience, 37*, 407–412.

Costanza, R. (1989). What is ecological economics? *Ecological Economics, 1*, 1–7.

Costanza, R. (2000). Visions of alternative (unpredictable) futures and their use in policy analysis. *Conservation Ecology, 4*(1), 5. [online] URL: http://www.consecol.org/vol4/iss1/art5

Costanza, R., & Maxwell, T. (1994). Resolution and predictability: An approach to the scaling problem. *Landscape Ecology, 9*, 47–57.

Day, R. H. (1989). *Dynamical systems, adaptation and economic evolution* (MRG Working Paper No. M8908). Los Angeles: University of Southern California.

Day, R. H., & Groves, T. (Eds.). (1975). *Adaptive economic models*. New York: Academic.

Delcourt, H. R., Delcourt, P. A., & Webb, T. (1983). Dynamic plant ecology: The spectrum of vegetation change in space and time. *Quaternary Science Reviews, 1*, 153–175.

Ehleringer, J. R., & Field, C. B. (Eds.). (1993). *Scaling physiological processes: Leaf to globe*. New York: Academic, 388 pp.

Eriksson, K.-E. (1991). Physical foundations of ecological economics. In L. O. Hansson & B. Jungen (Eds.), *Human responsibility and global change* (pp. 186–196). Göteborg: University of Göteborg Press.

Gallopin, G. C. (1989). Global impoverishment, sustainable development and the environment: A conceptual approach. *International Social Science Journal., 121*, 375–397.

Gardner, R. H., Cale, W. G., & O'Neill, R. V. (1982). Robust analysis of aggregation error. *Ecology, 63*(6), 1771–1779.

Garrett, M. J. (1993). A way through the maze: What futurists do and how they do it. *Futures, 25*, 254–274.

Gibson, C., Ostrom, E., & Ahn, T. (2000). The concept of scale and the human dimensions of global change: A survey. *Ecological Economics, 32*(2), 217–239.

Grigg, N. (1995). Case method for teaching water-resources management. *Journal of Professional Issues in Engineering Education and Practice, 121*, 30–36.

Günther, F., & Folke, C. (1993). Characteristics of nested living systems. *Journal of Biological Systems, 1*(3), 257–274.

Hirata, H., & Ulanowicz, R. E. (1985). Informational theoretical analysis of the aggregation and hierarchical structure of ecological networks. *Journal of Theoretical Biology, 116*, 321–341.

Holland, J. H., & Miller, J. H. (1991). Artificial adaptive agents in economic theory. *American Economic Review., 81*, 365–370.

Holling, C. S. (1987). Simplifying the complex: The paradigms of ecological function and structure. *European Journal of Operational Research, 30*, 139–146.

Ijiri, Y. (1971). Fundamental queries in aggregation theory. *JASA, 66*, 766–782.

Jarvis, P. G., & McNaughton, K. G. (1986). Stomatal control of transpiration: Scaling up from leaf to region. *Advances in Ecological Research, 15*, 1–49.

Kaitala, V., & Pohjola, M. (1988). Optimal recovery of a shared resource stock: A differential game model with efficient memory equilibria. *Natural Resource Modeling, 3*, 91–119.

Kauffman, S. A., & Johnson, S. (1991). Coevolution to the edge of chaos: Coupled fitness landscapes, poised states, and coevolutionary avalanches. *Journal of Theoretical Biology, 149*, 467–505.

Kay, J. J. (1991). A nonequilibrium thermodynamic framework for discussing ecosystem integrity. *Environmental Management, 15*, 483–495.

Kouzes, J. M., & Posner, B. Z. (1996). Envisioning your future: Imagining ideal scenarios. *The Futurist, 30*, 14–19.

Lindgren, K. (1991). Evolutionary phenomena in simple dynamics. In C. G. Langton, C. Taylor, J. D. Farmer, & S. Rasmussen (Eds.), *Artificial life* (SFI studies in the sciences of complexity, Vol. X, pp. 295–312). Boston: Addison-Wesley.

Lines, M. (1990). Stochastic stability considerations: A nonlinear example. *International Review of Economics and Business, 37*, 219–233.

Livingston, R. J. (1987). Field sampling in estuaries: The relationship of scale to variability. *Estuaries, 10*, 194–207.

Mandelbrot, B. B. (1977). *Fractals. Form, chance and dimension*. San Francisco: Freeman.

Mandelbrot, B. B. (1983). *The fractal geometry of nature*. San Francisco: Freeman.

Maxwell, T., & Costanza, R. (1993). An approach to modeling the dynamics of evolutionary self organization. *Ecological Modeling, 69*, 149–161.

McCoy, R. (1994). *The best of Deming*. Knoxville: SPC Press.

Meadows, D. (1996). Envisioning a sustainable world. In R. Costanza, O. Segura, & J. Martinez-Alier (Eds.), *Getting down to earth: Practical applications of ecological economics* (pp. 117–126). Washington, DC: Island Press.

Meentemeyer, V., & Box, E. O. (1987). Scale effects in landscape studies. In M. G. Turner (Ed.), *Landscape heterogeneity and disturbance* (pp. 15–34). New York: Springer.

Nagpal, T., & Foltz, C. (Eds.). (1995). *Choosing our future: Visions of a sustainable world*. Washington, DC: World Resources Institute.

Nicolis, G., & Prigogine, I. (1977). *Selforganization in non-equilibrium systems*. New York: John Wiley & Sons, 491 pp.

Nicolis, G., & Prigogine, I. (1989). *Exploring complexity*. New York: W. H. Freeman, 313 pp.

O'Neill, R. V., DeAngelis, D. L., Waide, J. B., & Allen, T. F. H. (1986). *A hierarchical concept of ecosystems*. Princeton: Princeton University Press.

O'Neill, R. V., Johnson, A. R., & King, A. W. (1989). A hierarchical framework for the analysis of scale. *Landscape Ecology, 3*, 193–205.

Prigogine, I. (1972). Thermodynamics of evolution. *Physics Today, 23*, 23–28.

Quinn, D. (1992). *Ishmael*. New York: Bantam/Turner, 266 pp.

Rastetter, E. B., King, A. W., Cosby, B. J., Hornberger, G. M., O'Neill, R. V., & Hobbie, J. E. (1992). Aggregating fine-scale ecological knowledge to model coarser-scale attributes of ecosystems. *Ecological Applications, 2*, 55–70.

Razak, V. M. (1996). From the canvas to the field: Envisioning the future of culture. *Futures, 28*, 645–649.

Robinson, J. B. (1991). Modelling the interactions between human and natural systems. *International Social Science Journal, 130*, 629–647.

Rosser, J. B. (1991). *From catastrophe to chaos: A general theory of economic discontinuities*. Amsterdam: Kluwer.

Rosser, J. B. (1992). The dialogue between the economic and ecologic theories of evolution. *Jour. of Economic Behavior and Organization, 17*, 195–215.

Rosswall, R., Woodmansee, R. G., & Risser, P. G. (1988). *Scales and global change: Spatial and temporal variability in biospheric and geospheric processes*. New York: Wiley. 355 pp.

Salthe, S. N. (1985). *Evolving hierarchical systems: Their structure and representation*. New York: Columbia University Press.

Schneider, E. D., & Kay, J. J. (1994). Life as a manifestation of the second law of thermodynamics. *Mathematical and Computer Modelling, 19*, 25–48.

Schumpeter, J. (1954). *History of economic analysis*. London: Allen & Unwin, 1260 pp.

Scott, W., & Oulton, C. (1999). Environmental education: Arguing the case for multiple approaches. *Educational Studies, 25*, 89–97.

Slaughter, R. A. (1993). Futures concepts. *Futurist., 25*, 289–314.

Steyn, D. G., Oke, T. R., Hay, J. E., & Knox, J. L. (1981). On scales in meteorology and climatology. *Climatological Bulletin, 39*, 1–8.

Thiel, H. (1967). *Economics and information theory*. Amsterdam: North-Holland.

Troncale, L. R. (1985). On the possibility of empirical refinement of general systems isomorphies. *Proceedings of the Society for General Systems Research, 1*, 7–13.

Turner, M. G., Costanza, R., & Sklar, F. H. (1989). Methods to compare spatial patterns for landscape modeling and analysis. *Ecological Modelling., 48*, 1–18.

Vitousek, P., Ehrlich, P. R., Ehrlich, A. H., & Matson, P. A. (1986). Human appropriation of the products of photosynthesis. *BioScience, 36*, 368–373.

Weisbord, M. (Ed.). (1992). *Discovering common ground*. San Francisco: Berrett-Koehler, 442 pp.

Weisbord, M., & Janoff, S. (1995). *Future search: An action guide to finding common ground in organizations and communities*. San Francisco: Berrett-Koehler.

Wheeler, K., & Lewis, L. (1997). School-community links for environmental health: Case studies from GREEN. *Health Education Research, 12*(4), 469–472.

Williams, N. (1997). Biologists cut reductionist approach down to size. *Science, 277*, 476–477.

Wilson, E. O. (1998). *Consilience: The unity of knowledge*. New York: Knopf, 332 pp.

Chapter 2
Millennium Alliance for Humanity and the Biosphere (MAHB): Integrating Social Science and the Humanities into Solving Sustainability Challenges

Ilan Kelman, Eugene A. Rosa, Tom R. Burns, Paul Ehrlich, Joan M. Diamond, Nora Machado, Donald Kennedy, and Lennart Olsson

2.1 Introduction

2.1.1 Dealing with Scientific Silos and Uncertainties

Comprehensive assessments have shown the wide variety of severe environmental problems facing and caused by humanity (e.g. Ehrlich and Ehrlich 2013; IPCC (Intergovernmental Panel on Climate Change) 2007; MEA (Millennium Ecosystem Assessment) 2005; Mitchell et al. 2006). These problems result largely from the activities of a human population whose consumptive patterns have already exceeded the long-term capacity of the Earth to support that population (Rees 2006, 2013).

In memory of Gene Rosa who passed away in February 2013.

I. Kelman (✉)
Institute for Risk & Disaster Reduction (IRDR) and Institute for Global Health (IGH), University College London (UCL), Gower Street, London WC1E 6BT, UK
e-mail: ilan_kelman@hotmail.com

E.A. Rosa
510 East C Street, Moscow, ID 83843, USA

T.R. Burns
Department of Sociology, University of Uppsala, Box 624, 751 26 Uppsala, Sweden

P. Ehrlich
Department of Biological Sciences, Stanford University, Stanford, CA 94305-5020, USA

J.M. Diamond
Nautilus Institute for Security and Sustainability,
Shattuck Ave. #300, Berkeley, CA 94710, USA

N. Machado
Center for Research and Studies in Sociology, Lisbon University Institute,
Avenida Forças Armadas, 1649-026 Lisbon, Portugal

M.J. Manfredo et al. (eds.), *Understanding Society and Natural Resources*,
DOI 10.1007/978-94-017-8959-2_2, © The Author(s) 2014

The problems fit a general pattern of diminishing marginal returns (Klare 2012) that Tainter (1988) saw as indicative of the coming collapse of complex societies. Despite the physical science knowledge establishing the current, threatened state of the Earth, concerted action on this knowledge is lacking.

Irrespective of the lack of substantive, concerted action, many examples exist of improvements. One instance has been policies and pressures to reduce the use of leaded gasoline to cut the amount of lead contamination in our bodies. Thomas et al. (1999) conducted a meta-analysis of nineteen studies on blood lead levels across all six inhabited continents. Seventeen of the studies measure blood lead levels before and after major reductions in the use of leaded gasoline. The remaining two studies surveyed populations with limited exposure to gasoline. They conclude that reducing lead in gasoline reduces the amount of lead in people's bodies.

Another example of an environmental improvement relates to acid rain. Acid rain refers to emissions of sulfur and nitrogen compounds, such as from coal-fired electricity generation plants, reducing the pH of rain. When acidic rain falls, it harms ecosystems such as by reducing the pH of soils and lakes, among other effects. Legislation to limit sulfur and nitrogen emissions in places such as North America and Europe reduced acid rain, permitting the ecosystems to recover (e.g. Reis et al. 2012).

Nevertheless, at least two overarching sustainability challenges remain. First, new environmental problems have emerged. For instance, recent research on endocrine-disrupting chemicals has highlighted humanity's ignorance of both their direct effects on human and environmental health and the myriad of potential synergisms among these toxins (Vandenberg et al. 2012). Another poignant example concerns negative, unintended consequences of the otherwise major achievement of *The Montreal Protocol on Substances that Deplete the Ozone Layer* from 1987. This protocol phased out the production and use of a list of chemicals which, when vented into the atmosphere, depleted the stratospheric ozone layer. Many were also greenhouse gases. Ironically, the substitutes for the ozone-depleting chemicals are also significant greenhouse gases, although it is hard to determine which chemicals are worse because complete life cycle analyses are needed (Velders et al. 2009). A "solution" to one environmental problem can cause or exacerbate other ones.

The second overarching sustainability challenge is that major differences in environmental conditions are evident based on location. For example, the UK has significantly reduced urban air pollution leading to an improvement in human health (Seaton et al. 1995) in contrast to Beijing where air pollution and associated human health impacts are staggering (Zhang et al. 2007). Acid rain also continues to be a major problem in China (Zhang et al. 2012), compared to the improvements in Europe and the USA mentioned above. Similarly, forestry regulation for multiple

D. Kennedy
The Center for Environmental Science and Policy (CESP), Stanford University,
Encina Hall E401, Stanford, CA 94305, USA

L. Olsson
Lund University Centre for Sustainability Studies (LUCSUS), Lund University,
P.O. Box 170, SE-221 00 Lund, Sweden

uses including logging is detailed and is enforced in Oregon leading to intensive management of forestry ecosystems (Boyle et al. 1997), compared to rampant unregulated and highly destructive deforestation in Papua New Guinea (Bryan et al. 2010).

These examples illustrate a remarkable contrast. The threats to humanity's future are often clear from the scientific evidence. Environment and sustainability problems can be solved and have been solved in some locations. Other locations do not apply the available knowledge for action while some new problems continue to emerge. Overall, the conclusion is that society has been unable or unwilling to take comprehensive steps to address the well-documented and continuing environmental and sustainability challenges, including with respect to resource management.

One difficulty in making sense of the scientific evidence and applying it for concerted action is the large degree of disciplinary silos. Plenty is published on, for instance, factors influencing pollutant transport to the Arctic (e.g. Downie and Fenge 2003; Eckhardt et al. 2003), but the work has varying levels of engagement with different disciplines and varying levels of resultant action from the knowledge. Sometimes, publications provide only a physical or chemical description without connection to any form of social science or policy. That is not inherently detrimental, since the physical science is a needed input and deserves publication in its own right. Nevertheless, much more than physical science is needed to understand society's interaction with resources and the environment—and how to inspire and formulate action addressing the problems identified.

Often, caught in their disciplinary silos, physical scientists will aim for full and comprehensive knowledge of a problem before being willing to recommend any form of action. Social science indicates that is not necessary, since techniques for decision-making under uncertainty exist alongside approaches for selecting action pathways which are likely to be beneficial over the long-term irrespective of the uncertainties and irrespective of what is not known. In fact, many positive examples exist of tackling sustainability problems without full physical science knowledge. These examples emerge from recent history, such as cleaning up Lake Erie and *The Montreal Protocol on Substances that Deplete the Ozone Layer* mentioned above.

Current initiatives exist as well. As a prominent example, little scientific doubt exists regarding observations about contemporary climate change and the human influences on it (IPCC 2007). Much work remains to be completed regarding, amongst other physical science challenges, feedback mechanisms from clouds (Dessler 2010) and the impact of climate change on tropical cyclones (Knutson et al. 2010). Supporting such physical science research would not only better understand the ultimate consequences of climate change, but would also highlight the importance of supporting curiosity-driven research with its unknown, and often spectacular, gains for humanity. While that research is ongoing, many communities are nonetheless taking action on their own, based on what is known, irrespective of the uncertainties and any knowledge limits.

Despite, or perhaps because of, any scientific uncertainties regarding climate change, Transition Towns (Barry and Quilley 2009) and relocalization movements (Kelman 2008) aim to transform entire cities toward pathways that are sustainable, irrespective of the climate pathway which emerges. Sector-specific approaches include "guerilla gardening" to use open space for food (Reynolds 2008) and community teams to reduce disaster vulnerability and to improve disaster response (Flint and Brennan 2006). These initiatives accept the physical science description of the problems, including the uncertainties and unknowns. They nevertheless aim to act on the basis of social sciences and humanities knowledge that exists, in order to help society to effect change irrespective of which pathway the climate pursues.

2.1.2 Solving These Challenges

Even with these polycentric examples, a significant need remains for a larger effort to fuse knowledge about physical and social systems into blueprints for action that are acceptable. Civilization needs to be "rescaled" to stabilize the population while reducing the average individual impact on the planet (Ehrlich et al. 2012). This means focusing on behavior and the reasons for behavior to understand better why we tend to ignore the high, destructive, and known human impact on the planet (Ehrlich et al. 2012). Humanity can no longer avoid dramatic change to society or the environment at global scales, but can potentially do a much better job of managing such change. Social science and humanities skills, interests, knowledge, and wisdom need to be mobilized and integrated into solving sustainability challenges, taking into consideration human behavior and values.

That does not need to come at the expense of tackling the many remaining fascinating scientific problems across the physical and social sciences. That does mean joining physical sciences, social sciences, humanities, and other fields to embrace as much knowledge as possible in order to break down the silos.

As one contribution towards that goal, this chapter highlights the importance of understanding and influencing human behavior: actions of individual and collective actors. The focus on human behavior, its causes, and mechanisms for influencing it is examined in the context of integrating physical sciences, social sciences, and the humanities to ensure that all available scientific knowledge contributes to action for sustainability.

To contribute to identifying the current status of integration, one initiative for doing so is presented: the Millennium Alliance for Humanity and the Biosphere (MAHB, pronounced "mob"). The next section defines and describes MAHB, including research and application. Then, a research agenda regarding resource management for and from MAHB is offered to pursue future opportunities for social science and humanities integration with physical sciences and policy-making. The key is not to await full knowledge and limited uncertainty before acting on any sustainability challenge. Instead, it is about using multiple disciplines in science to monitor and evaluate any measure implemented as an ongoing process, to ensure that actions do not exacerbate the existing problem or cause new problems.

2.2 Millennium Alliance for Humanity and the Biosphere (MAHB)

2.2.1 MAHB's Mission and Structure

Extensive literature (e.g. Brown et al. 1987; Gatto 1995; Santillo 2007) compiles and critiques definitions of "sustainability" and "sustainable development". While recognizing the importance of definitional discussions, MAHB adopts a comparatively generic and succinct definition. Paraphrasing the Oxford English Dictionary, "sustainability" is societal processes (e.g. livelihoods and governance) that are maintained and continued without the long-term depletion of human or natural resources. Based on this definition, MAHB's mission is to foster, fuel, and inspire global conversations and actions to shift human cultures and institutions toward sustainable practices, through dealing with the drivers of environmental degradation, yielding an equitable and satisfying future.

These conversations and actions are conducted through three connected activities on human behavior for sustainability (Ehrlich and Kennedy 2005; Rosa et al. 2011):

1. Knowledge generation, i.e. producing new science.

MAHB facilitates and supports research which integrates physical sciences, technological knowledge, social sciences, and the humanities to better understand human sustainability-related behavior. One example of ongoing work is Ehrlich and Ehrlich's (2012) analysis going beyond the standard mantra that perpetual economic growth is the antithesis of sustainability in order to demonstrate how it is "biophysically impossible" (pp. 558–559; see also Bartlett 2004). For a sector-based approach, food is a theme with Jerneck and Olsson (2014) seeking to understand poverty-agroforestry connections in Kenya so that the poorest people could have better opportunities to improve their situation without harming their land's sustainability.

As another example, in September 2012, MAHB opened the Institute of Foresight Intelligence at the Center for the Advanced Study in the Behavioral Sciences at Stanford University, California. The ethos is that, like "emotional intelligence", a set of human characteristics exists producing "future smart" individual and institutional actions. Future smart actions are concerned with the gap between our understanding of the threats to humanity and effective action. Why does society know so much about what is coming in the future, yet fails to act in ways that will result in a more equitable and sustainable future for all?

As such, the knowledge generated through MAHB is both theoretical and empirical, as well as connecting the two. Frameworks are being developed for determining how and why human behavior does and does not aim for sustainability, but then those frameworks are focused for on-the-ground analysis in specific locations and specific sectors. Other, specific practical studies which are ongoing to test and refine the theory include small island sustainability and energy use for transportation.

2. Knowledge dissemination in scientific and popular science venues.

Generating new knowledge in the form of scientific publications is important, but MAHB's approach to knowledge supply does not stop with academic publishing. Videos are part of the outreach effort, such Ehrlich's efforts for academic audiences through the Jack Beales Lecture on the Global Environment (http://www.youtube.com/watch?v=YHc7-275h0Y) and videos aimed at more general audiences such as radio interviews (http://www.youtube.com/watch?v=XGoG3fD7_GQ) and clips on climate change (http://www.youtube.com/watch?v=HE4xsgz5uew).

Speaking about the potential collapse of global civilization and what could be done to avert it (Ehrlich and Ehrlich 2013) naturally draws media attention which helps to engage those beyond the scientific world, including non-Anglophone audiences (e.g. Foucart 2013). MAHB also runs an online library (http://mahb.stanford.edu/library/mahb-library) for members to share relevant material in any media. That covers scholarly work alongside children's books, popular media, and public lecture notes. The criterion for selecting library material is fact-based presentation of the sustainability challenges emphasizing solutions related to human behavior.

Using multimedia approaches does not preclude face-to-face contact. Several MAHB workshops have been organized, in locations including Stanford University, Gothenburg and Lund in Sweden, and Lisbon, focusing on fostering collaboration among social and natural scientists as well as humanists while engaging with concerned citizens, including those with policy- and decision-making power. Topics have including environmental modeling, governance, and risk analysis; sustainability in island communities; new forms of governance, especially when government is an inhibitor to sustainability processes; and business pathways to sustainability. In larger academic settings, MAHB members presented MAHB's work at conferences including the Ecological Society of America, the World Congress of Sociology, and the American Sociological Association.

Sustainability Summits in Oslo have also been a core venue for MAHB since 2007. The Sustainability Summits are designed to accomplish three purposes. First, to support the social sciences and humanities as global players regarding environment and sustainability topics. Second, to provide a platform engaging the greatest diversity of people to establish dialogue and mutual challenges among different sectors that often do not communicate. Social scientists, natural scientists, and humanists interact with non-scientists, including leaders of business, non-governmental organizations, and government—plus concerned citizens who attend. Third, to bring university students from around the world to formulate questions and to propose conceptualizations and strategies that are alternatives to those presented by the researchers and leaders at the summits. These "young challengers" collaborate to prepare their questions, arguments, and proposals, with the aim of positioning themselves as the new generation of leaders and researchers who will achieve a sustainable future.

3. Knowledge brokering, i.e. engaging non-scientists in sustainability-related action based on science.

Knowledge dissemination cannot just be one way, from the ostensible "expert" to the masses of the public. Instead, MAHB further serves as an intermediary matching up those seeking sustainability knowledge with those who have ideas and actions to offer. The key is bringing together scientists and non-scientists to provide desired science to those without a background to or in science, as well as to indicate to the scientists the form of knowledge which is requested in order to act.

As an example linked to the knowledge dissemination workshops mentioned above, in January 2013, MAHB hosted a meeting at Stanford University titled "Can Foresight Intelligence Prevent the Collapse of Civilization?" A diverse group met for a 2-day conversation on the psychological, economic, historical, sociological, and natural science dimensions of that question. The aim was to inspire discussion on all sectors acting against such a collapse.

Another example of MAHB's engagement beyond the scientific community is the work of Bob Horn from Stanford University who collaborated with the World Business Council for Sustainable Development to assess the business vision for a sustainable world in 2050 (WBCSD 2010). A major feature of this project is assessments of where we are now, with a significant component of assessing human behavior—of businesses, governments, non-governmental organizations, communities, and everyday citizens—with respect to sustainability. Strategists from more than two dozen companies went through an 18-month process of setting 350 milestones and 70 metrics for achieving a sustainable planet. They distilled these lists into 40 "must-haves" that would be essential to achieve the sustainability vision, indicating how each of the next four decades needs to look like to reach the 2050 goal. For example, for materials, during the 2010s, new legislation is needed to reduce dependence on landfills and to encourage reduction, reuse, and recycling. For agriculture in the 2030s, productivity in Brazil will need to be double the current levels while in Africa, it will need to have increased five-fold.

A major goal of MAHB is reaching out to its members to help those involved in social action to have access to understandable information for their work on sustainability. MAHB membership includes a substantial list of "concerned citizens", including those from business, religion, non-governmental organizations, youth groups, and home-makers of many socio-economic statuses. The key is to go beyond knowledge generation and dissemination towards knowledge-based action. MAHB's network and media offer opportunities for members to share ideas and experiences, with discussion and exploration of ideas being an important component but also ensuring that action results. To facilitate this, MAHB is developing a set of measurable impact goals which focus on working with scholars and concerned citizens to include the primary drivers of environmental degradation—namely inequality, population, and over-consumption by the wealthy—in their literature, public outreach, and activism.

To enact these activities of knowledge production, dissemination, and brokering, MAHB's basic structure is an informal, international network of social scientists, physical scientists, humanists, academics from professions, and other engaged scholars, alongside members of business, political, and civic communities. The openness ensures that anyone who joins the mission can do so on their own terms and contribute in the way in which they feel most comfortable. For that, MAHB uses various media: a website with blogging, a facebook group generating debates, seminars, and workshops—all with academic material as well as popular science content. As such, MAHB media and venues serve as meeting and interaction points between scientists and non-scientists.

MAHB members can also create their own meeting and interaction points, to pursue MAHB's mission in their own location, by setting up a "node". The term "node" implies a connection point within a network or a vertex where several lines or vectors intersect in a large graph. MAHB's nodes are semi-autonomous groups to bring people together locally for pursuing the overarching MAHB goal. They draw on common MAHB information sources and goals to pursue actions locally pertinent to each node and each node's location.

MAHB nodes and partners exist across the globe. The first node was set up from 2007–2008 at Stanford University, California. The Stanford node focuses on a program of seminars and workshops designed to draw in researchers from across Stanford's academic strengths, such as energy engineering, climate change science, and environmental sociology.

2.2.2   MAHB's Research Approach

To understand pathways towards sustainability, MAHB specifically adopts an approach of use-inspired/problem-driven research that does not rely on a single discipline or single set of disciplines (e.g. see Clark 2007; NRC 1999). The research is use-inspired, because it aims at a practical application where policy-makers and decision-makers need the science and wish to use it for their policy and decisions. The research is problem-driven in that a practical problem is identified and research is used for tackling that problem, irrespective of the academic origins of the research approaches selected.

An example from MAHB is Hilary Schaffer Boudet's post-doctoral project at Stanford University. The U.S. Department of Energy is interested in how households decide to reduce their household energy use and so they funded a project to contribute toward solving this problem. Boudet and her team developed two curricula for children, based on the tenets of social cognitive theory (Bandura 1986), to teach the advantages and implementation of sustainable energy behavior. The curricula have been tested via a randomized controlled trial with 30 California Girl Scouts to determine their effectiveness in changing behavior. Thus far, results look promising.

Meanwhile, following on from WCED (1987), Burns and Witoszek (2012) outline a humanistic agenda for integrating humanist knowledge into global sustainability research. They provide a baseline for understanding the institutional and cultural barriers to accomplishing more sustainable processes within society. That is the problem driving the research. They also go further, suggesting several steps and strategies which can help to bring humanities concepts into, for instance, resource management in order to improve the economic, education, governance, and culture systems which favor unsustainable approaches. This work demonstrates that large-scale societal transformations are one way of effecting behavioral change—as are less dramatic approaches such as viewing society as a learning system where a multitude of small actions can add up to a major difference (Burns 2012).

Where a specific discipline can contribute to such research, its theories, literature, and methodologies are applied—and mixed with other disciplines to build on each discipline's strengths while shoring up any limitations. This research is not just for the pursuit of new knowledge, but is also about catalyzing and creating appropriate action based on the sound scientific knowledge produced. The focus of MAHB is not to displace other disciplines with social sciences and humanities inputs, nor is it to make social science and humanities inputs dominate. Instead, it is to ensure that all disciplinary voices, as well as inter-disciplinary voices, are heard and that they work together with mutual respect.

MAHB creates the space whereby such interaction can occur. By signing up to MAHB's mission, members accept the need for working with other disciplines. By attending a MAHB workshop, members accept that different disciplinary perspectives and approaches need to be respected, while also pushing themselves to think beyond specific disciplines. For example, at a MAHB workshop in Sweden in 2010 on risk, discussions ranged across different interpretations and understandings of risk in order to compare various disciplinary perspectives without becoming entrapped by one. One aim was to compile methods for influencing people's risk-related behavior in order to inform risk reduction policy and practice.

The space created by MAHB for dialogue amongst all disciplines is furthermore about encouraging scientists to interact with policy and decision-makers. That means that policy- and decision-makers understand more about the scientific process. Meanwhile, scientists are encouraged to work with policy and decision-makers to produce science that can be used. Ehrlich, for instance, started his policy-influencing career with *The Population Bomb* (Ehrlich and Ehrlich 1968) and continues to discuss the population-resource nexus with policy makers, highlighting that the sustainability challenges, according to him, are population, overconsumption, and inequality.

Ensuring this two-way exchange has various precedents, such as 'people's science' (Wisner et al. 1977) and 'useable science' (Glantz et al. 1990). Too often, science is seen as a linear process whereby knowledge is produced and then it might or might not be sent to policy- and decision-makers in a form which the policy and decision-makers might or might not be able to use. Among many others, Martin (1979) undercuts the myth of the objectivity of physical science results, using pollution and resource examples. One consequence is that environmental science cannot be assumed to stand alone from its policy and decision arena. MAHB therefore brings together the scientific and application arenas by focusing on a problem which parties wish to solve, recognizing that different skills and knowledge bases are needed to solve the problem and to use the results. That improves over many past endeavors, supporting better social science integration, because policy- and decision-makers are involved from the beginning—usually helping to define the problem to be solved—and by avoiding one discipline dominating others.

An example is Bob Horn's work with the World Business Council for Sustainable Development described earlier (WBCSD 2010). Because the business strategists were involved from the beginning and helped to define the problem to be tackled along with the tasks, they had an incentive to complete the work fully and then to consider how to apply the results for themselves. An added advantage is that the policy- and decision-makers involved gain an indication of the intricacies and uncertainties of scientific investigation, while educating scientists about the needs of the policy- and decision-makers.

The key area highlighted by MAHB for social science integration in the context of resource and sustainability challenges is showing how social science and humanities research draws attention to behavioral influences other than economic and technological considerations and framings. The latter often dominate discussions and assumptions regarding sustainability-related decisions, leading to the state of the world witnessed today. That does not denigrate the importance of material and economic interests, especially since they often operate in tandem with cultural and institutional factors. Indeed, the effectiveness of many economic incentives and technical innovations first requires major behavioral changes.

But moving beyond purely economic and technological considerations means recognizing that social sciences and humanities have much more to offer than understanding human behavior and perceptions to increase the uptake of advertised products (e.g. Mela et al. 1997) or indicating how people respond to economic incentives for risk reducing behavior (Kane et al. 2004). MAHB's framing and research approach treats the cause of the lack of sustainability (human behavior) rather than the symptoms (the physical indicators of resource depletion and environmental contamination). This does not denigrate or eliminate the past work or other framings. It builds on them, embraces them, and uses them as a springboard to understand more about the underlying drivers of poor sustainability-related behavior.

One example of needed behavioral change is travelling less in order to save energy. For example, using e-based (e.g. skype) meetings, learning, and conferences tends to save money, is more environmentally friendly than travelling, and is becoming increasingly easier due to technological developments. It does, though, require users to accept that forum of interaction rather than the expectation of more personal face-to-face approaches.

Much of that acceptance or rejection is cultural and people have different levels of comfort for "Personal connections in the digital age" (the title of Baym 2010). Shea (2005) points out that informing and dealing with climate change in the Pacific islands requires "Establishing and sustaining 'eyeball-to-eyeball' contact" (p. 4). That is notwithstanding the PEACESAT operation which, for over three decades while based in Honolulu, has used remote education through video and then the internet for training and education on development and sustainability topics, including climate change adaptation and resource management. No studies have yet examined the elements of PEACESAT which build up long-distance trust and credibility, compared to the cultural desire for the "eyeball-to-eyeball" contact. Understanding these dimensions of human behavior with respect to sustainability would be a research and application project directly in line with MAHB's aims and approach.

2.3   A Research Agenda for and from MAHB

Ehrlich and Kennedy (2005, p. 563) earlier defined a five-point research agenda for MAHB to highlight social sciences and humanities integrating with physical sciences:

> (i) what social scientists and others know about mechanisms of cultural evolution and how changes in direction might be steered democratically; (ii) how scarce and unevenly distributed non-renewable resources are used and some of the ethical connections between distribution, economic opportunity, and access; (iii) ethical issues related to the world trade system; (iv) conflicts between individual reproductive desires and environmental goals; and (v) economic, racial, and gender inequity as contributors to environmental deterioration.

This section specifies expanded research questions for MAHB within re-configured categories, as future opportunities for social science integration for resource management and sustainability.

2.3.1   Socio-cultural Change for Sustainability

Behavioral factors for socio-cultural change are frequently missed from technical perspectives of meeting sustainability challenges. In particular, for problems such as resource management, technical solutions are often imposed without considering how the problem might be overcome through behavior or how the solution itself might unintentionally alter behavior (e.g. Wilde 1994). By integrating social science and the humanities into physical sciences and policy processes, the risk of unintended consequences is diminished and policy making is likely to be more effective.

For example, in speaking with people owning hybrid cars, the authors have noticed a tendency for the owners to assume that driving is not a problem because their car is a hybrid. In fact, they often drive more than before. A useful solution to reduce fossil fuel consumption exists, through hybrid cars, leading to increased driving which counteracts some of the gain while increasing congestion and the need for road maintenance. Further investigation of two key classes of behavioral factors in conjunction with the technical solutions could assist in overcoming such challenges.

The first class is socio-cultural mechanisms of re-framing, re-definition, and other cognitive shifts so that problems are seen from new perspectives and new solutions are envisioned. That includes developing new narratives and discourses which play a major role in many cognitive shifts. One classic baseline is 'paradigm shifts' (Kuhn 1962), a concept which has been critiqued (Toulmin 1972) with the debate raging ever since, but nonetheless applied to public policy paradigm shifts (e.g. Carson et al. 2009). As is usual, reality seems to display both sudden and evolutionary changes in ideas, thoughts, and actions. Much more work is needed to understand the traits of changes at different time scales and how the time scale of behavioral change could be influenced. For example, some attribute a sudden change in U.S. forest management towards wildfire suppression as a result of the movie *Bambi* (Nash 1985). That compares to a later, much more gradual shift towards different regimes of managed burns (North et al. 2012).

The second class of key behavioral factors is understanding the main players who influence behavioral change with respect to policy and institutional shifts. The categories players which are particularly underrepresented in studies are:

(a) Social movements, because they raise awareness and play critical roles in cognitive changes and the development of new identities (Carson et al. 2009). Examples are "Corporate (Social) Responsibility" and "Green Citizens".

(b) Institutions exercising social power which may facilitate or constrain sustainability-related behavior. Advertising plays a key role. An example is airlines and car companies using environmental imagery and identities to sell their products. Another example is brandjacking, such as an environmental organization hijacking a corporate brand as epitomized by Greenpeace mocking Shell's "Arctic Ready" campaign (http://arcticready.com).

(c) Specific champions or icons, individual and organizational, in promoting a new sustainability ethos and new sustainability practices. Performing artists and sports stars often play key roles. United Nations agencies, for instance, use celebrities as Goodwill Ambassadors and Special Envoys. Midttun (2013) highlights the role of "cultural educators and protagonists" in developing a sustainability ethos and sustainability practices.

Within socio-cultural change for sustainability, several MAHB members have embarked on a study of island communities. Many island communities seek socio-cultural change because they are now highly vulnerable to the forms of social and environmental disasters which will be expected to affect most of humanity in the future, unless sustainable pathways are chosen. This research focuses on innovative responses, particularly to climate change challenges, from technical, economic, governance, and cultural perspectives. For example, innovation in energy technologies and policies, which often need to be self-sufficient for isolated island communities, are described by Baumgartner and Burns (1984) and Woodward et al. (1994). A related research program involving MAHB partners identifies ways in which human agents (individuals and collectives) bring about technical, economic, governance, and cultural innovation in response to climate change through case studies of cultures and institutions in Scandinavia, China, and Ghana (Midttun 2009).

2.3.2   Population and Sustainability

Malthusian and neo-Malthusian debates focusing on population numbers permeate sustainability research, policy, and practice. Few claim that population numbers are the only factor causing resource problems, just as few claim that population numbers are irrelevant for analyzing and solving resource problems. Reality is persistently complicated, as shown by relationships between population size and carbon dioxide emissions (Jorgenson and Clark 2013) and between population density and agro-diversity (Conelly and Chaiken 2000).

As such, MAHB's research agenda for population and sustainability embraces parameters such as population numbers, population densities, consumption rates, waste rates, affluence, and technology. Analyzing these various factors and the circumstances under which they contribute more to a specific resource problem, or less, is MAHB's research agenda.

For example, a small island such as Malé, the capital of the Maldives, is 100 % urbanized. Building further high-rises is not straightforward because the island's land, effectively at sea-level, has the potential of sinking with such added weight. There is an upper limit to how many people can live on the island without land reclamation. Conversely, the suburbs of Los Angeles are a clear example of urban sprawl in which long, wide streets and large plots for big houses epitomize high resource consumption per capita. What are the behavioral factors drawing different classes of people to these different urban environments? How could behavior be influenced to reduce population density in Malé and to reduce resource consumption in Los Angeles? Both locations display a combination of technical and social challenges. Neither can be solved without the social sciences and the humanities and neither can be solved with only the social sciences and the humanities. Instead, a combination of disciplines working in tandem to solve the place-specific problem is needed, exactly in line with MAHB's ethos.

Another layer can be added to these questions: How can researchers, policy makers, and practitioners focus on the fundamental population-related factors based on science? When population numbers are raised as a specter, the debate often leads to accusations of advocating reproductive control, perhaps through forced sterilization or forced abortion. Such unethical measures are supported by only an extremist minority, yet they often dominate the debate. That is the case even though social science provides details on how raising people's education and affluence levels, especially in terms of giving women reproductive-related education and choices, tends to lead to smaller families, higher infant survival, and better educated children (e.g. Martin 1995). Solving the challenge within MAHB's work is two-fold: Ensuring that scientific arguments dominate debates and keeping the discussion on the fundamental factors rather than having to defend against extremist arguments.

2.3.3 Environmental Governance for Sustainability

MAHB researchers have been contributing to bringing social sciences and the humanities into environmental governance regimes—including the governance of risk and using democratic change to achieve sustainability processes. Midttun (2010) edited a special issue of *Corporate Governance*, called "Rethinking Governance for Sustainability". Carson et al. (2009) investigated public policy paradigm shifts in the EU's management of asbestos, chemicals, climate change, and gas markets. Other resource-related studies from MAHB on environmental governance for sustainability include Baltic fisheries (Burns and Stohr 2011) and tropical forests (Nikoloyuk et al. 2010).

This governance research has been identifying and analyzing a variety of mechanisms of "soft means" for advancing public policy. "Soft means" stress non-economic and non-coercive incentives and pressures. Of particular interest for further investigation are how issues are framed politically (such as defining a policy issue as "European"); how data are selected, collected, and distributed; standardization of measurements and classification schemes; monitoring of opinion and behavior; and support for forming informed opinions and mutual learning processes. This wide variety of means shows that, even though managing resources such as forests and fish might have traditionally been seen as pursuits in ecology or biology, integrating social science (e.g. governance, individual attitudes, and education) is needed to achieve effective public policy and action.

2.3.4   Inequity and Sustainability

When determining how to use and misuse resources, many discussions within sustainability refer to resource distribution, access, and choices. People's individual and collective behavior is often attributed to political ideology, whether it be the approach epitomized by the legend (and likely reality) of Robin Hood, through stealing from the rich in order to give to the poor, or through modern-day unchecked capitalism, often interpreted as being as much short-term profit as feasible. Yet empirical evidence suggests that the links between values or ideology and behavior are rarely linear or straightforward (Osbaldiston and Schott 2012; Schultz et al. 2005).

MAHB aims to contribute to research on this topic by trying to understand more about how and why inequalities are created and perpetuated for resource distribution, access, and choices. "Selfishness", "greed", "ignorance", or "egoism" are answers which are too simplistic in themselves, because these characteristics, amongst many others, tend to be present to different degrees.

For instance, in terms of ignorance, commendable efforts to tackle deforestation in less affluent countries, such as by celebrities including Harrison Ford (http://www.youtube.com/watch?v=r87wJ1QmyYw), do not necessarily acknowledge that the deforestation is driven primarily by large-scale agriculture for markets in more affluent countries (Butler and Laurance 2008). That is, affluent consumers desire products which are cheap to produce through rain forest destruction. The affluent consumers then blame those working on the land which used to be rain forest. Those with the power and resources to change are blaming those without the power and resources to change for sustainability problems.

How could such inequalities of power and perception be overcome? Does the disparity between the thoughts and actions of the affluent consumers emerge from ignorance, greed, or other characteristics? Could consumer behavior be changed to reduce inequalities even if product costs increase (although life-cycle costs might decrease due to less environmental destruction)? These are questions on MAHB's research agenda regarding inequality and sustainability.

This topic connects back to topic (1) in terms of socio-cultural mechanisms of re-framing, re-definition, and other cognitive shifts. Ethical and value systems play an important role, which influence and are influenced by political ideology. That requires further work into how ethical systems such as "do no harm", "risk/benefit analysis", and "utilitarianism" view inequalities and overcoming inequalities both theoretically and operationally. Some also differentiate between equity, equality, and egalitarianism (e.g. Espinoza 2007). None of that addresses the fundamental challenge with respect to inequalities and sustainability: understanding and overcoming the disconnect between beliefs and actions so that certain sectors or institutions do not hoard or dominate control of available (and always constrained) resources (including information and knowledge).

In fact, one common thread through the above themes is that simple conceptual models of influencing behavior, and of understanding the root impetus of action, rarely manifest in reality, even when they appear in the literature. The reason is that these simple models are usually for highly specific cases in highly specific contexts, often with many variables controlled for the study which could not be controlled in reality. For instance, one model of behavioral change applies ABC referring to first influence Attitude which affects Behavior leading to the Change sought (Kumar 1996). Empirical evidence is not always supportive of the ABC sequence for sustainability behavior. Ample studies indicate that, even when people have an appropriate attitude, such as wishing to be environmentally friendly, and even when they identify the appropriate behavior, such as flying less to save fossil fuels, they do not always change in order to implement what they know (McKercher et al. 2010). Environmental scientists are a poignant example (Stohl 2008).

Whether with respect to socio-cultural change, ethics, population, or inequity, the fundamental objective within MAHB's research is to determine the underlying motivations to sustainability decision-making leading to successful action, rather than just attitudes and behavioral awareness. Part of that is drilling deeper than the simpler models which often do not work in practice, such as ABC. In particular, differentiating and conceptualizing values, attitudes, knowledge, and behavior is often poorly effected in studies. Overall, there is a dearth of research in determining how and why information and knowledge are and are not converted into behavioral changes and action.

The current status of integrating social science into understanding sustainability behavior has not yet fully described the links amongst values, attitudes, and knowledge—or how those lead to influencing behavior and action. MAHB, amongst other initiatives, contributes to engaging all science and other knowledge forms to build on and support ongoing work and to more fully engage everyone in addressing the challenges to the planet and humanity.

2.4 Concluding Reflections

1. MAHB is part of a global development which is forging links among researchers in the physical sciences, social sciences, and humanities—as well as with non-scientists.
2. MAHB stresses the necessity of behaviorally-focused approaches to achieving sustainability processes. One challenge is to identify and develop the kind of social science and humanities information, knowledge, and wisdom which could play a useful, even if not decisive, role in policy- and decision-making. Areas which social scientists have shown to be important include: (a) cognitive and framing concepts; (b) social networks; (c) social movements; (d) social power; (e) social change and evolution; and (f) methods and theoretical frameworks encompassing systems analysis, social ecology, human interaction and agency.
3. In spite of considerable progress in science, it seems that policy and strategy development for sustainable resource management is not informed enough by or through the social sciences and humanities. Social scientists and humanists need to learn from physical scientists who have become increasingly skillful in reformulating scientific knowledge into everyday language and communicating with concerned citizens who seek such knowledge.
4. The social sciences and humanities may identify policy openings and unseen opportunities as well as policy and institutional barriers.
5. All in all, the social sciences and the humanities have had rich and productive histories providing a substantial scholarly base upon which to draw for sustainable resource management. Integrating that knowledge means systematically applying it for encouraging behavior that will support sustainability processes.

MAHB is a unique initiative, establishing a permanent arena for dialogue and collaboration amongst all scientists, humanists, and non-scientists in the context of public policy engagement and outreach.

References

Bandura, A. (1986). *Social foundations of thought and action*. Englewood Cliffs: Prentice-Hall.
Barry, J., & Quilley, S. (2009). The transition to sustainability: Transition towns and sustainable communities. In L. Leonard & J. Barry (Eds.), *The transition to sustainable living and practice* (pp. 1–28). Bingley: Emerald.
Bartlett, A. A. (2004). *The essential exponential!* Lincoln: University of Nebraska.
Baumgarter, T., & Burns, R. R. (1984). *Transitions to alternative energy systems: Entrepreneurs, new technologies, and social change*. Boulder: Westview Press.

Baym, N. K. (2010). *Personal connections in the digital age*. Cambridge: Polity.

Boyle, J. R., Warila, J. E., Beschta, R. L., Reiter, M., Chambers, C. C., Gibson, W. P., Gregory, S. V., Grizzel, J., Hagar, J. C., Li, J. L., Mccomb, W. C., Parzybok, T. W., & Taylor, G. (1997). Cumulative effects of forestry practices: An example framework for evaluation from Oregon, U.S.A. *Biomass and Bioenergy, 13*(4–5), 223–245.

Brown, B. J., Hanson, M. E., Liverman, D. M., & Meredith, R. W. (1987). Global sustainability: Toward definition. *Environmental Management, 11*(6), 713–719.

Bryan, J., Shearman, P., Ash, J., & Kirkpatrick, J. B. (2010). Estimating rainforest biomass stocks and carbon loss from deforestation and degradation in Papua New Guinea 1972–2002: Best estimates, uncertainties and research needs. *Journal of Environmental Management, 91*(4), 995–1001.

Burns, T. R. (2012). The sustainability revolution: A societal paradigm shift? *Sustainability, 4*(6), 1118–1134.

Burns, T. R., & Stohr, C. (2011). Power, knowledge, and conflict in the shaping of commons governance. The case of EU Baltic fisheries. *International Journal of the Commons, 5*(2), 233–258.

Burns, T. R., & Witoszek, N. (2012). Sustainability: A humanistic agenda. *Journal of Human Ecology, 39*(2), 155–170.

Butler, R. A., & Laurance, W. F. (2008). New strategies for conserving tropical forests. *Trends in Ecology & Evolution, 23*(9), 469–472.

Carson, M., Burns, T. R., & Calvo, D. (2009). *Public policy paradigms: Theory and practice of paradigm shifts in the European Union*. Frankfurt/New York/Oxford: Peter Lang.

Clark, W. C. (2007). Sustainability science: A room of its own. *Proceedings of the National Academy of Sciences, 104*(6), 1737–1738.

Conelly, W. T., & Chaiken, M. S. (2000). Intensive farming, agro-diversity, and food security under conditions of extreme population pressure in Western Kenya. *Human Ecology, 28*(1), 19–51.

Dessler, A. E. (2010). A determination of the cloud feedback from climate variations over the past decade. *Science, 330*(6010), 1523–1527.

Downie, D. L., & Fenge, T. (Eds.). (2003). *Northern lights against POPs: Combating threats in the Arctic*. Montreal: McGill-Queen's University Press.

Eckhardt, S., Stohl, A., Beirle, S., Spichtinger, N., James, P., Forster, C., Junker, C., Wagner, T., Platt, U., & Jennings, S. G. (2003). The North Atlantic oscillation controls air pollution transport to the Arctic. *Atmospheric Chemistry and Physics, 3*(5), 1769–1778.

Ehrlich, P. R., & (uncredited) Ehrlich, A. (1968). *The population bomb*. New York: Sierra Club/Ballantine Books.

Ehrlich, P. R., & Ehrlich, A. H. (2012). Solving the human predicament. *International Journal of Environmental Studies, 69*(4), 557–565.

Ehrlich, P. R., & Ehrlich, A. H. (2013). Can a collapse of civilization be avoided? *Proceedings of the Royal Society B, 280*(1754), 20122501. doi:10.1098/rspb.2012.2501.

Ehrlich, P. R., & Kennedy, D. (2005). Millennium assessment of human behavior. *Science, 309*, 562–563.

Ehrlich, P. R., Kareiva, P. M., & Daily, G. C. (2012). Securing natural capital and expanding equity to rescale civilization. *Nature, 486*, 68–73.

Espinoza, O. (2007). Solving the equity–equality conceptual dilemma: A new model for analysis of the educational process. *Educational Research, 49*(4), 343–363.

Flint, C., & Brennan, M. (2006). Community emergency response teams: From disaster responders to community builders. *Rural Realities, 1*(3), 1–9.

Foucart, S. (2013, February 9). Notre civilization pourrait-elle s'effondrer? Personne ne veut y croire. *Le Monde*, Culture & Idées, p. 4.

Gatto, M. (1995). Sustainability: Is it a well defined concept? *Ecological Applications, 5*(4), 1181–1183.

Glantz, M. H., Price, M. F., & Krenz, M. E. (1990). *Report of the workshop "On assessing winners and losers in the context of global warming, St. Julians, Malta, 18–21 June 1990"*. Boulder: Environmental and Social Impacts Group, National Center for Atmospheric Research.

IPCC. (2007). *IPCC fourth assessment report*. Geneva: Intergovernmental Panel on Climate Change.

Jerneck, A., & Olsson, L. (2014). Food first! Theorising assets and actors in agroforestry: Risk evaders, opportunity seekers and 'the food imperative' in sub-Saharan Africa. *International Journal of Agricultural Sustainability, 12*(1), 1–22.

Jorgenson, A. K., & Clark, B. (2013). The relationship between national-level carbon dioxide emissions and population size: An assessment of regional and temporal variation, 1960–2005. *PLoS ONE, 8*(2), e57107. doi:10.1371/journal.pone.0057107.

Kane, R. L., Johnson, P. E., Town, R. J., & Butler, M. (2004). A structured review of the effect of economic incentives on consumers' preventive behavior. *American Journal of Preventive Medicine, 27*(4), 327–352.

Kelman, I. (2008). Relocalising disaster risk reduction for urban resilience. *Urban Design and Planning, 161*(DP4), 197–204.

Klare, M. T. (2012). *The race for what's left: The global scramble for the world's last resources*. New York: Metropolitan Books.

Knutson, T. R., McBride, J. L., Chan, J., Emanuel, K., Holland, G., Landsea, C., Held, I., Kossin, J. P., Srivastava, A. K., & Sugi, M. (2010). Tropical cyclones and climate change. *Nature Geoscience, 3*, 157–163.

Kuhn, T. (1962). *The structure of scientific revolutions*. Chicago: Chicago University Press.

Kumar, S. (1996). ABC of PRA: Attitude and behaviour change. *PLA Notes: Participation, Policy and Institutionalisation, 27*, 70–73.

Martin, B. (1979). *The bias of science*. Canberra: Society for Social Responsibility in Science.

Martin, T. C. (1995). Women's education and fertility: Results from 26 demographic and health surveys. *Studies in Family Planning, 26*(4), 187–202.

McKercher, B., Prideaux, B., Cheung, C., & Law, R. (2010). Achieving voluntary reductions in the carbon footprint of tourism and climate change. *Journal of Sustainable Tourism, 18*(3), 297–317.

MEA. (2005). *Millennium ecosystem assessment, ecosystems and human well-being*. Washington, DC: Island Press.

Mela, C. F., Gupta, S., & Lehmann, D. R. (1997). The long-term impact of promotion and advertising on consumer brand choice. *Journal of Marketing Research, 34*(2), 248–261.

Midttun, A. (Ed.). (2009). *Stakeholders on climate change. North and South perspectives* (Report No. 3). Oslo: CERES21—Creative Responses to Sustainability.

Midttun, A. (2010). Rethinking governance for sustainability. *Corporate Governance, 10*(1), 6–109.

Midttun, A. (2013). *The anatomy of green transition*. London: Palgrave Macmillan.

Mitchell, R. B., Clark, W. C., Cash, D. W., & Dickson, N. M. (2006). *Global environmental assessments: Information and influence*. Cambridge, MA: MIT Press.

Nash, R. (1985). Sorry, Bambi, but man must enter the forest: Perspectives on the old wilderness and the new". Section 7, "Banquet Address". In J. E. Lotan, B. M. Kilgore, W. C. Fischer, & R.W. Mutch (Technical coordinators), *Proceedings—Symposium and workshop on wilderness fire, 1983 November 15–18, Missoula MT* (pp. 264–268). General Technical Report INT-182 from U.S. Department of Agriculture, Forest Service, Intermountain Forest and Range Experiment Station.

Nikoloyuk, J., Burns, T. R., & de Man, R. (2010). The promise and limitations of partnered governance: The case of sustainable palm oil. *Corporate Governance, 10*(1), 59–72.

North, M., Collins, B. M., & Scott, S. (2012). Using fire to increase the scale, benefits, and future maintenance of fuels treatments. *Journal of Forestry, 110*(7), 392–401.

NRC. (1999). *Our common journey: A transition towards sustainability*. Washington, DC: National Academy Press.

Osbaldiston, R., & Schott, J. P. (2012). Environmental sustainability and behavioral science meta-analysis of pro-environmental behavior experiments. *Environment and Behavior, 44*(2), 257–299.

Rees, W. E. (2006). Ecological footprints and biocapacity: Essential elements in sustainability assessment. In J. Dewulf & H. V. Langenhove (Eds.), *Renewables-based technology: Sustainability assessment* (pp. 143–157). Chichester: John Wiley & Sons.

Rees, W. E. (2013). Ecological footprint, concept of. In S. Levin (Ed.), *Encyclopedia of biodiversity* (2nd ed.). San Diego: Academic.

Reis, S., Grennfelt, P., Klimont, Z., Amann, M., ApSimon, H., Hettelingh, J.-P., Holland, M., LeGall, A.-C., Maas, R., Posch, M., Spranger, T., Sutton, M. A., & Williams, M. (2012). From acid rain to climate change. *Science, 338*(6111), 1153–1154.

Reynolds, R. (2008). *On guerrilla gardening: A handbook for gardening without boundaries.* London: Bloomsbury.

Rosa, E. A., Kennedy, D., Ehrlich, P., Burns, T. R., Kelman, I., Midttun, A., & Witoszek, N. (2011). The millennium assessment of human behavior—5+ years later. *Mother Pelican: A Journal of Sustainable Human Development, 7*(8), 2.

Santillo, D. (2007). Reclaiming the definition of sustainability. *Environmental Science and Pollution Research, 14*(1), 60–66.

Schultz, P. W., Gouveia, V. V., Cameron, L. D., Tankha, G., Schmuck, P., & Franěk, M. (2005). Values and their relationship to environmental concern and conservation behavior. *Journal of Cross-Cultural Psychology, 36*(4), 457–475.

Seaton, A., MacNee, W., Donaldson, K., & Godden, D. (1995). Particulate air pollution and acute health effects. *The Lancet, 345*, 176–178.

Shea, E. (2005, January 9–13). *Living with a climate in transition: Pacific Islands experience.* Paper 14.3 at the 16th conference on climate variability and change, San Diego, CA, USA.

Stohl, A. (2008). The travel-related carbon dioxide emissions of atmospheric researchers. *Atmospheric Chemistry and Physics, 8*, 6499–6504.

Tainter, J. A. (1988). *The collapse of complex societies.* Cambridge: Cambridge University Press.

Thomas, V. M., Socolow, R. H., Fanelli, J. J., & Spir, T. G. (1999). Effects of reducing lead in gasoline: An analysis of the international experience. *Environmental Science and Technology, 33*(22), 3942–3948.

Toulmin, S. (1972). *Human understanding: The collective use and evolution of concepts.* Princeton: Princeton University Press.

Vandenberg, L. N., Colborn, T., Hayes, T. B., Heindel, J. J., Jacobs, D. R., Jr., Lee, D. H., Shioda, T., Soto, A. M., Vom Saal, F. S., Welshons, W. V., Zoeller, R. T., & Myers, J. P. (2012). Hormones and endocrine-disrupting chemicals: Low-dose effects and nonmonotonic dose responses. *Endocrine Reviews, 33*(3), 378–455.

Velders, G. J. M., Fahey, D. W., Daniel, J. S., McFarland, M., & Andersen, S. O. (2009). The large contribution of projected HFC emissions to future climate forcing. *Proceedings of the National Academy of Sciences, 106*(27), 10949–10954.

WBCSD. (2010). *Vision 2050: The new agenda for business.* Geneva: WBCSD.

WCED (World Commission on Environment and Development). (1987). *Our common future.* Oxford: Oxford University Press.

Wilde, G. J. S. (1994). *Target risk.* Toronto: PDE Publications.

Wisner, B., O'Keefe, P., & Westgate, K. (1977). Global systems and local disasters: the untapped power of people's science. *Disasters, 1*(1), 47–57.

Woodward, A., Ellig, J., & Burns, T. R. (1994). *Municipal entrepreneurship and energy policy: A five nation study of politics, innovation, and social change.* New York: Gordon and Breach.

Zhang, M., Song, Y., & Cai, X. (2007). A health-based assessment of particulate air pollution in urban areas of Beijing in 2000–2004. *Science of the total environment, 376*(1–3), 100–108.

Zhang, X., Jiang, H., Jin, J., Xu, X., & Zhang, Q. (2012). Analysis of acid rain patterns in northeastern China using a decision tree method. *Atmospheric Environment, 46*, 590–596.

Part II
Topics in Integration

Chapter 3
Science During Crisis: The Application of Interdisciplinary and Strategic Science During Major Environmental Crises

Gary E. Machlis and Kristin Ludwig

3.1 Introduction

On the French Caribbean island of Martinique in late April 1902, *La Commission Sur le Volcan* (Commission on the Volcano) met to decide a course of action. The island's Mt. Pelée was sending steam and smoke skyward, the smell of sulfur was in the air, and swarms of insects were moving down the mountain into neighboring cane fields. Frequent earthquakes and a thin layer of ash had set the population (particularly in the coastal city of St. Pierre) on edge and created a sense of crisis. The Commission included doctors, pharmacists, and science teachers, all appointed by the Governor. They discussed the potential of an eruption and what precautions, including evacuation, should be considered. The island was in the midst of general elections, complicating a response. After several meetings, the Commission made its decision, and announced "There is nothing in the activity of Pelée that warrants a departure from St. Pierre…the safety of St. Pierre [is] absolutely assured." Posters were placed throughout the town announcing the public's safety.

On May 8 Mt. Pelée erupted with an incandescent, high-velocity ash flow, associated hot gases, and dust – a pyroclastic flow of great destructive power. The cloud of hot ash and gases raced into St. Pierre at an estimated speed of 160 km/h (Fig. 3.1). Approximately 30,000 residents (including all members of the Commission) died within minutes, leaving only two survivors. One eyewitness described the scene:

> The whole side of the mountain seemed to gape open, and from the fissure belched a lurid whirlwind of fire, which wreathed itself into vast masses of flame as, with terrible speed, it

G.E. Machlis (✉)
School of Agricultural, Forest, and Environmental Sciences, Clemson University,
260 Lehotsky Hall, Clemson, SC 29634-0735, USA
e-mail: machlis@clemson.edu

K. Ludwig
Natural Hazards Mission Area, U.S. Geological Survey,
12201 Sunrise Valley Drive – MS905, Reston, VA 20192, USA

M.J. Manfredo et al. (eds.), *Understanding Society and Natural Resources*,
DOI 10.1007/978-94-017-8959-2_3, © The Author(s) 2014

Fig. 3.1 Photograph of Mt. Pelée May 7, 1902 (Photograph by Angelo Helprin, survivor. St. Pierre, Martinique, French West Indies. 1902 collection, Prints & Photographs Division, Library of Congress, LC-USZ62-47617. http://www.loc.gov/pictures/item/2006689820/)

descended on the doomed town. Before the true extent of the peril could be grasped, the fiery mass swept like a river over the town, and thrusting the very waters of the sea before it, set the ships ablaze. (Fermor 1950)

Environmental crises require decisions, and such fateful decisions require science. The distinctive and increasingly critical role of interdisciplinary science – including the physical, biological, and social sciences – during environmental crises is the topic of this chapter.

The structural processes of science have long been studied and debated (see for example Kuhn's *The Structure of Scientific Revolutions*, 1962, and commentary by Sarder 2000). However, the distinctive context of science during crisis events – and how best to conduct and deliver "crisis" science – has largely been left to historians (such as Richard Rhodes in *The Making of the Atomic Bomb*, 1986), scientists engaged in such work (Freudenburg and Gramling 2011; Machlis and McNutt 2011; Lubchenco et al. 2012), and critics focused on specialized or unusual cases (Taleb 2007). Crises vary in intensity, consequence, and scope – and range from events of war and security to health and public safety. They are often reflected in the strange and vivid metaphors surrounding crisis management: "black swans," "wicked problems," "acute events," and so forth (Rittel and Webber 1973; Taleb 2007; Brown et al. 2010).

Historical and contemporary experience suggests that science – including the physical, biological, and social sciences – plays an increasingly critical role in governmental and institutional responses to major environmental crises such as those caused by natural hazards or man-made disasters. Recent examples include major western US wildfires (2009), the Deepwater Horizon oil spill (2010), the Fukushima nuclear plant failures (2011), and Hurricane Sandy (2012).

Understanding the structural processes of science during environmental crises may have considerable value in developing best practices for the conduct and delivery of science during crisis. In addition, focusing on the potential role of social science during these events is critical to social science practitioners and the broader community of scientists, decision makers, and emergency responders who use social science to inform crisis response. There is also a substantial need to better define the roles of *strategic* and *tactical* science during crises. While tactical science focuses on immediate challenges and technical solutions, *strategic* science focuses on the longer-term issues of response and recovery, and considers longer chains of cascading consequences than is typical in tactical approaches (Machlis and McNutt 2010, 2011).

In this chapter, we explore the role and significance of science – including all disciplines and focusing attention on the social sciences – in responding to the needs of emergency response and recovery during major environmental crises. First, we examine the role of science during two recent major environmental crisis events – the Deepwater Horizon oil spill (2010) and Hurricane Sandy (2012). Second, we briefly review several specific examples of social science applied to environmental crisis events – Cyclone Sidr in Bangladesh (2007), the Puerto Aysen earthquake in Chile (2007), and Hurricane Katrina in the US (2005). Third, we identify several distinctive characteristics of strategic science during environmental crises. Finally, we describe a modest research agenda to advance the role of science during environmental crises.

3.2 Science During Crisis: Two Examples

3.2.1 2010: Deepwater Horizon Oil Spill

On April 20, 2010, the *Deepwater Horizon* drilling platform catastrophically exploded and later collapsed into the sea, killing 11 men and spilling over 4.9 million barrels of oil into the Gulf of Mexico, making it one of the worst man-made environmental disasters in US history (Mabus 2010; McNutt et al. 2012). Compared to other oil spills, Deepwater Horizon was unprecedented in its complexity and impact. At its peak, oil and tar balls contaminated the coastlines of all five Gulf states and led to the closure of 229,271 sq. km of federal waters to fishing (Mabus 2010). Response efforts included more than 47,000 personnel, 7,000 vessels, 120 aircraft, and the participation of scores of federal, state, and local agencies, universities, and non-governmental organizations (Mabus 2010).

In contrast to surface spills such as Exxon Valdez in Alaska (1989) or the Santa Barbara oil spill (1969), the Deepwater Horizon spill occurred at depth – crude oil flowed from a broken drill pipe approximately 1,500 m below the surface of the water. The extreme depth of the spill introduced new challenges in both engineering and environmental conditions that had to be overcome. Response crews needed ships with remotely operated vehicles equipped with sophisticated sensors, cameras, and robotic arms to navigate the wreckage and access the well. Engineers had to rapidly devise new capping devices to kill the well, which were thwarted by the formation of gas hydrates – crystals of methane ice that only form at depth – clogging the devices during several deployment attempts. Oil spilled into the Gulf continuously for nearly three consecutive months, polluting a three-dimensional area that extended vertically from the seafloor to the surface, and laterally across the Gulf, impacting the people, the environment, and the economy of the region.

Science played a vital role in stopping and responding to the spill. Because of the extreme complexity of the disaster, researchers and engineers from across academia, the federal government, and the private sector were called on to contribute their expertise in fields such as oceanography, geology, underwater engineering, physics, public health, and ecology (Lubchenco et al. 2012). Teams of scientists and the leaders of major federal science agencies including the Department of Energy, the US Geological Survey, and the National Oceanic and Atmospheric Administration (NOAA) were stationed at or near Incident Command centers established throughout the Gulf. Tactical science response efforts included geochemical "fingerprinting" of the oil, calculating the rate of flow from the broken pipe, and modeling the surface migration of oil using information on currents in the Gulf. The National Science Foundation awarded over 11 million dollars through its Rapid Response grants to research the spill.

Social science research was ongoing during the spill (April–September 2010), though it was fragmented, sometimes *ad hoc*, and largely peripheral to the engineering, toxicology, and ecological research that formed the core of the scientific response. While the Natural Resources Damage Assessment (NRDA) mandated the documentation of human health, social impacts, economic impacts, and cultural resource damage, this work often lagged behind other NRDA needs. Later, in a post-incident review of science conducted during the crisis, Lubchenco et al. (2012) called for a "greater emphasis on social science data collecting including adequate baselines, to understand costs to the region and the nation of oil spill disasters" in the future.

During the crisis, the unplanned and sporadic nature of on-the-ground social science led to specific topics receiving significant attention. An example is the research on the psychological impacts of the spill. Grattan et al. (2011) and Morris et al. (2013) used a community-based participatory model to perform standardized assessments of psychological distress, comparing populations in communities directly and indirectly impacted by the spill. They found no significant differences: residents in both communities displayed clinically significant depression and anxiety. Abramson et al. (2010) focused on the impact of the spill on children in the region, and found heightened mental health distress. Lee and Blanchard (2012) found, interestingly, that community attachment associated with higher levels of anxiety and fear, based on data collected in three Louisiana parishes during the spill.

During the spill, there were numerous calls for interdisciplinary approaches for dealing with the spill, its environmental and socioeconomic impacts, and the need to bolster resilience of affected communities (see for example Levy and Gopalakrishnan 2010). One significant response was scenario-building conducted by the Department of the Interior's (DOI) experimental Strategic Sciences Working Group (SSWG),

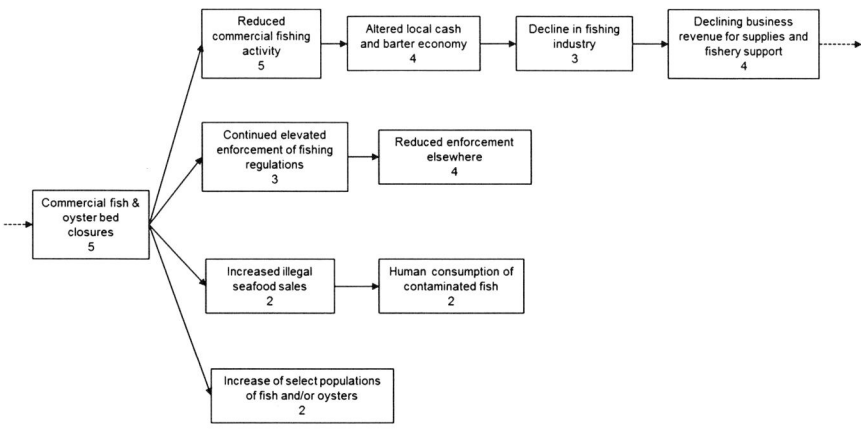

Fig. 3.2 A segment of one of the scenarios developed by the SSWG for the Deepwater Horizon oil spill. This segment shows the cascading effects of commercial fish and oyster bed closures (Department of the Interior 2012)

which analyzed the cascading consequences of the spill to inform decision-makers on near-term and long-term impacts (Machlis and McNutt 2010, 2011).

The SSWG was established quickly and included both federal and non-federal ecologists, social scientists, oceanographers, and other disciplinary experts. The SSWG worked extensively to create "chain of consequences" scenarios that included both biophysical and socioeconomic impacts (Department of the Interior 2010). Using the human ecosystem model (Machlis et al. 1997) as an organizing framework, and qualitatively assessing uncertainties, the SSWG created several scenarios and briefed DOI leadership on findings several times during the crisis. Figure 3.2 illustrates a small segment of one of the scenarios, focusing on commercial fishing and oyster bed closures. The numbers in the figure reflect the uncertainties associated with each consequence, with 5 being certain and lower numbers reflecting less certainty.

In September 2010, the spill officially ended when two relief wells enabled the well to be sealed. British Petroleum (BP), which had contracted the *Deepwater Horizon* platform, later committed $500 million in research funds to be spent over a 10-year period to study the aftermath of the spill. An additional $350 million from the $4 billion settlement between BP and the federal government was given to the National Academy of Sciences to establish a new program focused on human health and ecosystem science of the Gulf of Mexico to be spent over a 30-year period (Shen 2012).

Even with the tremendous efforts of the scientific community to deliver critical information to the response, the Deepwater Horizon oil spill highlighted the need to improve coordination between agencies and the scientific community for ensuring efficient, innovative, and thoughtful response to environmental crises. This necessarily includes coordinated social science. As one report stated, "there is no national lead entity coordinating the mobilization of science assets across federal agencies and within the broader science community"(Consortium for Ocean Leadership 2010). While the National Response Framework defines the responsibilities of each federal agency for responding to a disaster, lessons learned from Deepwater Horizon suggest that new and/or improved organizational structures are necessary to facilitate the mobilization of the scientific community to aid response, and this continues to be a fertile area for innovations in science policy (e.g., Nature 2010).

3.2.2   2012: Hurricane Sandy

In October 2012, Hurricane Sandy advanced toward the eastern seaboard of the United States. At the time of landfall near Atlantic City, New Jersey (NJ) on October 29, Hurricane Sandy measured over 1,770 km in wind field diameter and was classified as a post-tropical storm (NOAA 2012; Blake et al. 2013). Combining with a nor'easter, Hurricane Sandy affected 17 states, producing storm surges of up to 2.6 m, high precipitation including nearly 1 m of snow in areas of Maryland and West Virginia, and over 8.5 million households without power (Department of Energy 2012; Blake et al. 2013; US Geological Survey 2013).

Multiple dimensions of Sandy have required – and continue to require – tactical interdisciplinary science to support response efforts. Atmospheric scientists and meteorologists played a critical role in monitoring and assessing the formation and evolution of Sandy as it moved through the Caribbean, making landfall in Cuba before slowly progressing northward to pick up speed again before making its second landfall in New Jersey (Blake et al. 2013). Hydrologists deployed over 150 stream gauges to monitor storm surge while oceanographers evaluated potential damage to protective dunes and barrier islands (US Geological Survey 2013). In the aftermath of the storm, engineers were called upon to assess structural damage caused by flooding and wind. Public health experts, toxicologists, and chemists continue to assess health threats posed by mold in flooded houses, asbestos released from destroyed buildings, and other contaminants mobilized during fires that broke out during the storm.

Beginning days before Sandy's landfall and during the storm, social science efforts focused on providing necessary psychological and mental health services to the affected region. FEMA and American Red Cross deployed mental health professionals to the area days before the storm in preparation for supporting the citizens of the affected area. In the aftermath of the storm, multiple organizations launched social science studies to assess different dimensions – ranging from post-traumatic stress to the use of social media – of the storm's impacts on the social fabric of the region. FEMA awarded $82 million to the state of New York to "deliver immediate mental health outreach, crisis, and education services" to 200,000 individuals in the region through its Immediate Services Crisis Counseling Assistance and Training Program (Sederer 2012). The Pew Research Center Project for Excellence in Journalism analyzed the public's use of social media from October 29–31, 2012 to examine how individuals interacted with one another and with news and information. The study found that "fully 34 % of the Twitter discourse about the storm involved news organizations providing content, government sources offering information, people sharing… eye witness accounts, and still more passing along information posted by others" (Pew Research Center's Project for Excellence in Journalism 2013). At the organizational scale, one study examined the development of new partnerships in disaster relief operations, using Sandy as a case study and showing that 66 % of the partnerships that were relied on during Sandy response were new (Coles and Zhuang 2013). Other ongoing social science studies have examined how volunteer organizations have played a critical role in stabilizing communities, and how the mental and physical stress of disruption and displacement may impact local citizens and health care providers.

In the aftermath of Hurricane Sandy, strategic science was used to support recovery efforts. In January 2013, the Secretary of the Interior directed the Strategic Sciences Group (SSG, formerly the Strategic Sciences Working group described above) to stand up a crisis science team to support the Department's role on the cabinet-level Hurricane Sandy Rebuilding Task Force. In response, the SSG assembled a team of experts from government, academia, and non-governmental institutions to develop scenarios for the Task Force. The team was to examine the short- and long-term impacts of Hurricane Sandy and future major storms (such as another major hurricane) on the ecology, economy, and people of the affected New York/New Jersey region.

The SSG's Operational Group Sandy identified 13 primary or "first-tier" consequences of Sandy on coastal communities and ecosystems – from ecological change and changes in coastal geomorphology to altered storm preparedness and response activity and altered perception of risk (Department of the Interior 2013). Together, these consequences and their cascading consequences span a broad and complex range of environmental, economic, and social effects. Similar to the work completed during the Deepwater Horizon oil spill, the Hurricane Sandy scenarios used the human ecosystem model as an organizing framework; the scenarios are interdisciplinary and the impacts on the environment, infrastructure, and society are integrated throughout the scenarios.

One example of the SSG's work is shown in Fig. 3.3, which illustrates the cascading consequences resulting from Hurricane Sandy's flood damage to the built environment. This chain of consequences shows multiple dimensions of this damage, including the creation of hazardous and non-hazardous debris, new challenges in transportation, and downstream impacts to the local economy.

Using the results of the scenario, the SSG identified potential interventions, defined as institutional actions that support recovery and increase the resilience of the coupled human-natural system to future storms. The 17 interventions included several recommendations to bolster research in different areas, including ecosystem services, environmental contamination, social services, and risk education and communication (Department of the Interior 2013).

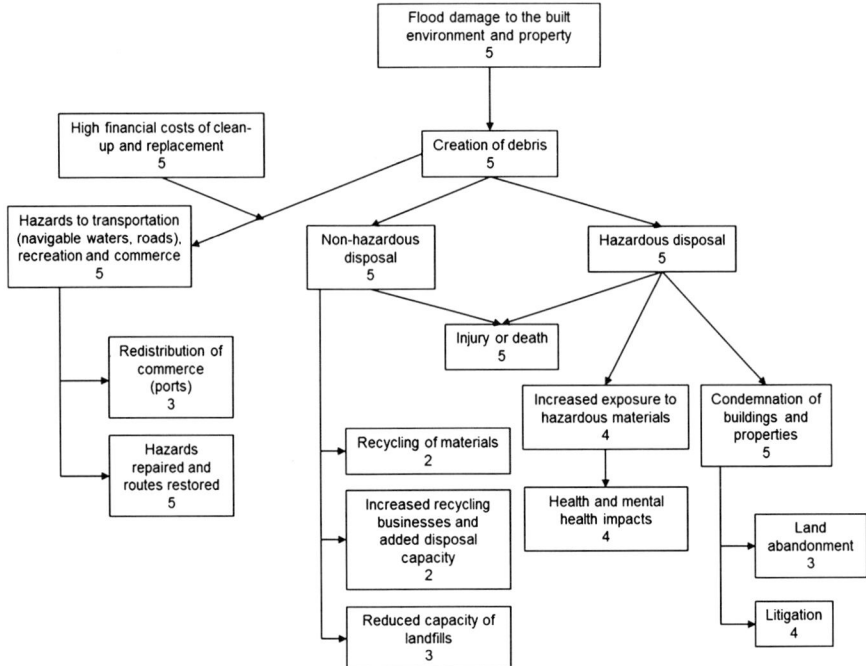

Fig. 3.3 Example chain of consequences from the SSG's Hurricane Sandy scenarios, showing the cascading consequences resulting from flood damage to the built environment in coastal communities. The numbers in the figure reflect the uncertainties associated with each consequence, with 5 being certain and lower numbers reflecting less certainty (Department of the Interior 2013)

3.3   Examples of Social Science During Environmental Crisis Events

Social science played a role in both of the major environmental crises described above. In these events and other crises, the social sciences (particularly but not uniquely sociology of risk, decision sciences, and community sociology) have contributed to both science during the crises and in the immediate aftermath as emergency response and recovery is underway. In many cases the research led to specific recommendations for action to improve crisis response. Several examples follow.

In 2007, Bangladesh was devastated by Tropical Cyclone Sidr, a Category 4 storm that killed 3,406 people and displaced or affected 27 million persons. A study by Paul (2012) examined factors that led to the resident's response to evacuation orders; the range of response is compelling, with only one-third of respondents evacuating to shelters. Policy recommendations to improve response during the crisis include expanded outreach programs, additional shelters, and evacuation drills.

From December 2006 to April 2007, the small Chilean town of Aysen (population 15,000) experienced intense seismic activity. Thousands of tremors were detected in the area, building up to a large earthquake on April 21, 2007. Though the earthquake itself was not a major event (magnitude 6.2), it did cause massive landslides, which in turn created localized tsunami waves that led to the death of 10 residents. During the seismic activity period, a parallel socio-political crisis developed, as controversy arose regarding the "decisions and best-suited measures required to prevent a potential disaster." A study by Soule (2012) reported on the factors that led to the socio-political crisis, which was centered on the perception (and eventual realization) of an imminent disaster, and makes recommendations for improving risk management as a result. Soule concludes that these findings "must be connected to a broader tendency to reject technocratic and centralized risk management" and calls for the incorporation of social science information during the assessment and decision making stages of risk management.

Hurricane Katrina made US landfall in Louisiana on August 29, 2005 as a large Category 3 hurricane. The levee system protecting New Orleans was breached, and over 75 % of the city and nearby parishes were flooded. The storm (and subsequent flooding) led to 1,833 deaths and over $81 billion dollars in property damage. A relatively large social science literature has emerged about the event, much of it based on data collected during or immediately after the crisis. For example, Millin et al. (2006) examined disaster medical assistance in both Mississippi and a volunteer site near New Orleans. Treatment of chronic disease, primary health care and routine emergency care not related to the hurricane were the most common needs. The authors suggested that in addition to acute medical needs, "disaster planners should prepare to provide primary health care, administer vaccinations, and provide missing long-term medications."

Vu and Van Landingham (2012) took advantage of survey work done just weeks prior to Hurricane Katrina, and were able to conduct a pre- and post-disaster assessment of physical and mental health consequences for working-age Vietnamese immigrants to New Orleans. The researchers located and re-assessed more than two-thirds of the original study cohort. They found statistically significant declines in physical and mental health status after the first anniversary of the storm, and substantial recovery by the second anniversary. Recovery varied by a number of key sociological variables (such as occupational type and marital status), and the authors suggested the results "present clear opportunities for targeted interventions."

3.4 Distinctive Characteristics of Science During Environmental Crises

The examples presented reveal the importance of science during environmental crises. In addition to traditional discipline-focused tactical research, the need and opportunity for interdisciplinary strategic science is intensified during such crises: decision makers need to quickly understand the impacts on coupled natural-human systems, the uncertainties and limitations that surround findings and analysis, the cascading consequences of the event, and an accurate sense of place that links the science to "on-the-ground" (or in the water) realities associated with a specific crisis event, time, and place.

Hence, the application of strategic science during environmental crises has several distinctive characteristics that are essential requirements if it is to be useful to decision makers. Many of these characteristics may also be relevant to crises other than environmental. With all of these, it is critical to stress that science during crisis can only be effective when *all* relevant disciplines of science – the physical, biological, and social – are fully integrated and actively engaged. Six key characteristics are described below.

3.4.1   The Importance of Coupled Human-Natural Systems

Science during environmental crises benefits from recognizing the need to evaluate and respond to the crisis using a systems approach, where consequences such as dune erosion during a hurricane are not just interpreted as an environmental change and loss of habitat, but as a storm consequence that may also compromise the safety of houses (and thereby households) that rely on these natural storm buffers for protection. Models of coupled human-natural systems are especially valuable to such strategic science.

One example (among many) is the human ecosystem model (Machlis et al. 1997, see Fig. 3.4) applied during the Deepwater Horizon oil spill (Department of the Interior 2010, 2012) and most recently Hurricane Sandy. It describes a reasonably detailed coupled human-natural system, including both biophysical and sociocultural variables, as well as flows of individuals, energy, nutrients, information, materials, capital, and information. The human ecosystem model originated in the 1997 paper entitled "The human ecosystem as an organizing concept" that was published in two parts in the journal *Society and Natural Resources*. A modest commentary has appeared (see for example Rudel 1999), and applications have included the National Science Foundation's Long Term Ecological Research Program in Baltimore, MD, a National Oceanic and Atmospheric Administration training program, and the United Nations Environmental Program on Sustainability. Models like these are essential for achieving a holistic approach to assessing impacts and anticipating cascading consequences, particularly during crises where the full range of consequences is both unknown and uncertain.

3.4.2   The Challenge of Collaboration and Interdisciplinary Teams

During non-crisis times, scientific research is conducted by individual principal investigators and/or teams of scientists. Research teams are often multi-institutional and in most cases, researchers collaborate with colleagues they have worked with in the past or with whom they have some pre-existing relationship. New collaborations are often formed through the long-term exchange of knowledge and ideas at regularly scheduled workshops, academic conferences, and peer-reviewed publications.

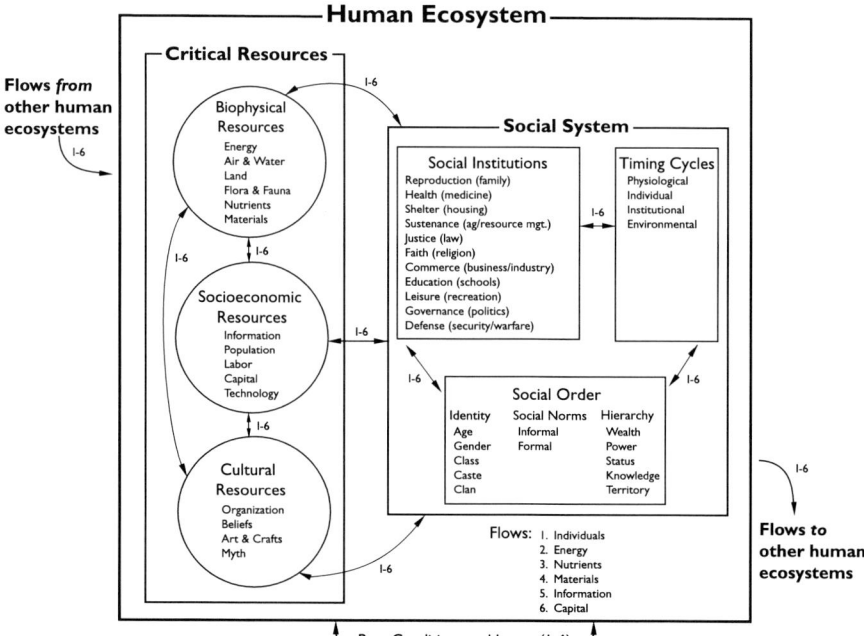

Fig. 3.4 A framework for the coupled human-natural system, showing the interconnectedness of critical resources and the social system is useful for guiding science during environmental crises (Adapted from Machlis et al. 1997)

By necessity, science during crisis is also often conducted by multidisciplinary teams where these teams are often formed quickly in response to the event. For environmental crises, members may represent fields ranging from the physical and natural sciences to human biology and social sciences. These teams are also multi-institutional and include scientists from the academic, government, non-profit, and private sectors. In many cases, the individuals in crisis science teams have not previously worked together before (see Fig. 3.5). Examples include the nuclear physics theorists and weapons engineers of Manhattan Project during WWII, teams of engineers from manufacturer plants and universities working together to solve the Apollo 13 crisis, and academic and federal geoscientists working with oil industry engineers to address the Deepwater Horizon oil spill.

The urgency of the task, compression of time available for research, and lack of previous collaboration can add additional challenges in communication among scientists, as well as issues of trust and collaboration styles. At the same time, a shared and critical mission can promote cooperative behavior and remove traditional barriers to collaboration by establishing common ground, focus upon mission rather than process, and recognition of expertise rather than representation of organizations, institutions, and academic pedigree or rank.

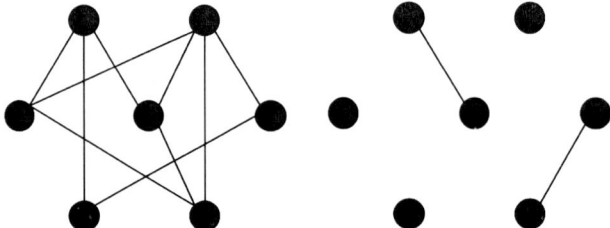

Fig. 3.5 Schematic collaboration patterns of science teams during non-crisis and crisis times. On the *left* is a schematic network diagram of a science team operating during non-crisis times. Nodes represent individual scientists and links represent previous collaboration (e.g., via co-investigators on a grant proposal or co-authoring a publication). Members of this team have worked with one another before. On the *right* is a schematic of a crisis science team, showing that only a few of the members have previously collaborated (After Börner 2010, 2011)

3.4.3 The Importance of Uncertainties and Limitations

During an environmental crisis, conditions can rapidly change: for example, an earthquake may trigger a tsunami, which may cause a nuclear emergency; evolving weather conditions may intensify an approaching hurricane, divert a storm track, and complicate evacuations. Working with limited knowledge and operating with uncertainty is inherent to responding to – and making decisions during – a crisis. For science during crisis to be useful to decision makers, it is essential to establish and explicitly state levels of uncertainty and knowledge limitations. For example, during the Deepwater Horizon oil spill, multiple studies produced different evaluations for the volume of oil leaking from the broken pipe on the seafloor (McNutt et al. 2012). These assessments had important and immediate implications for determining the best technical solution to capping the wellhead, determining the amount of chemical dispersant to be applied, planning for containment of oil once it reached the surface, and evaluating the extent of damage to the environment (McNutt et al. 2012). Determining and communicating scientific uncertainty with the flow rate estimates was essential to guiding sound decision-making during the spill, and retrospective analysis of these estimates have provided valuable lessons learned for responding to future deep sea blowouts (McNutt et al. 2012).

Similarly, the scenarios built by the DOI Strategic Sciences Group included formal evaluations of scientific uncertainty for each consequence in a chain of consequences; the evaluation (made using expert opinion and following the precautionary principle) was adapted from the Intergovernmental Panel on Climate Change's uncertainty scale associated with climate change, and other scales (see Weiss 2003).

3.4.4   The Value of Cascading Consequences and Assessing Impacts

To be effective during emergency response, recovery, and restoration, science during environmental crises often requires the examination of cascading consequences over both short- and long-term time scales. "Chains of consequences" illustrate changes, effects, or impacts resulting from an event. A chain of consequences begins with an event – such as a major oil spill or hurricane – and branches out, like a flow chart or tree diagram, showing possible cascading events. Each consequence in the chain has the potential to lead to other consequences. Each consequence in the chain can be assigned a level of scientific uncertainty – an assessment used to communicate the certainty or likelihood of a consequence.

Chains of consequences can reveal unanticipated effects of different events. For example, during Hurricane Sandy, storm surge caused severe flooding in homes and businesses across the affected region. In the community of Breezy Point, New York, a flooded electrical system led to fires that destroyed more than 120 homes (New York 1 News 2012), leading to the potential release of lead-based paints and the threat of additional health risks to first responders and the community (Plumlee et al. 2012).

Examining such chains of consequences is an area of science during crisis where strategic science can be extremely valuable. While essential tactical science, such as analyzing contaminated flood sediments from a hurricane or monitoring radioactivity in local water supplies, can be on-going during and after an environmental crisis, strategic science can complement tactical efforts by evaluating the cascading effects of an event across the coupled human-natural system.

This approach is illustrated by the work of the Department of the Interior's (DOI) Strategic Sciences Working Group (SSWG) during the Deepwater Horizon oil spill. The SSWG convened two scenario-building sessions (the first just days after the start of the event, the second while the wellhead had not yet been capped) to build scenarios analyzing the cascading consequences of the spill. Defining boundary conditions such as a flow rate estimate, geographic extent, and time horizons, the SSWG assessed short- and long-term consequences such as the effects of chemical dispersants, damage to wetlands, and impact to the local economy (Department of the Interior 2010, 2012; Machlis and McNutt 2010).

3.4.5   The Need for Sense of Place

Every environmental crisis is different from the last or the next: a major earthquake in southern California will require different response than an oil spill in the Gulf of Mexico or a severe tornado in Oklahoma. Even similar crises have place-based differences: an Alaskan Arctic oil spill differs from a Gulf of Mexico oil spill in technology, impacts, response, and restoration/recovery strategies. Depending on where and when the crisis occurs and the severity of damage, people in the affected region are impacted differently – the result of factors including geographic location, socio-economic status, and sociocultural traits. Hence, science during environmental crises must work with an accurate sense of place. Crisis science teams must rely heavily on members with local knowledge who can provide place-specific information on communities, cultures, values, history, and environment that can be essential to assessing risk and responding to unfolding events during a crisis.

3.4.6   The Demands of Communicating Science During Crisis

Effectively communicating science is essential if the scientific information is to be used under the rapidly changing conditions, constrained time frames, multiple demands on decision makers, and limited resources that are typical during a crisis. First and foremost, scientific information must be communicated with extraordinary clarity and conciseness. Because the information may be used by non-scientific audiences, technical terms should be well defined if they must be used. Explanation of results, findings, uncertainties and implications must take priority over descriptions of background, relevant literature, or methods.

Communicating science during crisis can also benefit from the use of compelling visualization. An example emerged during the Deepwater Horizon oil spill when graphic artists developed schematic diagrams of the broken pipe on the seafloor. Derived from observations made with remotely operated vehicles, these visualizations aided scientists, responders, decision makers, and the public in understanding the complexity of the damaged riser pipe over 1,500 m below the surface of the Gulf. Presentation tools – ranging from sketchpads to visualization software and mobile tablets – can be useful for translating scientific information quickly and efficiently during a crisis.

To be effective in supporting decision making, science (and scientists) during crisis should have the capacity to speak "truth to power," delivering difficult or unpopular findings or analyses. Direct access to decision-makers is essential. Access requires trust. As shown in many of the previous examples, scientific information can be pivotal for decision making during a crisis. It must be delivered directly to decision makers unfettered by layers of bureaucracy and/or the public diversion of "science by interview" (whereby competing scientists present their personal views to reporters and/or commentators) now fashionable in the contemporary media.

Issues of transparency and public right-to-know are considerable, and must be adjudicated carefully. Post-crisis publication through peer reviewed literature and third-party evaluation are both potential solutions. However, while science during crisis should have access to decision-makers (and vice versa), it is the responsibility of the scientists involved to maintain their independence and credibility and role as "honest brokers" (Pielke 2007) by presenting information rather than attempting to make policy or response decisions. For trust (and thereby access), it is essential that this distinction be maintained in communications between scientists and decision makers during crisis.

3.5   A Modest Research Agenda

Clearly, the application of science during crisis is not novel: it has been used to monitor and respond to events ranging from epidemics and terrorist attacks to man-made disasters and natural hazards. However, there has been little coordinated effort to formally characterize science (including social science) during crisis and to identify ways in which it can be improved for responding to future crisis events. This is particularly true for environmental crises, with the oft-repeated pattern of multiple jurisdictions, overlapping responsibilities, a traditional focus on tactical rather than strategic science, and high levels of uncertainty.

There are ample opportunities for improvement. New organizational frameworks could streamline the use and application of science during crisis. New technologies could improve visualization, communication, and the sharing of information among scientists, emergency responders, and the public. Advanced training, simulations, and workforce development could improve the preparation of the next generation of scientists needed to respond to future crisis events. Preparing decision makers to use science during crises and to make science-informed decisions is equally important.

While the role of science during crises – war, natural disasters, industrial accidents, pandemics, and more – has increased significantly in contemporary times, there has been little scholarly attention devoted to the distinctive character of science during crisis and how such science can most effectively be planned, conducted, examined, communicated, and applied to decision-making. This is particularly true for interdisciplinary and strategic science. Organizational frameworks for science during crisis have not been described, best practices have not been systematically identified (Machlis and McNutt 2011; Machlis and Kooistra 2012), and a research agenda for understanding and improving science during crisis has not been proposed or implemented.

We suggest a modest first step is to examine several essential questions:

1. Is science during crisis different than science practiced in non-crisis periods, and if so, how?
2. If it is different, how do these differences affect the management, design, conduct, analysis, application, and dissemination of science?
3. How can science during crisis be improved and made more useful?
4. How can the workforce and scientific community be better prepared?
5. What are the most appropriate organizational frameworks and best practices for science during crisis?
6. What role can interdisciplinary and strategic science play in responding to major crises?

A range of disciplines including sociology, anthropology, economics, organizational and management science, as well as policy studies can be fruitful partners in answering these questions. Historians of science can provide thoughtful guidance based on the role of science in historical and recent past events. For environmental crises, professionals in hazards management, emergency response, risk assessment, and resources management can be vital contributors, both as end-users of strategic sciences and as "first responders" responsible for emergency and recovery. Results should be shared broadly and thoughtfully converted to usable knowledge. The result would be improved science during crisis.

3.6 Conclusion

In early 2009, tremors and foreshocks were increasing in the Abruzzo region of central Italy. The swarms of small quakes concerned local citizens, and Italian science technician Giampaolo Giulian was predicting a major quake, only to be reported to the police. A select group of Italian scientists, all members of the National Commission for the Forecast and Prevention of Major Risks, met on March 31, to assess the situation, and decide on a course of action. A press conference was held after the meeting, led by the technical head of Italy's Civil Protection Agency. He announced,

> the scientific community tells me there is no danger because there is an ongoing discharge of energy. The situation looks favorable. (Nosengo 2010)

Many citizens of the mountainous region were relieved, and evacuation or precautionary pre-positioning of emergency supplies did not occur.

On April 6 a significant (magnitude 6.3) earthquake epicentered near the town of L'Aquila, the capital of the Abruzzo, struck the region. It was at relatively shallow depth (8.8 km), and the region's soil structure amplified the seismic impact. Nearly

Fig. 3.6 Damage from 2009 L'Aquila Earthquake (Website of the Italian Civil Protection Department – Presidency of the Council of Ministers, http://www.protezionecivile.gov.it/jcms/en/descrizione_sismico.wp;jsessionid=6EED29F25DA52C422634EE009FC67CAE?pagtab=3)

70 % of the buildings in L'Aquila were severely damaged or destroyed (Fig. 3.6). Over 300 persons died, 1,500 were injured, and thousands were left homeless (Kaplan et al. 2010). In a trial watched by the global scientific community with alarm, six of the scientists were convicted of manslaughter, for giving falsely assuring advice on possibility of a major and devastating quake. The convictions are under appeal.

Severe environmental crises disrupt multiple dimensions of social, economic, and environmental systems over both short- and long-term time scales. It is likely that the complexity and impact of such crises will increase as human population continues to rise, technology becomes more complex and vulnerable, climate change acts as a driving force and/or accelerant for many environmental crises, and as local, regional, and national economies become more globalized and interdependent. Fatalities will likely increase in the future due to more people living in hazard-prone areas (e.g., Holzer and Savage 2013). The insurance industry has shown that the cost of property damage from natural hazards is increasing and even single events "can greatly strain a nation's ability to deal with direct damage costs and indirect economic, social, and cultural losses" (American Geosciences Institute 2012).

Because of this growing cost and complexity, it is likely that science will play an increasingly significant role in supporting response to and preparation for future environmental crises. Scientists, emergency managers, business leaders, educators, and local, state, and federal decision makers will have to cooperate to ensure public safety and to develop solutions to mitigating and adapting to risk. Beyond these challenges, the scientific community – including the social sciences and its practitioners – must grapple with the responsibility of science and scientists during crisis, and the implications of events on the island of Martinique and the Italian region of Abruzzo.

References

Abramson, D. M., Redlener, I. E., Stehling-Ariza, N. A., Sury, J., Banister, A. N., Park, Y. S. (2010, August). *Impact on children and families of the Deepwater Horizon oil spill: Preliminary findings of the Coastal Population Impact Study*. National Center for Disaster Preparedness, Research Brief 2010:8.

American Geosciences Institute. (2012, October 17). *Critical needs for the twenty-first century: The role of the geosciences* [Internet]. Alexandria: American Geosciences Institute. Available from: http://www.agiweb.org/gap/criticalneeds/hazards.html. Accessed 20 Mar 2012.

Blake, E. S., Kimberlain, T. B., Berg, R. J., Cangialosi, J. P., & Beven, J. L., II. (2013). *Tropical cyclone report Hurricane Sandy (AL182012) 22–29 October 2012*. Miami: National Hurricane Center.

Börner, K. (2010). *Atlas of science: Visualizing what we know*. Cambridge, MA: The MIT Press.

Börner, K. (2011). Network science: Theory, tools, and practice. In B. William Sims (Ed.), *Leadership in science and technology: A reference handbook*. Thousand Oaks: Sage.

Brown, V. A., Harris, J. A., & Russell, J. Y. (2010). *Tackling wicked problems throughout the transdisciplinary imagination*. London: Earthscan.

Coles, J. B., & Zhuang, J. (2013). Partnership behavior in disaster relief operations: A case study of the response to Hurricane Sandy in New Jersey (Atlantic City, New Jersey, 2013), Field Report, University at Buffalo

Consortium for Ocean Leadership. (2010). *Deepwater horizon oil spill: Scientific symposium meeting summary*. Washington, DC: Consortium for Ocean Leadership.

Department of Energy. (2012, November 7). *Responding to Hurricane Sandy: DOE situation reports* [Internet]. Washington, DC: Department of Energy. Available from: http://energy.gov/articles/responding-hurricane-sandy-doe-situation-reports. Accessed 20 Mar 2013.

Department of the Interior. (2010). *DOI Strategic Sciences Working Group Mississippi Canyon 252/Deepwater Horizon oil spill progress report 1, 9 June 2010*. Washington, DC: Department of the Interior.

Department of the Interior. (2012). *DOI Strategic Sciences Working Group Mississippi Canyon 252/Deepwater Horizon oil spill progress report 2, May, 2012*. Washington, DC: Department of the Interior.

Department of the Interior. (2013). *DOI Strategic Sciences Group Operational Group Sandy technical progress report*. Washington, DC: Department of the Interior.

Fermor, P. L. (1950). *The traveler's tree: A journey through the Caribbean Islands*. London: John Murray Press.

Freudenburg, W. R., & Gramling, R. (2011). *Blowout in the Gulf: The BP oil spill disaster and the future of energy in America*. Cambridge, MA: MIT Press.

Grattan, L. M., Roberts, S., Mahan, W. T., Jr., McLaughlin, P. K., Otwell, W. S., & Morris, J. G., Jr. (2011). The early psychological impacts of the Deepwater Horizon oil spill on Florida and Alabama communities. *Environmental Health Perspectives, 119*(6), 838.

Holzer, T. L., & Savage, J. C. (2013). Global earthquake fatalities and population. *Earthquake Spectra, 29*(1), 55–175.

Kaplan, H., Bilgin, H., Yilmaz, S., Binici, H., & Oztas, A. (2010). Structural damages of L'Aquila (Italy) earthquake. *Natural Hazards and Earth System Sciences, 10*, 499–507.

Kuhn, T. (1962). *The structure of scientific revolutions*. Chicago: University of Chicago Press.

Lee, M. R., & Blanchard, T. C. (2012). Community attachment and negative affective states in the context of the BP Deepwater Horizon disaster. *American Behavioral Scientist, 56*(1), 24–47.

Levy, J., & Gopalakrishnan, C. (2010). Promoting ecological sustainability and community resilience in the US Gulf Coast after the 2010 Deepwater Horizon oil spill. *Journal of Natural Resources Policy Research, 2*(3), 297–315.

Lubchenco, J., McNutt, M. K., Dreyfus, G., Murawski, S. A., Kennedy, D. M., Anastas, P. T., Chu, S., & Hunter, T. (2012). Science in support of the Deepwater Horizon response. *Proceedings of the National Academies of Sciences, 109*(50), 20212–20221.

Mabus, R. (2010). *America's Gulf Coast: A long term recovery plan after the Deepwater Horizon oil spill*. Washington, DC: RestoretheGulf.gov.

Machlis, G. E., & Kooistra, C. (2012). Science during crisis: The DOI strategic sciences group and its OSS inspiration. *The OSS Society Journal, Fall*, 48–50.

Machlis, G. E., & McNutt, M. K. (2010). Scenario-building for the Deepwater Horizon oil spill. *Science, 329*(6007), 1018–1019.

Machlis, G. E., & McNutt, M. K. (2011). Ocean policy: Black swans, wicked problems, and science during crises. *Oceanography, 24*(3), 318–320.

Machlis, G. E., Force, J., & Burch, W. R. (1997). The human ecosystem part I: The human ecosystem as an organizing concept in ecosystem management. *Society and Natural Resources, 10*(4), 347–367.

McNutt, M. K., Camilli, R., Crone, T. J., Guthrie, G. D., Hsieh, P. A., Ryerson, T. B., Savas, O., & Shaffer, F. (2012). Review of flow rate estimates of the Deepwater Horizon oil spill. *Proceedings of the National Academies of Sciences, 109*(50), 20260–20267.

Millin, M., Jenkins, L., & Kirsch, T. D. (2006). A comparative analysis of two external healthcare disaster responses following Hurricane Katrina. *Prehospital Emergency Care, 10*(4), 451–456.

Morris, J. G., Jr., Grattan, L. M., Mayer, B. M., & Blackburn, J. K. (2013). Psychological responses and resilience of people and communities impacted by the Deepwater Horizon oil spill. *Transactions of the American Clinical and Climatological Association, 124*, 191.

Nature. (2010). All at sea. *Nature, 465*, 397–398.

New York 1 News. (2012, December 25). *Fire officials determine origin of Breezy Point fire caused by Sandy*. [Internet]. New York: NY1.com. Available from: http://www.ny1.com/content/top_stories/174510/fire-officials-determine-origin-of-breezy-point-fire-caused-by-sandy. Accessed 21 Mar 2013.

NOAA. (2012, October 29). *NOAA Hurricane Report 29* [Internet]. Miami: National Hurricane Center. Available from: http://www.nhc.noaa.gov/archive/2012/al18/al182012.discus.029.shtml. Accessed 23 Feb 2013.

Nosengo, N. (2010). Italy puts seismology in the dock. *Nature, 465*, 992.

Paul, B. K. (2012). Factors affecting evacuation behavior: The case of 2007 Cyclone Sidr, Bangladesh. *Professional Geographer, 64*(3), 401–414.

Pew Research Center's Project for Excellence in Journalism. (2013). *Hurricane Sandy and Twitter*. Available from: http://www.journalism.org/index_report/hurricane_sandy_and_twitter. Accessed 5 Sept 2013.

Pielke, R. A., Jr. (2007). *The honest broker: Making sense of science in policy and politics*. London: Cambridge University Press.

Plumlee, G., Morman, S. A., & Cook, A. (2012). Environmental and medical geochemistry in urban disaster response and preparedness. *Elements, 8*(6), 451–457.

Rhodes, R. (1986). *The making of the atomic bomb*. New York: Simon and Schuster.

Rittel, H. W. J., & Webber, M. M. (1973). Dilemmas in a general theory of planning. *Policy Sciences, 4*, 155–169.

Rudel, T. K. (1999). Critical regions, ecosystem management, and human ecosystem research. *Society and Natural Resources, 12*(3), 257–260.

Sarder, Z. (2000). *Thomas Kuhn and the science wars*. Cambridge: Icon Books UK.

Sederer, L. (2012, November). FEMA approves $8.2 million for post-Sandy mental-health outreach. *The Atlantic*.

Shen, H. (2012, November 15). *Historic Gulf oil spill settlement to bolster US research* [Internet]. London: Nature.com. Available from: http://blogs.nature.com/news/2012/11/historic-gulf-oil-spill-settlement-to-bolster-us-research.html. Accessed 15 Nov 2012.

Soule, B. (2012). Coupled seismic and socio-political crises: The case of Puerto Aysen in 2007. *Journal of Risk Research, 15*(1), 21–37.

Taleb, N. N. (2007). *The black swan: The impact of the highly improbable*. New York: Random House.

US Geological Survey. (2013). *Response to Hurricane Sandy* [Internet]. Reston: US Geological Survey. Available from: http://coastal.er.usgs.gov/hazard-events/sandy. Accessed 21 Mar 2013.

Vu, L., & VanLandingham, M. J. (2012). Physical and mental health consequences of Katrina on Vietnamese immigrants in New Orleans: A pre-and post-disaster assessment. *Journal of Immigrant and Minority Health, 14*(3), 386–394.

Weiss, C. (2003). Expressing scientific uncertainty. *Law, Probability, and Risk, 2*, 25–46.

Chapter 4
Who's Afraid of Thomas Malthus?

Jörg Friedrichs

4.1 Introduction

The main impediment to science integration in the study of resource management, not only between various social scientific disciplines but also between the social and the physical sciences more generally, is a refusal of social scientists to appreciate how deeply the societal sphere is embedded in wider biophysical and social-ecological systems. Recently, however, researchers working at the intersection between human and natural systems have come to acknowledge that society is inextricably embedded in, and constrained by, wider ecological systems including the earth system as a whole. This research program is commonly called the social-ecological, socio-metabolic, or earth-systems perspective (Berkes et al. 2003; Walker et al. 2004; Haberl et al. 2011; Bierman et al. 2012), and it undeniably holds significant promise for the study of resource management.

It is important to note, however, that integrating a social with a biophysical perspective is not new if we take the long view of the history of science. This is not a problem in and of itself, as science is always a kind of palimpsest. But since amnesia can also hamper the development of new ideas, it is worthwhile for those interested in a social-ecological systems perspective and other related research programs to scrutinize earlier traditions for potentially useful contributions.

Indeed, the linkages between natural resources and social change were studied long before the separation between physical and human sciences, and the subsequent specialization of social science into various academic disciplines. Take for example the *physiocrats* of the eighteenth century, who emphasized that all economic

This contribution contains excerpts totaling ca. 2,500 words from Chapters 1 and 3 of my book *The Future Is Not What It Used to Be: Climate Change and Energy Scarcity* (MIT Press, 2013).

J. Friedrichs (✉)
Oxford Department of International Development, Queen Elizabeth House,
3 Mansfield Road, Oxford OX1 3TB, UK
e-mail: joerg.friedrichs@qeh.ox.ac.uk

M.J. Manfredo et al. (eds.), *Understanding Society and Natural Resources*,
DOI 10.1007/978-94-017-8959-2_4, © The Author(s) 2014

wealth is ultimately derived from a land base. In the present chapter, I focus on another early integrated framework, namely the tradition founded by the enlightenment polymath Thomas Malthus (1766–1834). Malthus was versed in an impressive array of areas, from theology to philosophy and from population analysis to the emerging field of political economy. He integrated all of these disparate fields of knowledge in order to study the interaction between population dynamics and food production, including the social consequences of that interaction.

Today, Malthus' determination to integrate whatever field of knowledge had something to contribute to the issues under study is a source of inspiration to all those who want to take a genuinely integrated look at resource management. As we will see, modified Malthusian theories constitute a uniquely promising bid for grounding the study of resource management on science integration, not only between various social scientific disciplines but also between the social and the lages between population dynamics, food production, and social change, modified Malthusian theories go beyond his original framework. The most sophisticated models are equipped to consider *any* kind of resource constraint and incorporate *any* challenge to the ecosphere, from biodiversity loss to climate change.

Despite the considerable potential of modified Malthusian theories, most social scientists have a hard time accepting that social change can be anything but endogenous. Physical scientists are more open to Malthusian hypotheses, but their social theorizing often lacks sophistication and is therefore duly criticized.

To overcome this unproductive state of affairs, I start from the classical Malthusian framework and gradually add complexity to it. After an introduction and discussion of classical Malthusianism I show how, despite the failing of Malthusian predictions, its logical structure is reproduced by simple neo-Malthusian theories that have been developed to account for contemporary global challenges. Subsequently, I show the potential of more sophisticated neo-Malthusian models and theories, from the iconic *Limits to Growth* study in the 1970s to the eco-scarcity theory of the 1990s and from climate-based eco-scarcity to Tainter's theory of diminishing returns on civilizational complexity. I conclude by pondering the prospects of modified Malthusian theories contributing to better science integration.

4.2   Classical Malthusianism

The original theory of Thomas Malthus is neatly summarized by an oft-quoted statement from the *Essay on the Principle of Population*: "Population, when unchecked, increases in a geometrical ratio. Subsistence increases only in an arithmetical ratio" (1798, 14). Population is assumed to grow exponentially, but the growth of a society's means of subsistence is assumed to be only linear. If this is so, exponential growth of population unavoidably outpaces the linear increase of subsistence. Alas, population levels are constrained by food supply as people need enough food. Tragically, food intake per capita shrinks as population grows faster than subsistence. Linear growth in food supply cannot make up for the skyrocketing needs of the exponentially growing population. At some point, population growth runs against the limit imposed by minimum food intake per capita.

In a society characterized by social inequality, the poorest of the poor will be the first to feel the looming food scarcity. As population levels rise and food per capita decreases, the food available to the poor will fall below the minimum intake that is necessary for their subsistence. Redistribution can keep the poor fed for a while, but this will not prevent more and more people from becoming destitute due to the inexorable fall of food per capita. In the end, the system is likely to be readjusted by brutal mechanisms such as famine, war, and pandemics.

Logically speaking, another solution would be to limit population growth to "arithmetical ratio" in line with the linear growth of food production. In practical terms, this would mean birth control. To Malthus, who was an Anglican country curate and a moralist, family planning and any kind of sex without the aim of reproduction came under the category of sinful behavior. He therefore advocated voluntary forms of "moral restraint", but at the same time believed that curtailing the reproductive instinct of the masses was simply not realistic.

During his lifetime, Malthus modified his theory several times: first in the two-volume version of the *Essay* (1803) and then in various further editions (Winch 1987). These modifications need not detain us here, as they left the basic theory in place. Nor is there any need to dwell on the finer points of the theory or its policy implications, which were important during the nineteenth-century debate about the poor laws. For our present purposes, we are only interested in the logical structure of the theory and its applicability to issues of resource management.

The enduring appeal of the theory is mostly due to its plausible assumptions and axiomatic elegance. It is indeed plausible to assume that population grows by an annual rate multiplied by current numbers—much like the stock on a bank account grows by the iterative application of an interest rate. The result of compound interest, or of children and children's children following the reproductive behavior of their forefathers, is exponential growth. Similarly, it appears plausible to assume linear growth for a population's means of subsistence because agricultural innovation and other improvements in food production tend to happen in an incremental fashion, suggesting linear progress rather than a self-reinforcing mechanism. This appears much more plausible than to assume that improvements in food production are like a compound interest rate applied over a stock.

There is an important element missing from the account, or rather implicit in it: namely the notion of *overshoot*. Overshoot means that a system can temporarily exceed its long-term limits. Malthus assumed that this was indeed possible. Otherwise, why did he assume that population levels would be readjusted through "vice and misery"—shorthand for famine, war, pandemics, and sinful behavior—rather than simply being limited by minimum food intake per capita? In fact, "vice and misery" are unavoidable only insofar as population can temporarily exceed subsistence. Plain commonsense has it that this can easily happen. Population levels may exceed agricultural yields during years of good harvest, but the famine bound to occur in a later year of bad harvest will then be even more catastrophic. Malthus

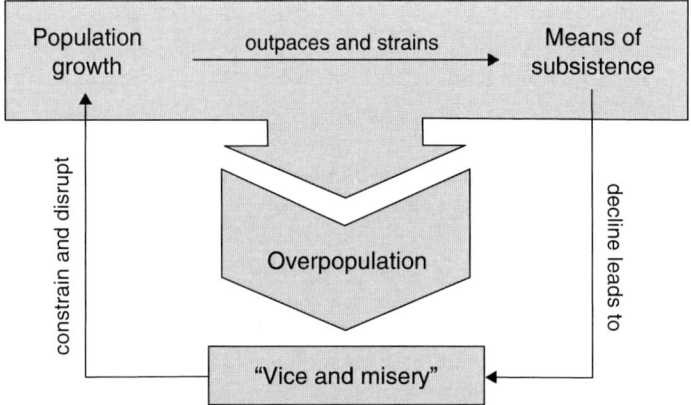

Fig. 4.1 Classical Malthusianism

assumed that such misery was likely to be accompanied by war and pandemics, as well as objectionable forms of non-reproductive sex, or "vice".

To illustrate the axiomatic elegance of the theory, consider Fig. 4.1.

4.2.1 The Logical Structure of Malthusianism

Malthusianism is more than simply a theory about the social interaction effects of population dynamics and food production. Logically speaking, it is the study of how different functions, which are all essential to social production and reproduction, enable and constrain each other. In abstract formal terms, this logical structure can be visually expressed by the following general scheme (Fig. 4.2).

At the heart of the model, there are two functions which are both vital to social production and reproduction. The first function (f_1) outpaces and strains the second one (f_2). For a while, this is obfuscated by the fact that time lags built into the system enable a temporary overshoot. In the long run, however, there is an inexorable mechanism by which the second function (f_2) constrains the first one (f_1). The way the mechanism operates is that the decline of f_2 leads to significant problems, which at the end of the day disrupt the unsustainable growth of f_1.

As we have seen, in classical Malthusianism population growth (f_1) outpaces and strains food supply (f_2) because the former function is exponential while the latter is only linear. Overshoot is possible for a while, for example due to a series of good harvests. In the long run, however, food supply (f_2) inexorably constrains population growth (f_1) because caloric intake per capita cannot fall below subsistence level. Famine and other calamities are then unavoidable. According to Malthus, "vice and misery" will ultimately bring population levels down.

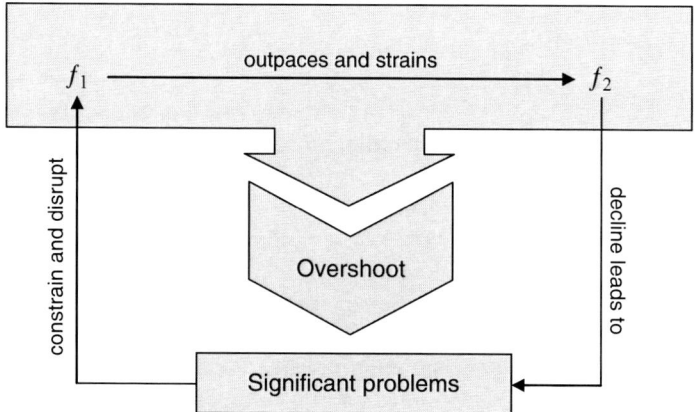

Fig. 4.2 The logical structure of Malthusianism

4.2.2 Why Malthus Was Wrong

The theory is axiomatically true if one assumes, with Malthus, that the growth of food production is at best linear while population growth is inherently exponential. Or, more mildly, if one assumes that population growth outpaces but is ultimately constrained by the means of subsistence. Quite obviously, this is not how modern history has unfolded. So far, overpopulation has neither led to mass starvation nor to planetary pandemics or other forms of catastrophic rebalancing.

With hindsight, there are four reasons why Malthus has not been vindicated. First, his assumption of exponential population growth was largely correct at the time but is less so today. As a result of the so-called demographic transition, world population is moving away from familiar patterns of exponential growth. It is still projected to grow by another two billion people, from around seven billion in 2011 to about nine billion in 2050. But, at the same time, population growth has started to level off in most parts of the world (Lutz and Samir 2010; UN 2011).

Second, growth in food production has been far more than linear. Since the nineteenth century, industrial inputs such as chemical fertilizer and motorized machinery have dramatically intensified agricultural productivity. Thanks to an abundant supply of such inputs, food production has been largely able to keep pace with population growth. For the last couple of centuries, agricultural innovation has eluded Malthusian predictions over and over again (Trewavas 2002).

Third, globalization has enabled an unprecedented growth of both world population and food production. In line with circumstances in the early modern period, Malthus saw population levels as constrained by food production at the local level. Over the last two centuries, however, mobility and trade have shifted the territorial frame of reference first from the local to the national level, then to the international, and finally to the global level. To begin with, Europeans were able to move to "underpopulated" landmasses such as America and Siberia and to import raw materials and foodstuffs from the colonies. Subsequently the globalization of trade, and more recently of aid, has had similar effects, although in the reverse direction, buttressing indigenous population levels in developing countries.

Fourth, vulgar forms of Malthusianism tend to assume that any given resource base can sustain only a fixed number of individuals of some species, commonly called *carrying capacity*. For example, wild deer can for some time overgraze the available herbs on an island, but their population level will inevitably be adjusted downward to carrying capacity after a period of overshoot. While this notion of carrying capacity is suitable for simple cases of population biology, for example algal growth constrained by the surface of a lake, it is far too static for the study of more complex constellations.[1] When applied to human populations, carrying capacity can only be understood as a dynamic cultural concept, depending *inter alia* on technological innovation and social choice (Cohen 1995; Seidl and Tisdell 1999). The carrying capacity for irrigation agriculture is higher than for rain-fed agriculture, and the carrying capacity for a population of vegans riding on bicycles is higher than for a population of meat lovers driving about in SUVs.

4.2.3 Why Malthus May Still Turn Out to Be Right

Today, industrial civilization is buttressing a globalized system that injects trade and aid to some of the most vulnerable parts of the world, which would otherwise suffer serious problems of overpopulation. In our globalized world, even the poorest countries are embedded in industrial civilization, both by virtue of transnational interdependence and through governmental links such as development aid and military intervention. This does not always apply to the extent desirable from a humanitarian viewpoint, but in most places and most of the time Malthusian scenarios are successfully prevented by world industrial civilization.

Alas, this applies only as long as world industrial civilization is in a position to bail out places afflicted by overpopulation. In a way, the industrial era with its enormous energy inputs and technological inventiveness may have created a "fool's paradise" which temporarily abrogates the worst effects of overpopulation. Once industrial civilization enters a terminal decline, Malthusian fears may still be vindicated after all (for the "worst case", see Duncan 1993, 2001, 2005, 2007).

---

[1] Even in the case of wild deer, overshoot may lead to a lowering of overall carrying capacity due to various forms of ecological damage. For example, after a cycle of overgrazing an island may be able to sustain fewer deer than previously.

4.2.4 Science Integration

For our present purposes, classical Malthusianism is interesting not only as an intuitively plausible and axiomatically elegant theoretical model to study important phenomena, but also as a paradigm case of science integration. At its core, classical Malthusianism deals with a wide array of vexing ethical and empirical questions pertaining to multiple areas of knowledge, connecting the physical and human sciences and spanning various social scientific disciplines.

Let me simply list a selection of the questions broached and scientific disciplines involved. What are the empirical patterns driving population growth, and how do they operate at the level of individual reproductive choices (population biology, human demography)? How is subsistence affected by various regimes of technological innovation and social distribution, and how is it impacted by a population's level of affluence and food habits such as meat consumption versus vegetarianism (agronomy; food studies)? At what point must a specific territory be considered overpopulated, taking account of the fact that trade and aid can support very high levels of population density in urban areas and countries receiving an inflow of food and other means of subsistence (economics; development studies)? Which social and political mechanisms are triggered by over-population, and under what circumstances (comparative sociology; political science)? When is there a serious risk of population pressure leading to a pandemic (epidemiology)?

4.3 Simple Neo-Malthusian Theories

Simple neo-Malthusian theories apply the logical structure of classical Malthusianism to other important issues of resource management. Like classical Malthusianism, they have a certain commonsensical appeal due to their plausible assumptions and axiomatic elegance. Simple neo-Malthusian theories therefore often play a powerful role in the popular imagination, although more often than not without any direct reference to classical Malthusianism as the source of the tradition.

4.3.1 Environmental Neo-Malthusianism

Environmental neo-Malthusianism is a typical case in point. According to this school, environmental impact (f_1) such as land degradation and biodiversity loss outpaces and strains nature's ability to provide ecosystem services (f_2) such as biomass production and carbon sequestration. The reason is that environmental impact constantly increases, while ecosystem services are either stagnant or declining. After a period of overshoot, the decline of ecosystem services inexorably leads to

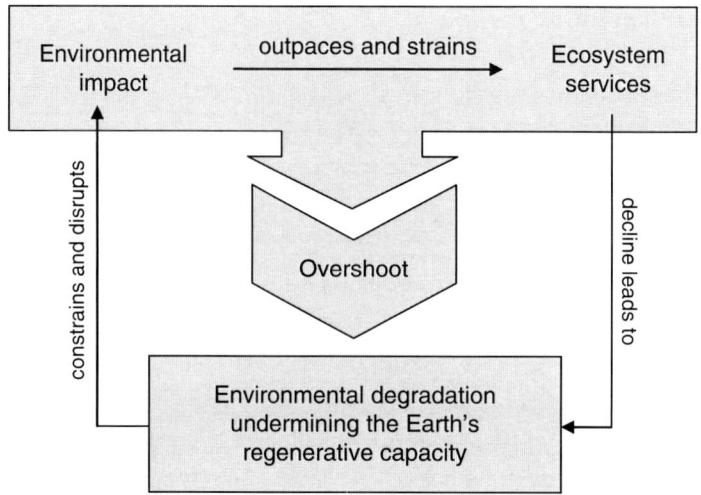

Fig. 4.3 Environmental neo-Malthusianism

environmental degradation, undermining the Earth's regenerative capacity. This must lead to catastrophic consequences, constraining humanity's ability to make further demands on ecosystems and ultimately rebalancing environmental impact with nature's ability to provide ecosystem services (Fig. 4.3).

Environmental neo-Malthusianism is neatly illustrated by ecological footprint analysis, as in the World Wildlife Fund's *Living Planet Report* (WWF 2012).[2] The report closely follows the neo-Malthusian template, with ecological footprint outpacing and straining biocapacity but ultimately constrained by it.

Ecological footprint (f_1) is a measure of environmental impact. It is understood as the land base that would be required to compensate for a given level of environmental impact, most notably greenhouse gas emissions. It is based on the so-called IPAT equation, which specifies environmental impact in terms of population, affluence, and technology (Ehrlich and Holdren 1971). The equation has seen many specifications over the years (Chertow 2000).[3] To cite just one prominent example, Ehrlich et al. (1999, 270) define environmental impact (I) as:

> a product of population size (P), per capita affluence (A) measured as per capita consumption, and the environmental impact of the technologies, cultural practices, and institutions through which that consumption is serviced (T), measured as damage per unit of consumption.

---

[2] Ecological footprint analysis goes back to Wackernagel and Rees (1996) and is also applied by the Global Footprint Network (Ewing et al. 2010).

[3] Most authors interpret IPAT as an equation [$I=P \times A \times T$] or even as an identity [$I = P \times \dfrac{GDP}{P} \times \dfrac{I}{GDP}$], although it is better understood as a complex function allowing for interaction effects between its variables [$I=F(P;A;T)$].

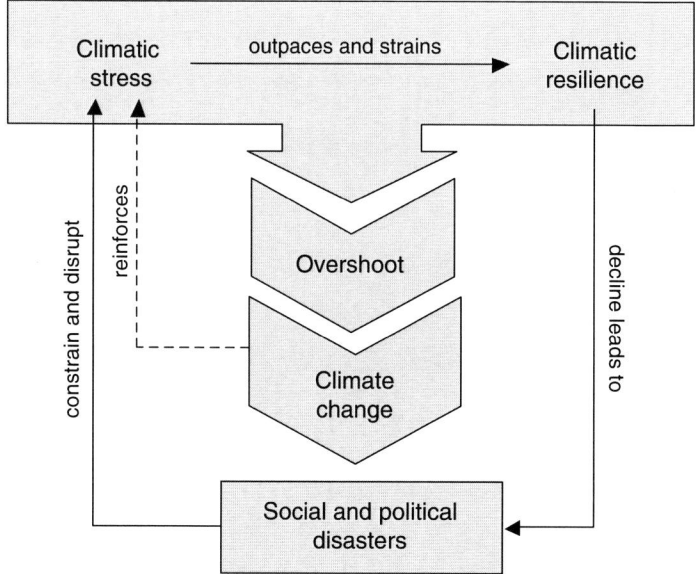

Fig. 4.4 Climate-based neo-Malthusianism

Biocapacity (f_2) is a measure of ecosystem services. It is defined as "[t]he capacity of ecosystems to produce useful biological materials and to absorb waste materials generated by humans" (WWF 2012, 146).[4] Ecological footprint constrains biocapacity insofar as, after a period of overshoot, the overburdening of biocapacity by ecological footprint must lead to dismal consequences such as land degradation and climate change, which in turn must lead to significant social calamities: environmental migration, resource wars, pandemics, and so on. Short of a sustainability transformation, such calamities may be the only way for ecological footprint and biocapacity to return to a long-term global equilibrium.

4.3.2 Climate-Based Neo-Malthusianism

In climate-based neo-Malthusianism (Fig. 4.4), climatic stress (f_1) is understood in terms of atmospheric concentrations of greenhouse gases, most notably CO_2. Climatic resilience (f_2) is almost impossible to measure, but it is usually understood as the ability of the climate system to absorb stresses without exceeding an envelope of change deemed acceptable to human society, such as a maximum global warming

---

[4] Biocapacity is understood here as a specific ecosystem service, namely the bioproductivity of the earth. It is operationalized as the average bioproductivity of a "global hectare", multiplied by the surface of the earth in hectares.

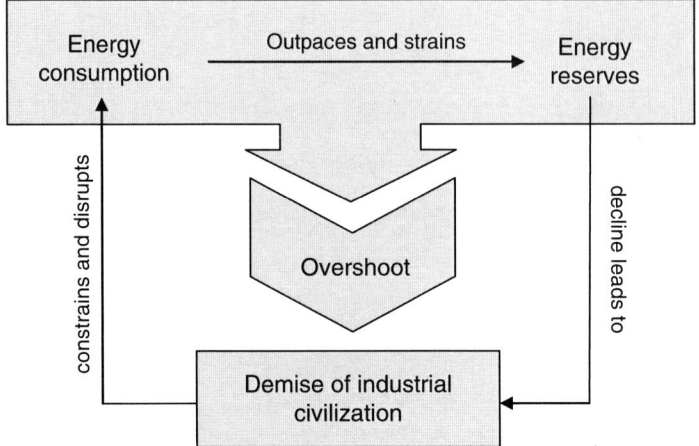

Fig. 4.5 Energy-based neo-Malthusianism

of 2 °C. To the extent that, at the global level, greenhouse gas emissions are an unavoidable collateral of economic growth (Peters et al. 2012), climatic stresses are destined to outpace climatic resilience. Despite considerable time lags in the climate system, climate-based neo-Malthusians warn that the erosion of climatic resilience is destined to operate as a constraint on tolerable levels of climatic stress. When unchecked, climate change is expected to lead to a variety of social and political disasters that may eventually force a concomitant reduction of climatic stress (Dyer 2010; Welzer 2011).[5]

4.3.3   Energy-Based Neo-Malthusianism

Energy-based neo-Malthusians (Fig. 4.5) emphasize, first, that energy consumption (f_1) is a fundamental precondition for economic growth. Second, they point out that energy consumption is constrained by the availability of energy reserves (f_2). They further claim that energy reserves are unavoidably depleted due to their finite nature. Therefore, energy consumption has an inherent tendency to outpace the ability to extract declining energy reserves, with the latter ultimately constraining the former. Insofar as economic growth and human subsistence are tightly linked with energy consumption, energy scarcity will ultimately reverse the growth trajectory and lead to the demise of industrial civilization. This in turn will lead to all sorts of social and political calamities while at the same time constraining future energy consumption (Hubbert 1993; Heinberg 2003; Kunstler 2005).

---

[5] As indicated by the dashed arrow, however, climate change itself may tragically reinforce climatic stress due to feedback mechanisms (Lenton et al. 2008).

4.3.4 Critique of Simple Neo-Malthusianism

Simple neo-Malthusian theories are problematic precisely because they are so simple. For example, it is only a half-truth that economic growth and CO_2 emissions, as well as economic growth and energy consumption, are inextricably linked, as technological innovation can weaken that link by reducing the carbon and energy intensity of GDP. Similarly, it is only a half-truth that energy production is inextricably linked to CO_2 emissions and resource depletion: this appears to be true in the case of non-renewable but not renewable sources of energy. Expanding the share of renewable energy such as solar and wind can weaken the link between economic growth, resource depletion, and climate change.

4.4 Complex Neo-Malthusian Theories

While simple Malthusian theories are limited to the examination of only a couple of functions and the way they outpace and constrain one another, more complex forms of neo-Malthusianism explore how a variety of different trajectories mutually enable and/or constrain each other. This is not to deny that Thomas Malthus has been so much discredited by his detractors that only few modified Malthusian theories openly claim a Malthusian lineage. Based on the logical structure of Malthusianism, however, it is easily possible to identify Malthusian theories even where their complexity goes beyond the original framework.

4.4.1 Limits to Growth

In 1972, a group of MIT researchers around Dennis Meadows applied a complex neo-Malthusian framework to the planetary level and used the emerging method of computer-driven system dynamics, developed by Jay Forrester, to examine the earth system as a whole. In their iconic study *The Limits to Growth* and its two sequels, they compellingly demonstrated that exponential growth on a finite planet is impossible in the long run (Meadows et al. 1972, 1992, 2004).[6]

Meadows and colleagues found that, for a while, the growth of various parameters such as world population, resource consumption, and environmental pollution may appear to defy physical limits, but only until the systemic feedbacks kick in. In the long run, as resource depletion and/or pollution exceed physical limits, an abrupt decline or indeed collapse of industrial society is the only way for the world system to return to equilibrium. The delay between temporary overshoot and ultimate collapse is due to the fact that there are various time lags between anthropogenic

---

[6] For a related warning, see Ehrlich and Ehrlich (2004); see also Bardi (2011).

State of the World

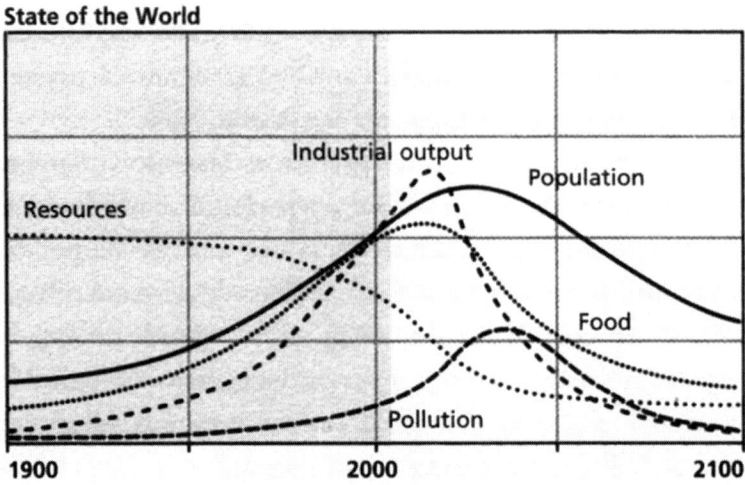

Fig. 4.6 World model standard run (Source: Meadows et al. (2004, 169). Despite some updating, Fig. 4.6 is remarkably similar to its precursor in Meadows et al. (1972, 124))

causes such as resource depletion and greenhouse gas emissions, and systemic outcomes such as energy scarcity and climate change.

The diagnosis of *The Limits to Growth* is a systemic pattern of exponential growth, overshoot, and collapse. Contrary to what their detractors sometimes surmise, Meadows and colleagues did not envision imminent doom. On the contrary, their baseline model, called "standard run", displays a continued pattern of exponential growth and overshoot until about 2010 or 2020, followed by the onset of systemic collapse between 2020 and 2050 (Fig. 4.6).[7]

The end result of the standard run scenario is a contraction of world population to the level of about 1960 by 2100.[8] Shockingly, this implies a dramatic decline by more than two billion people from current levels. However this decline would not happen by starvation alone, as it would occur over several generations and other demographic factors would also play a role: lower birth rates, pandemics, declining life expectancy driven by failing healthcare systems, and so on.

As the model suggests, it is perfectly possible for industrial civilization to "overshoot" and exceed planetary limits for a limited period of time. In the long run, however, no society, and much less the human race as a whole, can live beyond their means. No matter how recklessly we tap into the resources of the earth crust to sustain our unsustainable lifestyles, the improvement of our economic welfare and the increment on global carrying capacity are only temporary.

---

[7] The model is on track with historical data (Turner 2008; Hall and Day 2009).

[8] In the original version (1972, 124), the projected contraction of world population by 2010 was "only" to the level of about 1980.

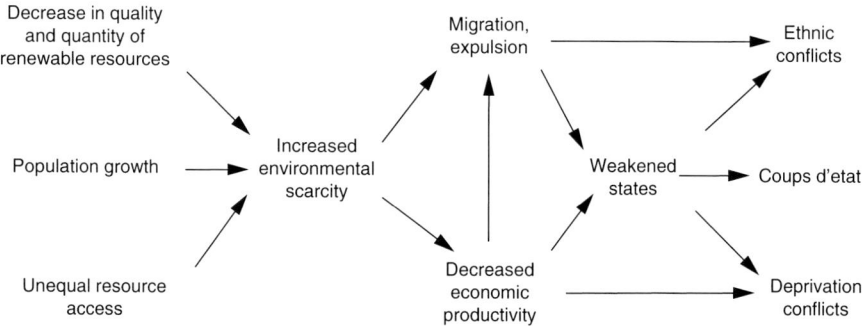

Fig. 4.7   Causal pathways from environmental scarcity to violent conflict

4.4.2   Eco-scarcity Theory

Another complex version of neo-Malthusianism is "eco-scarcity theory", whereby land degradation and other environmental strains combine with population pressure to unleash Malthusian scenarios of social conflict and political disorder.

Eco-scarcity began in the 1990s with conflict theorists suggesting complex causal links between environmental pressure, defined as scarcities of renewable resources, and the outbreak of violent conflict.[9] Their strategy was to collect case studies substantiating the claim that, particularly in overpopulated developing countries, environmental pressure can lead to the outbreak of violence. Two ample collections of case studies were produced roughly at the same time, one by a Canadian team (Homer-Dixon 1994, 1999) and the other by a team based in Switzerland (Bächler et al. 1996). Both of these teams focused on developing countries, and both had the aim of tracing the social processes leading from environmental scarcity, eventually combined with population pressure, to the outbreak of violent conflict. Thomas Homer-Dixon (1994, 31), the leader of the Canadian team, presented these "mechanisms" in a neat causal model (Fig. 4.7).[10]

According to the model, environmental scarcity is triggered by a combination of population growth and excessive strain on some dwindling renewable resource, typically exacerbated by unequal access to that resource. Together with the direct effects of the scarcity itself, the ensuing economic crisis engenders the forcible displacement of people and/or their voluntary emigration. The result is social segregation and a weakening of state structures, both in the country affected by the scarcity and in neighboring countries targeted by a massive inflow of migrants. In some cases this may lead to a coup d'état or even state collapse.

---

[9]For a recent survey, see Bernauer et al. (2012); see also Mildner et al. (2011).

[10]For a slightly modified version of the model, see Homer-Dixon (1999, 134); see also Kahl (2006, 59).

All of this increases the risk of conflict in two different ways. First, scarcity-driven migration may provoke violent clashes between the migrant population displaced by environmental pressure and the recipient population (ethnic conflicts). Second, the economic crisis in the area immediately affected by the scarcity, combined with a declining ability of the state to manage the crisis, can lead to an insurgency of citizens who feel deprived of the standard of living they either feel entitled to, or need in order to survive (deprivation conflicts).

4.4.3   Critique of Eco-scarcity Theory

Eco-scarcity theory is a logically sound extension of the original Malthusian framework which, at least sometimes and in some places, applied before the advent of industrial civilization (LeBlanc 2003); would apply in the absence of industrial civilization; and will again apply after its terminal demise. In the presence of industrial civilization, however, it is an easy target for empirical criticism. The reason for this is that, just as classical Malthusianism, eco-scarcity theory fails to account for the beneficial systemic effects of industrial civilization (see Sect. 4.2.2). Due to this failure, it is easy for critics to come up with countervailing case studies to "falsify" eco-scarcity theory (e.g. Peluso and Watts 2001).

For the same reason, eco-scarcity theory can also be undermined by the application of conventional statistical techniques. Here, the procedure is to collapse eco-scarcity models into bundles of causal factors, with violent conflict as the dependent variable and environmental pressure as the independent variable of interest. Factors intervening in eco-scarcity models, such as the strength of state institutions, are added to the list of independent variables as "controls". This reductive procedure makes it then possible to "test" via correlation analysis whether or not there is a connection between environmental pressure and violent conflict.

While early quantitative scholarship seemed to confirm the claim of a strong and significant causal relationship between environmental pressure and violent conflict, subsequent studies have undermined this belief.[11] Consider the fate of an early quantitative study that found a clear causal link between environmental pressures, such as land degradation and fresh water scarcity, and the risk of domestic armed conflict (Hauge and Ellingsen 1998). Ten years after its publication, the study was replicated by another scholar—and most of its findings turned out to be spurious (Theisen 2008). Overall, the balance of recent quantitative studies do not support the claim that environmental pressure has any statistically significant causal effect on violent conflict (Bernauer et al. 2012).

To be sure, the quantitative literature debunking eco-scarcity theory can itself be criticized. It is problematic to reduce complex social-ecological processes, with their multiple discontinuities and feedback mechanisms, to independent and dependent variables. Insofar as environmental strains and population pressure are remote

---

[11] See for example Urdal (2005); Binningsbø et al. (2007).

causes in complex social-ecological processes, it is unfair to place them alongside more proximate causes such as unequal distribution, ethnic hatred, or inadequate institutions. The danger of reductivism is occasionally recognized even by quantitative scholars: "Conventional statistical techniques run into problems when the relationships to be investigated are of a complex and interactive kind, which is exactly the case for eco-scarcity theory" (Theisen 2008, 814).

And yet, when measured against its own validity claims, eco-scarcity theory is in trouble. The absence of a strong and demonstrable statistical nexus linking environmental pressure with violent conflict questions the applicability of this complex neo-Malthusian school of thought to the analysis of conflict patterns.

That said, however, it is important to recall that the criticism applies only to the recent past. It does not alter the fact that eco-scarcity scenarios may yet be borne out in the near future if industrial civilization enters a terminal decline. Just as the neo-Malthusian proponents of eco-scarcity theory fail to acknowledge that we are still living in the industrial age, their critics fail to appreciate that the durability of industrial civilization cannot be taken for granted in a world entering various forms of geophysical turbulence. Climate change and energy scarcity, either to prevent catastrophic global warming or due to a terminal decline of global oil production, are dramatic game changers that may drive the world towards a post-industrial and post-global age where we may see precisely the complex neo-Malthusian scenarios that have so often been discarded (Friedrichs 2013).

4.4.4 Climate-Based Eco-scarcity

If eco-scarcity theory is a logical extension of classical Malthusianism, then climate-based eco-scarcity is in turn a logical extension of eco-scarcity theory. In essence, it explores the multiple ways by which climate change may lead to environmental scarcity and, thereby, affect the likelihood of violent conflict and other social problems through a variety of social mechanisms such as migration.

The academic debate about climate-based eco-scarcity is a kind of déjà vu in that it tracks the same trajectory as the previous debate about eco-scarcity theory. It started with some authors postulating a causal link between climate change and violent conflict. As is typical for eco-scarcity theory, environmental migration was considered as an important intervening factor (Barnett and Adger 2007; Reuveny 2007). The specific causal mechanisms under scrutiny are also similar to those previously considered by eco-scarcity theorists. Let us take as an example the model outlined in Fig. 4.8 (source: Buhaug et al. 2010, 82).

Like eco-scarcity more generally, climate-based eco-scarcity was countered by arguments based on the statistical analysis of recent events and highlighting the absence of a strong and significant causal link connecting climate change with violent conflict (Raleigh and Urdal 2007; Theisen et al. 2012). Also like in the case of eco-scarcity, even authors representing the variable-based approach sometimes acknowledge that statistical models based on recent historical events are unable to

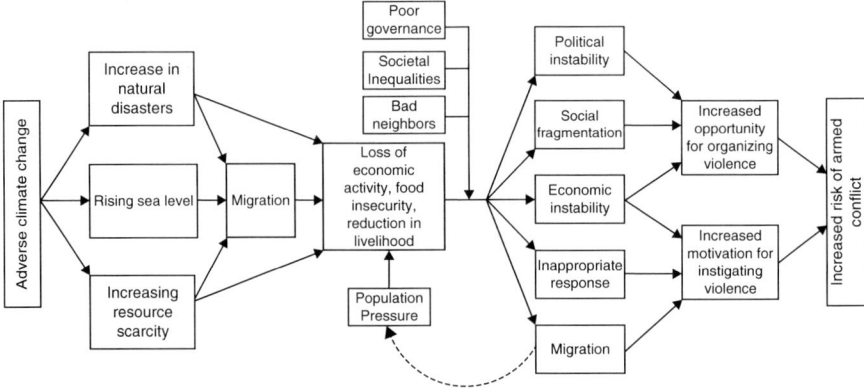

Fig. 4.8 Causal pathways from climate change to violent conflict

predict the conflict dynamics to be expected under abrupt climate change: "We are only beginning to experience the physical changes imposed by global warming […], so a lack of systematic association between the environment and armed conflict today need not imply that such a connection cannot materialize tomorrow" (Buhaug et al. 2010, 93–94).

In fact, climate change of a magnitude similar to what is currently underway has not happened for at least a couple of centuries. Therefore, the statistical analysis of recent events is not empirically adequate to understand the effects of future climate change. Instead, we need to hark back to earlier historical episodes when societies were actually confronted with comparable climatic stresses.

4.4.5   The Future in the Past

Climate-based eco-scarcity has been successfully applied in historical research. Most notably, Zhang and colleagues (2007, 2011) have looked at the period between 1500 and 1800 to understand the social and political effects of climate change. Based on time series from the Northern Hemisphere, especially from Europe but also from China, Zhang et al. (2011, 17298) have come up with a sophisticated causal model that is thoroughly grounded in empirical data (Fig. 4.9).

The model is neatly illustrated by Europe's "general crisis" of the seventeenth century. A drop in average temperature around 1560 was immediately followed by a reduction of bio-productivity, which negatively affected agricultural yields and thus food supply per capita. Over the next 30 years or so, this was followed by cascading escalations of social unrest, migration, famine, war, epidemics, and widespread malnutrition. From 1618, the crisis culminated in the Thirty Years War. Subsequent warfare, together with famines and epidemics, led to a considerable shrinkage of the European population (Zhang et al. 2011).

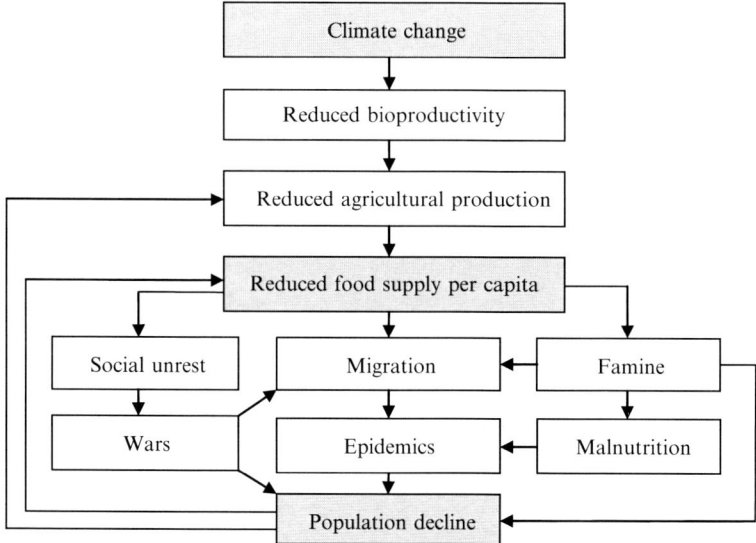

Fig. 4.9 Causal pathways from climate change to large-scale human crisis

When tested against data from the Northern Hemisphere more generally between 1200 and 1800, the expectations derived from the model are largely confirmed. The authors observe strikingly similar macro-patterns for regions as disparate as Europe and China, at a time when Europe and China were largely detached from one another both economically and politically. Zhang et al. (2007) suggest that this synchronicity can hardly be explained unless one assumes that similar social mechanisms were triggered by similar climatic stresses.[12]

4.4.6   Science Integration

While the insights of Zhang and colleagues are of a heuristic nature, the interdisciplinary nature of a research program such as that suggested by Fig. 4.9 is obvious. It takes climatologists, ecologists, and agricultural experts to trace the links between climate change, reduced bioproductivity, and agricultural shortfalls. The link between agricultural production and food supply per capita must be unpacked by social scientists sensitive to political inequality. One level further down, when it comes to the study of social unrest, migration, and famine, we are entering the

---

[12]While Zhang et al. have shown that social and political dislocations in the temperate regions of the Northern Hemisphere are mostly associated with climatic warming, others have demonstrated that the opposite holds for the tropics where warmer El Niño years have always been, and are still, associated with serious social and political trouble (Fagan 2009; Hsiang et al. 2011).

bailiwick of political scientists, economists, and sociologists. The study of war is the turf of international relations scholars, while epidemics and malnutrition are at the intersection of medical and social scientific disciplines. Demographers are competent to study the dynamic of population decline.

Systems scientists and people trained in advanced computer technology would be needed to further refine the operationalization of the model. Because the model is supposed to work across time and space, historians and area specialists would obviously have to actively contribute at every stage of the research cycle. Ironically, however, empirically oriented multidisciplinary papers such as those by Zhang et al. are hardly ever discussed by disciplinary social scientists.

Why do "hard" scientists such as Zhang et al. come up with deductive models, rather than social scientists developing them inductively? It is too comfortable and surely not helpful for social scientists to accuse those who develop complex models of "environmental determinism" while digging in behind disciplinary walls. Social scientists would not have to agree with every detail of such models, but they could make important contributions to improving and refining them.

4.4.7   Civilizational Neo-Malthusianism

Civilizational neo-Malthusianism is perhaps the most original modified Malthusian theory. It states that a civilization's problem solving capacity is depleted as social and technological complexity rises to unsustainable levels.

The classical statement is Joseph Tainter's theory of the emergence, survival, and collapse of complex societies (1988). According to this theory, the fate of societies depends on their ability to adapt to emerging challenges either by an upgrade or by a voluntary downgrade of their systemic complexity. In general, upgrades are obviously the preferred option. They are particularly rewarding at the early stages of civilizational development, when the marginal cost of higher complexity is still low. Later on, the growing marginal cost of complexification makes comparable upgrades gradually more expensive. The strategy of problem solving through complexification becomes entirely punitive at the final stages, when the return on investment in further complexity is negative. Tragically, however, the alternative option of voluntary simplification is hardly available because advanced civilizations are not "downward compatible". They are incapable of a planned reduction of their level of complexity because the existing complexity represents indispensable solutions to real problems. Consequently, involuntary collapse is often the only way for the fragments of the system to reach a new equilibrium.

The fundamental underlying point is that societies are always driven to respond to emerging problems (Wilkinson 1973). These problems can be either exogenous to the society in question, or they can be externalities produced by it. Either way, the logical answer is additional layers of complexity. Tragically, however, complexification has diminishing returns because the easy fixes are implemented first. Moreover, increasing complexity implies increasing costs for the maintenance of

Climatic stress	Dominant response	
10800–9500 BC Younger Dryas	More complexity	Agricultural revolution
3300–3000 BC	More complexity	Rise of urban culture
2500–1950 BC	Temporary increase in complexity, followed by systemic collapse	Northern Mesopotamia: rise and fall of the Akkadian empire Southern Mesopotamia: rise and fall of the Third Dynasty of Ur
1200–850 BC	Collapse	"Dark ages" all across the Old World

Fig. 4.10 The ancient Near East, 11000–1000 BC

that complexity (Homer-Dixon 2006). When the capacity for problem solving has been depleted due to the declining returns on complexification and the escalating cost for the maintenance of the existing level of complexity, only collapse remains because voluntary simplification is not a feasible option.

The framework has sometimes been applied to the rise and fall of civilizations in history. For example, archaeologists such as Weiss (2000) and Ur (2010) have explained the rise and fall of ancient civilizations in Mesopotamia by the initial ability of these civilizations to respond to climatic stresses with more complexity, followed by a later inability to avoid collapse in the face of otherwise similar stresses (Fig. 4.10, from Friedrichs 2013, 62).[13] The theory can be adapted for the diagnosis of current predicaments such as anthropogenic climate change, energy scarcity, or financial instability (Friedrichs 2013, Ch. 3; Korowicz 2010, 2012).

4.5 The Role of Social Science

While natural science is a main driver of unsustainable patterns of industrial development, it also acts as a catalyst for public awareness and political action to address the concomitant sustainability crisis (e.g. climate science). Social science, by contrast, more often than not plays a sedative role. For example, this is seen in energy studies where mainstream economists have largely defined away the problem of scarcity. Mainstream economists staunchly believe that the price mechanism invariably translates demand into supply. If a resource becomes more expensive, more of it

---

[13] See the interesting edited volumes by McIntosh et al. (2000) and Costanza et al. (2007). See also the work by climate historians (Lamb 1977; Fagan 2004, 2008, 2009), as well as Chew (2007, 2008) on the "recurring dark ages" and Greer (2009) on the "ecotechnic future".

will be produced—period. This axiomatic assumption is incompatible with the idea that there are physical limits to industrial growth.[14]

Even social scientific fields explicitly dedicated to environmental issues have a poor record when it comes to preparing the world for the possible demise of industrial civilization. For example, environmental sociology develops policy suggestions for mainstream environmental policy rather than addressing the fundamental unsustainability of industrial society. Similarly, the literature on ecological modernization and green growth pretends that industrial society can be made environmentally viable by technological innovation and incremental social and political reforms, while playing down the dreadful fact that the "treadmill of production" is going round and round while the planet is hopelessly in overshoot.[15]

Even worse, social scientists have been complicit in subverting the notion of sustainability. Originally, sustainability was about socio-political and socioeconomic regimes that are viable in the long run because they do not overstrain the environment. This is a vague regulative ideal that leaves many questions open, but it does imply that political and economic considerations ought to be subordinated to ecological concerns. But then the *Brundtland Report* introduced the notion of sustainable development, based on the optimistic assumption that sustainability and development go together rather than contradicting each other (World Commission on Environment and Development 1987). This has led some social scientists to claim that sustainability has three pillars: environmental, economic, and social (Littig and Grießler 2005).[16] The implication is that, insofar as any economic or social retrenchment is anathema to markets and citizens, suggestions for environmental sustainability that are not palatable to markets and societies must be seen as incompatible with the imperative of economic and social sustainability. This is exactly what the public and political decision makers like to hear, but as a result the original idea of environmental sustainability was turned on its head.

In principle, critical social scientists unsatisfied with the system-stabilizing role of mainstream social science can help us gain a better understanding of the current sustainability crisis and elucidate the moral dilemmas that make it so hard to address it. This does not automatically imply that the crisis can be overcome, but a better understanding of the predicament would be valuable in and of itself. Unfortunately, however, this is not how most critical social scientists are (re)acting. Instead, many have gone post-positivist. Rather than providing any guidance about the precise nature of the crisis and how it might be addressed, they develop sophisticated accounts of how industrial society engages in collective self-delusion (for a survey,

---

[14] Following pioneers such as Karl William Kapp, Nicholas Georgescu-Roegen and E. F. Schumacher, proponents of ecological economics such as Herman Daly, Kenneth Boulding, Robert Costanza, H.T. Odum, and David Pimentel have not been able to pose a significant challenge to mainstream economics. But note the important textbook by Ayres and Warr (2009).

[15] On ecological modernization, see Mol and Jänicke (2009); for a critical survey, see Warner (2010); on green growth, see Ekins (2000); on the treadmill of production, see Gould et al. (2004); see also Mol (2002).

[16] For an ambitious (and upbeat) attempt by a physicist-turned-development-economist to translate this into practice, see Munasinghe (2009).

see Blühdorn 2010). There is nothing fundamentally wrong with this, but it cannot replace a direct focus on the problems themselves.

4.6   Conclusion

Despite the considerable potential for science integration inherent in modified Malthusian theories, mainstream social scientists are generally reluctant to engage in, or even consider, such research programs. To put it in the words of the anthropologist Possehl: "We should stop thinking about the physical world and start looking at the fabric of society" (quoted in Lawler 2008).

Looking at the fabric of society is what social scientists have been doing all along, so what is the actual worry underlying Possehl's statement? Quite obviously, it is fear of transdisciplinary hybridization or bastardization. Indeed, integration with other disciplines may be less desirable to most social scientists than suggested by solemn calls for inter- or multi-disciplinary collaboration.

The main impediment is a refusal on the part of social scientists to accept that societal change can be anything but endogenous to the social sphere (for a critique, see Sørensen 2008). Mainstream social science follows an increasingly counterproductive division of labor whereby physical scientists study the physical world and social scientists study the social world—as if the two were separate and not interconnected. Natural scientists mirror this by a concentration on physical processes, although some are open to neo-Malthusian theories and models.

The self-encapsulation of the social sciences works reasonably well in times of resource abundance and material affluence. It is epitomized by economists reducing scarcity to a problem related to the allocation but not the physical availability of resources, and constructivists cordoning off their scholarship from the analysis of material factors and thus making social change endogenous to self-(re)producing patterns of human interaction. However, the separation between social and physical sciences rests on the cornucopian assumption that industrial society always expands and never contracts. Under conditions of abrupt climate change and looming energy scarcity, social scientists do themselves a disservice by dismissing "materialistic" theories as reactionary or deterministic.

Just like the intersubjective norms that are at the core of social constructivism, resources constrain and enable human action. Precisely for this reason, it is self-defeating for social scientific research to dismiss Malthusian hypotheses. Social scientists should seriously (re-)engage with modified Malthusian theories. As we have, seen some pioneering work has already been done at the fringes of social science, leading to remarkably sophisticated causal models belying knee-jerk allegations of "environmental determinism". Such research not only has the potential to better integrate the social and physical sciences, but it also provides a platform for better integration among social scientific disciplines. It is reasonable to assume that this would also make it easier to communicate the results to the public.

Despite the considerable promise of modified Malthusian theories, fundamental challenges remain. Most if not all existing Malthusian theories operate at the macro-level, whereas work on common-pool resources (Ostrom 1990) operates on a smaller scale. While work on common-pool resources can hardly be scaled up to the macro-level (Levin 2010), it is equally challenging to scale Malthusian theories down to the micro-level. Despite the best efforts made by the International Association for the Study of Society and Natural Resources, the greatest challenge remains to formulate convincing theories that work at an intermediate level, perhaps connecting Malthusian theories with work on common-pool resources.

Postscript

At a conference, one person from the audience objected that Malthusian theories were discredited because of repressive policies that had in the past been justified in their name. This is a serious objection. Nevertheless, the complex neo-Malthusian theories presented in this chapter are a far cry from the original theory formulated by Thomas Malthus. Moreover, shall we not ask the tough questions because we fear that we might not like some of the answers? Is it not better to intrepidly confront those questions, precisely in order to ensure the humane character of the policies and intellectual frameworks formulated in response to them?

References

Ayres, R. U., & Warr, B. (2009). *The economic growth engine: How energy and work drive material prosperity.* Cheltenham: Edward Elgar.
Bächler, G., Böge, V., Klötzli, S., Libiszewski, S., & Spillmann, K. R. (Eds.). (1996). *Kriegsursache Umweltzerstörung* (3 Vols.). Zürich: Rüegger.
Bardi, U. (2011). *The limits to growth revisited.* New York: Springer.
Barnett, J., & Neil Adger, W. (2007). Climate change, human security and violent conflict. *Political Geography, 26*(6), 639–655.
Berkes, F., Colding, J., & Folke, C. (Eds.). (2003). *Navigating social-ecological systems: Building resilience for complexity and change.* Cambridge: Cambridge University Press.
Bernauer, T., Böhmelt, T., & Koubi, V. (2012). Environmental changes and violent conflict. *Environmental Research Letters, 7*(1).
Bierman, F., Abbott, K., Andresen, S., Bäckstrand, K., Bernstein, S., Betsill, M. M., Bulkeley, H., Cashore, B., Clapp, J., Folke, C., Gupta, A., Gupta, J., Haas, P. M., Jordan, A., Kanie, N., Kluvánková-Oravská, T., Lebel, L., Liverman, D., Meadowcroft, J., Mitchell, R. B., Newell, P., Oberthür, S., Olsson, L., Pattberg, P., Sánchez-Rodríguez, R., Schroeder, H., Underdal, A., Camargo Vieira, S., Vogel, C., Young, O. R., Brock, A., & Zondervan, R. (2012). Navigating the anthropocene: improving earth system governance. *Science, 335*, 1306–1307.

Binningsbø, H. M., de Soysa, I., & Gleditsch, N. P. (2007). Green giant or straw man? Environmental pressure and civil conflict, 1961–99. *Population and Environment, 28*(6), 337–353.

Blühdorn, I. (2010, February 17–20). *The politics of unsustainability: Copenhagen, post-ecologism and the performance of seriousness*. Paper presented at the ISA 51st Annual Convention, New Orleans.

Buhaug, H., Gleditsch, N. P., & Theisen, O. M. (2010). Implications of climate change for armed conflict. In R. Mearns & A. Norton (Eds.), *Social dimensions of climate change: Equity and vulnerability in a warming world* (pp. 75–101). Washington, DC: World Bank.

Chertow, M. R. (2000). The IPAT equation and its variants: Changing views of technology and environmental impact. *Journal of Industrial Ecology, 4*(4), 13–29.

Chew, S. C. (2007). *The recurring dark ages: Ecological stress, climate changes, and system transformation*. Lanham: AltaMira.

Chew, S. C. (2008). *Ecological futures: What history can teach us*. Lanham: AltaMira.

Cohen, J. E. (1995). *How many people can the earth support*. New York: Norton.

Costanza, R., Graumlich, L. J., & Steffen, W. (Eds.). (2007). *Sustainability or collapse? An integrated history and future of people on earth*. Cambridge, MA: MIT Press.

Duncan, R. C. (1993). The life-expectancy of industrial civilization: The decline to global equilibrium. *Population and Environment, 14*(4), 325–357.

Duncan, R. C. (2001). World energy production, population growth, and the road to the Olduvai Gorge. *Population and Environment, 22*(5), 503–522.

Duncan, R. C. (2005). The Olduvai theory: Energy, population, and industrial civilization. *The Social Contract, 16*(2), 1–12.

Duncan, R. C. (2007). The Olduvai theory: Terminal decline imminent. *The Social Contract, 17*(3), 141–151.

Dyer, G. (2010). *Climate wars: The fight for survival as the world overheats* (2nd ed.). Oxford: Oneworld.

Ehrlich, P. R., & Ehrlich, A. H. (2004). *One with Nineveh: Politics, consumption, and the human nature*. Washington, DC: Island Press.

Ehrlich, P. R., & Holdren, J. P. (1971). Impact of population growth. *Science, 171*(3977), 1212–1217.

Ehrlich, P. R., Wolff, G., Daily, G. C., Hughes, J. B., Daily, S., Dalton, M., & Goulder, L. (1999). Knowledge and the environment. *Ecological Economics, 30*(2), 267–284.

Ekins, P. (2000). *Economic growth and environmental sustainability: The prospects for green growth*. London/New York: Routledge.

Ewing, B., Moore, D., Goldfinger, S., Oursler, A., Reed, A., & Wackernagel, M. (2010). *The ecological footprint atlas 2010*. Oakland: Global Footprint Network.

Fagan, B. (2004). *The long summer: How climate changed civilization*. London: Granta.

Fagan, B. (2008). *The great warming: Climate change and the rise and fall of civilizations*. New York: Bloomsbury.

Fagan, B. (2009). *Floods, famines and emperors: El Niño and the fate of civilizations*. London: Pimlico.

Friedrichs, J. (2013). *The future is not what It used to be: Climate change and energy scarcity*. Cambridge, MA: MIT Press.

Gould, K. A., Pellow, D. N., & Schnaiberg, A. (2004). Interrogating the treadmill of production: Everything you wanted to know about the treadmill but were afraid to ask. *Organization and Environment, 17*(3), 296–316.

Greer, J. M. (2009). *The ecotechnic future: Envisioning a post-peak world*. Gabriola Island: New Society Publishers.

Haberl, H., Fischer-Kowalski, M., Krausman, F., Martinez-Alier, J., & Winiwarter, V. (2011). A socio-metabolic transition towards sustainability? Challenges for another great transformation. *Sustainable Development, 19*(1), 1–14.

Hall, C. A. S., & Day, J. W. (2009). Revisiting the limits to growth after peak oil. *American Scientist, 97*(3), 230–237.

Hauge, W., & Ellingsen, T. (1998). Beyond environmental scarcity: Causal pathways to conflict. *Journal of Peace Research, 35*(3), 299–317.

Heinberg, R. (2003). *The party's over: Oil, war, and the fate of industrial societies.* Gabriola Island: New Society Publishers.

Homer-Dixon, T. F. (1994). Environmental scarcities and violent conflict: Evidence from cases. *International Security, 19*(1), 5–40.

Homer-Dixon, T. F. (1999). *Environment, scarcity, and violence.* Princeton: Princeton University Press.

Homer-Dixon, T. F. (2006). *The upside of down: Catastrophe, creativity and the renewal of civilization.* London: Souvenir Press.

Hsiang, S. M., Meng, K. C., & Cane, M. A. (2011). Civil conflicts are associated with the global climate. *Nature, 476,* 438–441.

Hubbert, M. K. (1993). Exponential growth as a transient phenomenon in human history. In H. E. Daly & K. N. Townsend (Eds.), *Valuing the earth: Economics, ecology, ethics* (pp. 113–126). Cambridge, MA: MIT Press.

Kahl, C. H. (2006). *States, scarcity, and civil strive in the developing world.* Princeton: Princeton University Press.

Korowicz, D. (2010). On the cusp of collapse: Complexity, energy, and the globalised economy. In R. Douthwaite & G. Fallon (Eds.), *Fleeing Vesuvius: Overcoming the risks of economic and environmental collapse.* Gabriola Island: New Society Publishers.

Korowicz, D. (2012). *Trade-Off: Financial system supply-chain cross-contagion: A study in global systemic collapse.* Bloomsday: Metis Risk Consulting and Feasta.

Kunstler, J. H. (2005). *The long emergency: Surviving the converging catastrophes of the twenty-first century.* New York: Atlantic Monthly.

Lamb, H. H. (1977). *Climate: Present, past, and future. Climatic history and the future* (Vol. 2). London: Methuen.

Lawler, A. (2008). Indus collapse: The end or the beginning of an Asian culture? *Science, 320,* 1281–1283.

LeBlanc, S. A. (2003). *Constant battles: The myth of the peaceful, noble savage.* New York: St. Martin's Press.

Lenton, T. M., Held, H., Kriegler, E., Hall, J. W., Lucht, W., Rahmstorf, S., & Schellnhuber, H. J. (2008). Tipping elements in the Earth's climate system. *Proceedings of the National Academy of Sciences of the United States of America, 105*(6), 1786–1793.

Levin, S. (2010). Crossing scales, crossing disciplines: Collective motion and collective action in the global commons. *Philosophical Transactions of the Royal Society B, 365*(1), 13–18.

Littig, B., & Grießler, E. (2005). Social sustainability: A catchword between political pragmatism and social theory. *International Journal of Sustainable Development, 8*(1), 65–79.

Lutz, W., & Samir, K. C. (2010). Dimensions of global population projections: What do we know about future population trends and structures? *Philosophical Transactions of the Royal Society B, 365,* 2779–2791.

Malthus, T. (1798). *An essay on the principle of population, as it affects the future improvement of society.* London: J. Johnson.

Malthus, T. (1803). *An essay on the principle of population, or, a view of its past and present effects on human happiness.* London: J. Johnson.

McIntosh, R. J., Tainter, J. A., & McIntosh, S. K. (Eds.). (2000). *The way the wind blows: Climate, history, and human action.* New York: Columbia University Press.

Meadows, D. H., Meadows, D. L., Randers, J., & Behrens, W. W., III. (1972). *The limits to growth: A report for the club of Rome's project on the predicament of mankind.* New York: Universe Books.

Meadows, D. H., Meadows, D. L., & Randers, J. (1992). *Beyond the limits: Confronting global collapse, envisioning a sustainable future.* Post Mills: Chelsea Green.

Meadows, D. H., Randers, J., & Meadows, D. L. (2004). *Limits to growth: The 30-year update.* White River Junction: Chelsea Green.

Mildner, S.-A., Lauster, G., & Wodni, W. (2011). Scarcity and abundance revisited: A literature review on natural resources and conflict. *International Journal of Conflict and Violence, 5*(1), 155–172.

Mol, A. P. J. (2002). Ecological modernization and the global economy. *Global Environmental Politics, 2*(2), 92–115.

Mol, A. P. J., & Jänicke, M. (2009). The origins and theoretical foundations of ecological modernization theory. In A. P. J. Mol, D. A. Sonnenfeld, & G. Spaargaren (Eds.), *The ecological modernization reader* (pp. 17–27). London/New York: Routledge.

Munasinghe, M. (2009). *Sustainable development in practice: Sustainomics methodology and applications*. Cambridge: Cambridge University Press.

Ostrom, E. (1990). *Governing the commons: The evolution of institutions for collective action*. Cambridge: Cambridge University Press.

Peluso, N. L., & Watts, M. (Eds.). (2001). *Violent environments*. Ithaca: Cornell University Press.

Peters, G. P., Marland, G., Le Quéré, C., Boden, T., Canadell, J. G., & Raupach, M. R. (2012). Rapid growth in CO_2 emissions after the 2008 global financial crisis. *Nature Climate Change, 2*(1), 2–4.

Raleigh, C., & Urdal, H. (2007). Climate change, environmental degradation and armed conflict. *Political Geography, 26*(6), 674–694.

Reuveny, R. (2007). Climate change-induced migration and violent conflict. *Political Geography, 26*(6), 656–673.

Seidl, I., & Tisdell, C. A. (1999). Carrying capacity reconsidered: From Malthus' population theory to cultural carrying capacity. *Ecological Economics, 31*(3), 395–408.

Sørensen, G. (2008). The case for combining material forces and ideas in the study of IR. *European Journal of International Relations, 14*(1), 5–32.

Tainter, J. A. (1988). *The collapse of complex societies*. Cambridge: Cambridge University Press.

Theisen, O. M. (2008). Blood and soil? Resource scarcity and internal armed conflict revisited. *Journal of Peace Research, 45*(6), 801–818.

Theisen, O. M., Holtermann, H., & Buhaug, H. (2012). Climate wars? Assessing the claim that drought breeds conflict. *International Security, 36*(3), 79–106.

Trewavas, A. (2002). Malthus foiled again and again. *Nature, 418*, 448–670.

Turner, G. M. (2008). A comparison of the limits to growth with 30 years of reality. *Global Environmental Change, 18*(3), 397–411.

UN. (2011). *World population prospects: The 2010 revision. Highlights and advance tables*. New York: United Nations.

Ur, J. A. (2010). Cycles of civilization in northern Mesopotamia, 4400–2000 BC. *Journal of Archaeological Research, 18*(4), 387–431.

Urdal, H. (2005). People vs. Malthus: Population pressure, environmental degradation, and armed conflict revisited. *Journal of Peace Research, 42*(4), 417–434.

Wackernagel, M., & Rees, W. E. (1996). *Our ecological footprint: Reducing human impact on earth*. Gabriola Island: New Society.

Walker, B., Holling, C. S., Carpenter, S. R., & Kinzig, A. (2004). Resilience, adaptability and transformability in social-ecological systems. *Ecology and Society, 9*(2).

Warner, R. (2010). Ecological modernization theory: Towards a critical ecopolitics of change? *Environmental Politics, 19*(4), 538–556.

Weiss, H. (2000). Beyond the younger Dryas: Collapse as adaptation to abrupt climate change in ancient West Asia and the Eastern Mediterranean. In G. Bawden & M. Reycraft (Eds.), *Environmental disaster and the archaeology of human response* (pp. 75–98). Albuquerque: Maxwell Museum of Anthropology.

Welzer, H. (2011). *Climate wars: Why people will be killed in the twenty-first century*. Cambridge: Polity.

Wilkinson, R. G. (1973). *Poverty and progress: An ecological model of economic development*. London: Methuen.

Winch, D. (1987). *Malthus*. Oxford: Oxford University Press.

World Commission on Environment and Development. (1987). *Our common future*. Oxford: Oxford University Press.

WWF. (2012). *Living planet report 2012: Biodiversity, biocapacity and better choices*. Gland: WWF.

Zhang, D. D., Brecke, P., Lee, H. F., He, Y.-Q., & Zhang, J. (2007). Global climate change, war, and population decline in recent human history. *Proceedings of the National Academy of Sciences of the United States of America, 104*(49), 19214–19219.

Zhang, D. D., Lee, H. F., Wang, C., Li, B., Pei, Q., Zhang, J., & An, Y. (2011). The causality analysis of climate change and large-scale human crisis. *Proceedings of the National Academy of Sciences of the United States of America, 148*(42), 17296–17301.

Chapter 5
A Conceptual Framework for Analyzing Social-Ecological Models of Emerging Infectious Diseases

Melissa L. Finucane, Jefferson Fox, Sumeet Saksena, and James H. Spencer

5.1   Introduction

Unraveling mechanisms that underlie new and reemerging infectious diseases (EID) requires exploring complex interactions within and among coupled natural and human (CNH) systems. This scientific problem poses one of the most difficult challenges for society today (Wilcox and Colwell 2005). EID are diseases that have recently increased in incidence or in geographic or host range (e.g., tuberculosis, cholera, malaria, dengue fever), and diseases caused by new pathogens and new variants assigned to known pathogens (e.g., HIV, SARS, Nipah virus, and avian influenza) (Morse 2005). Wilcox and Gubler (2005) and Wilcox and Colwell (2005) argue that transformations in ecological systems caused by multifaceted interactions with anthropogenic environmental changes such as urbanization, agricultural transformations, and natural habitat alterations produce feedbacks that affect natural communities and ultimately their pathogens, animal host, and human populations. These altered "host-pathogen" relationships facilitate pathogen spillover into "new" hosts, rapid adaptations by pathogens, more frequent generation of novel pathogen variants that result in new and reemerging infectious diseases, as well as range expansion and increasing epidemic intensity and frequency of existing diseases.

In this chapter we present a conceptual framework for examining the Wilcox-Gubler-Colwell hypothesis in the context of whether risks, and perceptions of risk,

M.L. Finucane (✉)
Behavioral and Policy Sciences, RAND Corporation,
4570 Fifth Ave, Suite 600, Pittsburgh, PA 15213-2665, USA
e-mail: finucane@rand.org

J. Fox • S. Saksena
The East-West Center, 1601 East-West Road, Honolulu, HI 96848-1601, USA

J.H. Spencer
Department of Planning, Development and Preservation, Clemson University,
Lee Hall, Box 340511, Clemson, SC 29634-0511, USA

M.J. Manfredo et al. (eds.), *Understanding Society and Natural Resources*,
DOI 10.1007/978-94-017-8959-2_5, © The Author(s) 2014

associated with highly pathogenic avian influenza (HPAI) caused by the H5N1 virus,[1] as measured in terms of poultry deaths, can be associated with anthropogenic environmental changes produced by urbanization, agricultural change, and natural habitat alterations. This is a novel way of looking at HPAI and other health risks like it, suggesting these risks are not an accident of time and place, but rather are the product of the modernization and urbanization transitions.

We present the conceptual framework in the context of Vietnam because (1) it has been one of the nations most affected by HPAI, (2) has a rapid rate of development, and (3) has very comprehensive secondary databases relevant to such studies. The emergence of the HPAI was first reported in Vietnam at the end of 2003 (Delquigny et al. 2004). Three major epidemic waves of HPAI have occurred in poultry, resulting in 45 million birds culled between December 2003 and August 2005, leading to a 0.5 % reduction in GDP in 2004. As of 2012, a total of 123 confirmed human cases and 61 deaths were recorded (World Health Organization 2012). The country has attempted to control the infection through massive, repeated vaccination campaigns in combination with other control measures (Gilbert et al. 2008a). Vietnam is a particularly useful country to examine development transitions and their associated environmental transformations because these processes are occurring both exceptionally rapidly and simultaneously as traditional agricultural lands are converted to intensified commercial farming or reshaped into urban settlements to meet the needs of the growing population attracted to cities for emerging job opportunities (Douglass et al. 2002; Spencer 2007). If development transitions do pose new challenges to governance, and in particular environmental health challenges, then one would expect to see more of these types of problems in transitional agricultural or peri-urban areas as distinct from both predominately urban and rural areas.

The Wilcox-Gubler-Colwell hypothesis of disease emergence was influenced by complexity theory, which argues that as complex adaptive systems (CAS) CNH systems exhibit far-from-equilibrium non-linear behavior often manifested as "surprise" as with the case of abrupt and unexpected epidemiological phenomena including the emergence of entirely new diseases. Parallel to this, in biological science a re-envisioning has occurred in which nature is no longer seen as consisting of balanced ecological systems made up of relatively linear processes. Rather, natural systems are now seen as hierarchical, self-organized, non-equilibrial and non-linear systems in which emerging diseases can themselves be seen as "emerging properties" of these CAS (Levin 1999). The traditional conception of the ecosystem, a fundamental paradigm in the ecological sciences, has thus been overturned (O'Neill 2001).

As such, ecosystems, including "social-ecological systems" are now understood as characteristically producing emergent phenomena like the unexpected appearance of new pathogens, inherently unpredictable by conventional approaches and theory (e.g., epidemiological models). Moreover, nearly all emerging diseases are vector borne or zoonotic (i.e., maintained in natural host-pathogen cycles that "spill over" to humans) (Woolhouse and Gowtage-Sequeria 2005). They, or their immediate progenitors, exist as part of naturally co-evolved host-parasite complexes

---

[1] In this chapter HPAI refers to HPAI caused by the H5N1 strain.

embedded within ecosystems, whose dynamics normally include non-linear, cross-scale behavior (Horwitz and Wilcox 2005). Based largely on complexity theory, social ecological systems and resilience (SESR) theory was developed to account for the non-linear dynamic behavior of CNH systems that result from intervention in "managed" ecosystems that unwittingly lead to unexpected and sometimes catastrophic outcomes, including disease re-emergence.

Holling (1973) introduced the concept of resilience into the ecological literature as a way of understanding nonlinear dynamics, such as the processes by which ecosystems maintain themselves in the face of change. In a resilient forest ecosystem, for example, four phases of change repeat themselves again and again. The first two phases, exploitation (the establishment of pioneering species) and conservation (the consolidation of nutrients and biomass), lead to an old growth forest or a climax community. But this climax community invites environmental disturbances such as fire or disease, and is more susceptible to disturbances than non-climax forests. When surprise or change occurs, the accumulated capital is suddenly released producing other kinds of opportunity, termed creative destruction. Release, a very rapid stage, is followed by reorganization in which, for example, nutrients released from the trees by fire will be fixed in other parts of the ecosystem as the renewal of the forest starts again (Berkes et al. 2003). Holling suggested that human societies also reproduce and reinvent in the process of cyclic transformations; he writes: "The bewildering, entrancing, unpredictable nature of nature and people, the richness, diversity and changeability of life come from the evolutionary dance generated by cycles of growth, collapse, reorganization, renewal and re-establishment" (Holling 2003: xv).

5.2 Integrating Social Science Theories Relevant to Development Transitions

Several social science theories relevant to the notion of types of development transitions are also relevant to both the Wilcox-Gubler-Colwell hypothesis and Holling's resilience theory. Our starting point is the theory of the Environmental Kuznet's Curve (EKC) from the discipline of development economics. Nobel Prize winner Simon Kuznets proposed that with respect to inequality, economic development is not linear but rather an inverted 'u' shape; economic development is a transition from an initial state of relative equality to an end state also of relative equality, but in the midst of economic development nation-states display high levels of economic inequality (line C in Fig. 5.1 is a Kuznets curve). While various scholars have shown that Kuznets curves are not universally applicable (see Park et al. 2007), Kuznets curves have become a simple but powerful method for empirically testing hypotheses about transitional states. The EKC hypothesizes that certain indicators of environmental degradation tend to get worse as modern economic growth occurs until average income reaches a certain point over the course of development (Grossman and Krueger 1995). Urban air pollution and deforestation have been cited as examples of environmental quality variables that follow the EKC. Recent

Fig. 5.1 Environmental risk
transition framework

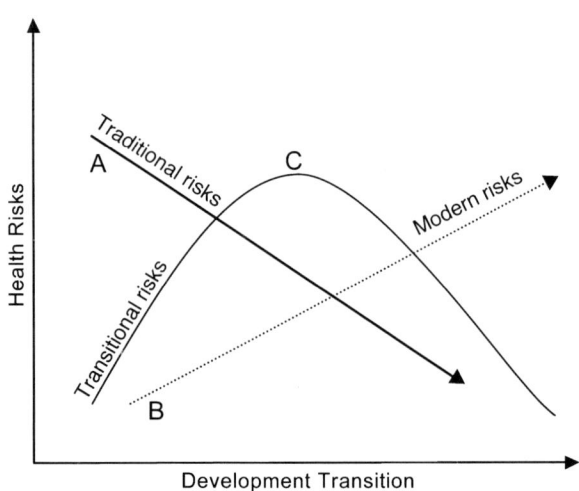

empirical work has applied this understanding to the issue of HPAI (Spencer 2013), indicating that disease occurrence may be most likely where settlements are undergoing the most intense transitions. Such evidence suggests that a deeper understanding of "transition" is warranted.

While some environmental quality indicators such as landfills and biodiversity do not seem to follow the EKC, supporters of the theory have argued that this may be more to do with issues of scale. Traditionally, most of the empirical work on EKC has been based on inter-country analysis of cross-sectional data. Often global regions (groups of countries) have been the unit of analysis. Most commonly, the metrics chosen for the predictor variable have been Gross Domestic Product (GDP) or GDP per capita, often adjusted by purchasing power parity or the Human Development Index (HDI). To address the scale issue we are examining whether the EKC is valid at the lowest level of government administration–the commune or municipal ward—for the entire country of Vietnam using readily available census data. The outcome variable is HPAI in domestic poultry. We are faced with the challenge that at the commune/ward level metrics such as GDP or HDI are not valid or are difficult to measure (e.g., because data are not available). Furthermore, using a binomial variable—whether a place is rural or urban (as classified by the government) has two problems: (1) one cannot test the non-linearity of the curve and (2) it contradicts our fundamental premise that a significant number of places are not easily classifiable as being either rural or urban. Thus, we are forced to find a new metric reflecting development, which is ordinal with at least three levels. The "urbanicity" method from the field of urban geography is useful for classifying place (Allender et al. 2008; Dahly and Adair 2007; McDade and Adair 2001; Vlahov and Galea 2002). Though most of the urbanicity metrics in the literature are continuous scale metrics, we adapt the principles to create an ordinal scale metric.

Another social science theory relevant to understanding the relationship between development transitions and EID comes from Smith (1990), who proposed an environmental risk transition where the environmental factors leading to ill health were categorized as traditional or modern. This categorization is based on the premise that the major environmental causes of traditional diseases are problems at the household level (e.g., water, sanitation, food availability and quality, ventilation and indoor air pollution). As these are addressed during development there is an increase in the relative importance of the major environmental causes of more modern diseases which operate at the community level (i.e., urban air quality, occupational hazards, toxic chemicals, and motorization). As these are addressed in richer societies a further transition occurs to increase the importance of environmental hazards at the global level (e.g., global warming, land-use change) (Holdren and Smith 2000; McGranahan et al. 2000; Smith and Akbar 2003).

Figure 5.1 shows the environmental risk transition framework in which traditional risks fall with social and economic development, transitional risks rise and then fall, and modern risks rise throughout the development process. Smith and Ezzati (2005) write that limited or no research has attempted to apply this framework to emerging and reemerging infectious diseases caused by evolving human activities such as those associated with trade, tourism, terrorism, and human interactions with natural environments.

Others have hypothesized forest (a natural habitat) (Grainger 1995; Mather 1992, 2007; Rudel 1998), agrarian (Hall 2004; Rigg 2005), and urban (Douglass 2000; Friedmann 2005) transitions. A forest transition occurs when an initial surge in economic activity spurs deforestation, but as economic activity continues to intensify and cities grow larger, a 'turnaround' occurs, and deforestation gives way to reforestation. The agrarian transition has been defined as a number of inter-related phenomena. These include agricultural extensification and intensification (the amount of agricultural land is hypothesized to follow a Kuznets curve as extensification precedes intensification, but intensification then leads to the abandonment of marginal land); increased integration of production into market-based systems of exchange; heightened mobility of populations both within and across national borders as people are attracted to opportunities both within and outside of the agricultural sector; and processes of environmental change that reflect new human impacts and new valuations of resources (Akram-Lodhi 2004; Rigg and Vandergeest 2012; De Koninck 2004). The urbanization transition includes two parallel processes: population concentration (population is hypothesized to increase linearly) and the development of socio-physical infrastructure to manage the inevitable conflicts and problems associated with higher density living (infrastructure is hypothesized to increase linearly). The urban transition in developing countries describes societies that have rapidly changed from rural to urban forms of social and physical organization in relatively short time periods (Douglass 2000; Montgomery et al. 2004) such as those found in Southeast Asia.

The broader implications of these simultaneous and related transitions remain unexplored in general, and more specifically, as Wilcox and Colwell (2005) argue, in relation to how they produce feedbacks that affect natural communities and ultimately their pathogens, animal host, and human populations. A better understanding of the relationship between development transitions and EID is critical for improving our ability to predict and respond to EID. This is particularly true in Vietnam where government policies have facilitated what can broadly be called a "transition to the market" (Arkadie and Mallon 2003; De Vylder 1990; Fforde and Vylder 1996). Economic policies have driven changes in the built environment that have created new ecological health risks (Oliveira et al. 2004; Smith 1997), migration to cities has simultaneously uprooted residents from local social networks and placed them into new neighborhood associations, water user-groups, and other forms of social organization (Crane 1994; Spencer 2007). These new socio-physical ecologies present new challenges that, in turn, require new forms of social organization and governance, many of which do not yet exist, to provide basic services such as water and sanitation, education, housing, and public health.

A final social science perspective that has received little attention to date in the developing world relates to theories of behavioral decision making and perceptions of risk. This field draws primarily from economics, psychology, philosophy, anthropology, and cognitive science. Researchers have developed tools such as the Social Amplification of Risk Framework (Kasperson et al. 2003; Pidgeon et al. 2003) to describe and explain the societal processing of risk signals, but tests of such frameworks are rare because of the difficulty in predicting when risk amplification conditions are likely to occur (Frewer et al. 2002). A central insight of decision theory is that risk responses are based on socially constructed perceptions of risk. That is, risk means different things to different people; it cannot be measured independent of our minds and cultures. People prioritize risks in different ways, depending on their beliefs about the need to try to reduce a risk (Douglas and Wildavsky 1982; Hofstede 1984; Park et al. 2007). Research conducted in the developed world suggests that perceived risk is related strongly to feelings of control and trust (Slovic 2000). Some authors suggest that people perceive low risk during modernization because they feel they are in control of technology, nature, or society and that regulatory authorities can be trusted (Bauman 1992; Beck 1999; Giddens 1992). Over time, however, perceived risk increases as feelings of control and trust are eroded. Some people will respond swiftly and comprehensively to a risk event and others will respond more slowly, depending on a range of psychological and socio-cultural variables and environmental conditions. However, the relative importance of various elements in CNH systems (i.e., socio-ecological and socio-psychological factors) in determining perceptions of and responses to the risk of EID in rapidly developing societies has not been examined.

To understand the relationships between characteristics of decision makers, their environmental context, and their risk responses, the field of decision science uses a wide range of methodological approaches and analytical tools. Qualitative interviews and focus groups permit an in-depth exploration of risk perceptions and responses, allowing participants to describe beliefs and experiences in their own words, rather than as a choice between predetermined survey responses (Pope and Mays 1995). These methods are useful in defining the range and variability of conceptualizations of risks such as disease outbreaks and how they might relate to environmental change (O'Brien 1993). Quantitative methods (e.g., decision analysis, process tracing, surveys) are more readily applied with larger samples and allow precise measurement of the information integration strategies decision makers use to determine their risk response. Applied to the problem of understanding the relationship between modernization and EID, we can test whether variation in individuals' risk responses is related to environmental change as represented in our degree of modernization metric.

Combining decision and risk research methods applied at the household level with environmental economics methods applied at the commune- and national-level analyses provides an opportunity to look for converging evidence for hypothesized relationships between constructs in our social-ecological model of EID. Identifying variation in risk responses to EID is only of theoretical interest if it furthers our understanding of the causal variables and structure in the coupled natural-human system that lead to the behavior. An understanding of the system is also crucial to making such results practically useful. For instance, observing variance in HPAI risk responses in Vietnam provides an empirical basis for making predictions about other diseases and other developing countries that have not been studied directly.

In sum, the conceptual framework introduced in this chapter relies heavily on the integration of multiple social science theories and methods from diverse disciplines (e.g., environmental economics, geography, decision and risk science, urban and regional development, and spatial information science). Identifying key components of these theories relevant to the notion of types of development transitions provides a coherent approach to analyzing the complex interactions among natural and human systems at diverse spatial, temporal, and organizational scales. In the next section we provide reasons for the choice of important elements to characterize CNH systems.

5.3   Anthropogenic and Ecological Determinants of HPAI in Southeast Asia

Kapan et al. (2006) hypothesized that the on-going process in Southeast Asia of replacing traditional farming methods such as multi-species livestock husbandry with industrial, mass-production-oriented operations poses significant environmental health risks (e.g., Mallin and Cahoon 2003) due to increases in livestock pools and thus opportunities for disease transmission. Simultaneously, rapid urban and peri-urban development in these countries has often been accompanied by more refuse, standing water, and animals in and around homes that have been correlated with environmental health risks (e.g., Graham et al. 2004). With respect to HPAI, expansion of the urban fringe has placed a larger proportion of the human population in contact with formerly dispersed farm environments that include potentially infected poultry and swine populations. Such urban–rural interfaces have been hotspots of other infectious diseases such as leishmaniasis (Oliveira et al. 2004).

An array of anthropogenic and ecological studies of the determinants of HPAI in Southeast Asia has supported these hypotheses. Gilbert et al. (2006, 2007) showed that the interaction of poultry and particularly domestic duck populations within the rice paddy production system was as an important factor for the maintenance and spread of HPAI virus in Thailand. Pfeiffer et al. (2007) showed that rice paddy production intensity and density of domestic chickens and water birds were also associated with a higher risk of HPAI outbreaks in Vietnam, lending support to the rice-duck-chicken hypothesis. The same study showed that increased distance from high density human population areas consistently decreased HPAI risk (Pfeiffer et al. 2007). The study finds support for the hypothesis of "the presence of a fairly widespread infection reservoir in Vietnam …, possibly in domestic and wild water birds" (Pfeiffer et al. 2007). Gilbert et al. (2008b) demonstrated that a few key factors such as human population density, rice cropping intensity, and to some extent poultry density, managed to explain a large proportion of the spatial variation in HPAI disease risk; the same study also notes that considerable variation remained unexplained, and suggests that other factors such as poultry production and marketing systems, agricultural seasonality, the potential for contacts between domestic and wild birds, and climatic and other conditions affecting the persistence of the virus in the environment should be considered. Fang et al. (2008) found the minimal distance to the nearest national highway, annual precipitation, and the interaction between minimal distance to the nearest lake and wetland, were important predictive environmental variables for the risk of HPAI in China. A study of post-vaccination outbreaks in southern Vietnam found poultry flock density, fraction of houses with electricity, rescaled Normalized Difference Vegetation Index, buffalo density and sweet potato yield to be significant risk factors (Henning et al. 2009).

Of particular interest to this study is the claim by Gilbert et al. (2014) that the highest risks of HPAI impact in Southeast Asia are to be expected where extensive and intensive systems of poultry production co-exist. The extensive systems allow virus circulation and persistence; the intensive systems promote disease evolution. A study in Thailand found differences in avian influenza risk rates across scale of operations (Otte et al. 2006), which was attributed to bio-security (waste management) features.

Spencer (2013) sought to establish whether bird deaths followed a Kuznets curve as settlement infrastructure patterns evolved. Vietnam's 1999 Census of Population and Housing provides counts of households by housing construction materials (traditional/temporary or modern), water supply (stream, rain, well, piped), and sanitation infrastructure (none, pit, composting, flush). Spencer converted each of these 4-category, ranked urbanization measures into four distinct measures of settlement "coherence". For each coherence measure, greater mixing (i.e. incoherence) of the four categories was set to center on a value of zero, with more "traditional" settlement a mixture dominated by the least sophisticated (e.g. no toilet) of each response category valued at (−1), and the most "modern" settlements a mixture dominated by the most sophisticated (e.g. running water) of each response category valued at (+1). Working at the district level, Spencer plotted these three coherence indices, as well as a composite index combining the three, against the probability that a district in any of Vietnam's provinces (including cities) had an outbreak of HPAI in 2004 or 2005. After accounting for a minimal threshold effect of development, Spencer (2013) found a distinct Kuznets relationship exists between settlement coherence and HPAI. In particular, the sanitation coherence index explained over one third of variance in outbreaks (R-square=0.37, bivariate), and the water supply coherence index explained over half (R-square=0.56, bivariate). Overall, the findings suggest that the urban infrastructure transition is associated with HPAI outbreaks in poultry and may be used as a general predictor of emerging infectious disease risk.

These initial findings illustrate the potential theoretical contributions of a transitional approach to the study of HPAI. This suggests that for the urbanization measure, at least those measures centered on water supply and sanitation, the basic function may be a Kuznets curve rather than a linear or a more complex curve. Our current project is conducting similar exercises for agricultural change and habitat alteration. We are developing transition indices for agricultural change and habitat alteration, plotting them against the probability of HPAI outbreak, and choosing the curve that best fits the data. A twice-changing slope as the best fit would suggest a more complex fluctuation of risk between traditional and modern landscapes, and a u-shaped curve would suggest that transitional landscapes are associated with reduced risk.

Lastly, we are examining how perceptions of HPAI risk vary with urbanization, agricultural change, and habitat alteration. Of the few studies that have examined determinants of HPAI risk perceptions, all have focused on perceived risk of HPAI to humans (rather than perceived risk to the health of poultry). Three studies conducted in Asian countries (de Zwart et al. 2007; Fielding et al. 2005; Figuié and Fournier 2008) showed perceived human risk was correlated with demographics (women and older people perceived more risk) and efficacy (perceived availability of protective actions and ability to engage in those actions led to lower perceived risk). In Laos, Barennes et al. (2007) reported that protective behavior was more likely with higher levels of education, urban living, knowledge of HPAI, and owning poultry. Only one of the above studies (Figuié and Fournier 2008) was conducted in Vietnam. No studies have examined the relative importance of socio-ecological variables (urbanization, agricultural change, habitat alteration) versus socio-psychological variables (efficacy, knowledge, affective response, risk avoidance, demographics) in determining perceptions of risk to the health of poultry. Moreover, no studies have examined whether risk perceptions and protective behaviors vary across traditional, transitional, and modern settings or with observed risk (poultry deaths).

The framework we are proposing is based on an assumption that risk management policies need to be derived from a broad-based understanding of how decision makers perceive, explain, and prioritize risk. An analysis of EID risk that focuses only on socio-ecological variables will not reveal the socio-psychological differentiation of individuals who are more or less successful in responding to and managing EID outbreaks. Currently, however, there exists a gap in knowledge about the underlying mechanisms that explain variation in perceptions of the risk of EID and how these perceptions vary with social-ecological transition. Furthermore, most research on perceived risk has been done in democratic, Western countries, not in a context where there is tight state control of key institutions that interpret and disseminate disease risk information. To further understand how EID risk signals are processed by individuals in the context of CNH systems in Vietnam, we need to examine in depth the processes through which people collect and integrate data from natural and human systems. This work will advance basic knowledge about the complex linkages among ecological, social, and psychological variables that amplify or attenuate the intensity and frequency of EID.

In sum, literature suggests an array of important elements characterize CNH systems. Using unidimensional measures, Spencer (2013) showed that the relationship between transition and disease can be examined empirically. However, more research is necessary to understand the complex interactions among natural and human elements at diverse spatial, temporal, and organizational scales, and how they relate to EID outbreaks.

5.4   Developing and Testing the Framework

Empirical studies on CNH system change and disease emergence depend on the assembly of a diverse set of independently generated neighborhood- and landscape-level data accurately matched with spatially aggregated or 'point' level data. In order to make such analyses possible we need an extensive spatial database that includes vector layers representing landscape characteristics (e.g., ecoregions, geology, soils, protected area boundaries, human settlements, road infrastructure) from available hardcopy maps, digital data, aerial surveys, global positioning system (GPS) data, and satellite image feature extraction, e.g., urban features (Zhang et al. 2002) and paddy fields (Xiao et al. 2005, 2006). We have also acquired national census data and other economic, demographic, institutional and cultural databases that describe socioeconomic variables at the household and commune level and that have been linked to biophysical data via a geographic information system (GIS) (Epprecht and Heinimann 2004; Epprecht and Robinson 2007). These databases can also provide information on the location and size of villages, roads, streams, and agricultural fields. Information collected on specific landscapes through interviews with farmers and key informants can be keyed to these databases using handheld GPS devices.

We can use these multivariate databases to test models of general relationships hypothesized between HPAI and urbanization, agricultural change (both crop land and poultry), and habitat alteration at the national level using commune level data. For instance, we can test an *a priori* specific structural equation model (SEM) based on extant literature. In Fig. 5.2, we present a model in which we hypothesize that outbreaks of HPAI in poultry are associated with variations in three latent constructs: (1) Urbanization, measured in terms of changes in quality of housing and drinking water supplies (Spencer 2013); proximity to cities (Pfeiffer et al. 2007) and major roads (Fang et al. 2008), and human population (Gilbert et al. 2008b); (2) Habitat Alteration, measured in terms of changes in wetlands (Fang et al. 2008; Gilbert et al. 2007), amount and diversity of natural water sources (Fang et al. 2008);

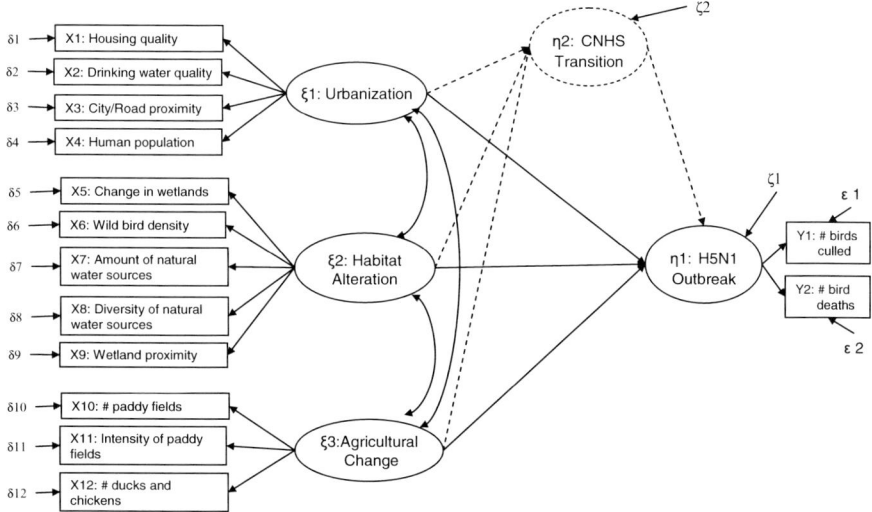

Fig. 5.2 Structural equation models (SEMs) of hypothesized relationships between HPAI outbreak and urbanization, habitat alteration, and agricultural change (both crop land and poultry) at the national level. Model 1 (*solid lines*) proposes an *a priori* SEM based on existing literature; model 2 (*dashed lines*) proposes an explanatory higher-order factor

changes in proximity to wetlands (Fang et al. 2008); and (3) Agricultural Change, measured in terms changes in terms of number and intensity of paddy fields (Gilbert et al. 2007, 2008b; Pfeiffer et al. 2007), and number of ducks and chickens (Gilbert et al. 2006, 2007, 2008b; Pfeiffer et al. 2007).

The second model shown in Fig. 5.2 (dashed lines) tests whether the relations among urbanization, habitat alteration, and agricultural change are attributable to a common higher order influence. While the first model acknowledges the existence of relations among the three latent constructs, it does not explicitly represent cause of covariation. The second model postulates that correlations among the latent constructs can be explained by a higher-order factor. We can thus examine direct and indirect pathways between the latent constructs and HPAI outbreak. The dashed lines in Fig. 5.2 represent the indirect pathways and higher-order factor, which we call CNH system Transition.

Evaluating these models is quantitatively challenging because the concepts of urbanization, agricultural change, and habitat alteration represent a complex multivariate response. Multiple regression analysis of these types of problems are subject to problems of interpretation that include covariances among interacting explanatory variables and an inability to assign unique explanatory capacity to individual factors (Grace and Bollen 2005; Laughlin and Abella 2007). To avoid these problems, techniques such as structural equation modeling (SEM) may be most useful. SEM allows researchers to theorize about why explanatory variables are correlated and to build directional relationships into their models of systems. Explanatory variables are often correlated because they have a common cause or because one factor influences the other (Laughlin and Abella 2007; Shipley 2000). These situations are common in observation studies of complex systems. Consequently a systems approach to the analysis and interpretation of composition in this CNH system may be optimal for explaining where driving forces interact to produce observed patterns of bird deaths across the landscape.

We can not only explore the relationship between HPAI and urbanization, agricultural change, and habitat alteration at the national level, but also examine whether this relationship exists at commune and household scales using focus groups, interviews, and a structured household survey. This is necessary because as numerous researchers have shown, complexity is scale sensitive (Fox 1992; Phillips 1999; Walsh et al. 1999). Processes that operate at one scale may not occur at other scales or resolutions.

5.5 Lessons Learned About Social Science Integration

Social science integration poses many challenges. First, there is a steep learning curve regarding terminology and methods for interdisciplinary research teams. For instance, "risk" may be expressed in monetary terms by an environmental economist, as probability by a statistician, or as a more qualitative and multi-dimensional construct by a decision theorist. These differences of course have important implications for choices about the measures collected and analyzed. Our team meets this challenge by holding frequent (often biweekly) meetings throughout proposal development and project implementation to identify and learn differences in our understandings and approaches.

Another challenge for social science integration is the need to measure complex phenomena such as "modernization," which often occur on very large scales. Our approach is to recognize the subjectivity and the value-laden judgments that scientists make about the validity of alternative measures. Accordingly, we recommend selecting variables directly from relevant theories and statistically testing (e.g., through factor analysis) the extent to which sets of variables combine in an internally consistent manner. The higher the consistency, the more confidence we have that the measures capture an underlying construct relevant to our model.

To illustrate, our primary goal for measuring the complex construct of "modernization" is to create a "degree of modernization" national map based on the smallest administrative unit as the unit of analysis and use secondary data sources to do so. We start by identifying three latent concepts for modernization—urbanization, agricultural intensification and land use changes. For each of these concepts, we identify theories and metrics from diverse fields and then select those variables which seem to be valid for the type of transition happening in Vietnam. This list is shortened simply by eliminating variables for which high quality secondary data do not exist. The next step in the process is to realize that we are creating a comparative rather than an absolute metric of modernization. Even though the metric helps in the comparative assessment of the level of modernization, there is still a need to validate it based on ground-truthing. So then we use a multi-disciplinary approach to ground truthing—walk through field surveys, random ground-level photographs, satellite imagery, commune data archives—to ensure that our classification was at least accurate on an ordinal scale.

A final key element facilitating the integration of diverse social sciences relates to the disciplines represented by the research team. Our core team is comprised of researchers from fields that already reflect an interdisciplinary approach (geography, urban planning, environmental science, and decision science). Also, each team member has experience working at different scales. We are fortunate to work in an institutional setting that encourages interdisciplinary work and multiply authored papers. Support for publishing in cross disciplinary journals helps generate recognition for the value of integrated work.

5.6   Conclusion

No single theory or method is sufficient to explain complex phenomena such as EID and the relationships between factors influencing disease outbreaks. Integrated approaches—bridging multiple social sciences and bridging social and non-social sciences—are time consuming and challenging enterprises, but arguably the most fruitful if they provide an in-depth description of and improved predictive capacity for a complex problem. The initial framework we present for the analysis of social-ecological models of EIDs is useful for scholars from diverse disciplines as a method for examining the relationships within and among multiple components of CNH systems. Given that other researchers have already identified the relevance of these components for explaining HPAI, we have some confidence in our model as a starting point. Future research will need to examine the extent to which relationships among these components meaningfully capture the construct of transition and explain HPAI outbreaks in Vietnam.

Once a model has been proven robust, we will be able to examine specific conditions and identify specific components of CNH systems that amplify or attenuate HPAI risk. More systematic analyses of CNH systems will improve our understanding of how transformations in social-ecological systems produce feedbacks that affect natural communities, their pathogens, animal host, and human populations at diverse spatial, temporal, and organizational scales.

Given its importance and difficulty, we conclude that social science integration requires a carefully considered theoretical rationale and a model-guided methodological approach. This approach will provide for cumulative results from multiple studies designed to investigate various aspects of the model. To test the robustness of this approach, interdisciplinary research teams will need to examine the consistency of results across independent data sets, ideally with different operationalizations of the relevant theoretical constructs.

By further developing and applying conceptual frameworks that take into account the complexity of real-world systems we can build the knowledge base necessary to advance our understanding in a manner meaningful to policy makers. Ultimately, such frameworks offer a flexible tool for diagnosing and dealing with the multiple challenges facing rapidly developing communities.

Acknowledgments Author order is alphabetical; the authors contributed equally to the development of this manuscript. The authors acknowledge Drs. Nancy Lewis, Bruce Wilcox, Michael DiGregorio, and Durrell Kapan for their contributions to conceptualization of the framework. We are grateful for support from the East-West Center, Honolulu HI, and the National Science Foundation through grant 0909410.

References

Akram-Lodhi, A. H. (2004). Are 'landlords taking back the land'? An essay on the agrarian transition in Vietnam. *European Journal of Development Research, 16*(4), 757–789.

Allender, S., Foster, C., Hutchinson, L., & Arambepola, C. (2008). Quantification of urbanization in relation to chronic diseases in developing countries: A systematic review. *Journal of Urban Health: Bulletin of the New York Academy Of Medicine, 85*(6), 938–951.

Arkadie, B. V., & Mallon, R. (2003). *Viet Nam: A transition tiger?* Canberra: ANU Press.

Barennes, H., Martinez-Aussel, B., Vongphrachanh, P., & Strobel, M. (2007). Avian influenza risk perceptions, Laos (letter). *Emerging Infectious Diseases, 13*(7), 1126.

Bauman, Z. (1992). *Modernity and ambivalence*. London: Blackwell.

Beck, U. (1999). *World risk society*. London: Blackwell.

Berkes, F., Colding, J., & Folke, C. (2003). *Navigating through social ecological systems: Building resilience for complexity and change*. Cambridge: Cambridge University Press.

Crane, R. (1994). Water markets, market reform, and the urban poor: Results from Jakarta, Indonesia. *World Development, 22*(1), 71–83.

Dahly, D. L., & Adair, L. S. (2007). Quantifying the urban environment: A scale measure of urbanicity outperforms the urban–rural dichotomy. *Social Science and Medicine, 64*(7), 1407–1419.

De Koninck, R. (2004). The challenges of the agrarian transition in Southeast Asia. *Labour, Capital and Society, 37*, 285–288.

De Vylder, S. (1990). *Towards a market economy? The current state of economic reform in Vietnam*. Stockholm: Stockholm School of Economics/SIDA, Planning Secretariat.

de Zwart, O., Veldhuizen, I. K., Elam, G., Aro, A. R., Abraham, T., Bishop, G. D., et al. (2007). Avian influenza, risk perception, Europe and Asia. *Emerging Infectious Diseases, 13*(2), 290–293.

Delquigny, T., Edan, M., Nguyen, D. H., Pham, T. K., & Gautier, P. (2004). *Evolution and impact of avian influenza epidemic and description of the avian production in Vietnam. Final report for FAO's TCP/RAS/3010 emergency regional support for post avian influenza rehabilitation*. Rome: FAO.

Douglas, M., & Wildavsky, A. (1982). *Risk and culture: An essay on the selection of technological and environmental dangers*. Berkeley: University of California Press.

Douglass, M. (2000). Mega-urban regions and world city formation: Globalization, the economic crisis and urban policy issues in Pacific Asia. *Urban Studies, 37*(12), 2315–2335.

Douglass, M., DiGregorio, M., Pichaya, V., & Boonchuen, P. (2002). *The urban transition in Vietnam*. Honolulu/Fukuoka/Hanoi: UNCHS/UNDP and University of Hawaii, Department of Urban and Regional Planning.

Epprecht, M., & Heinimann, A. (Eds.). (2004). *Socioeconomic atlas of Vietnam. A depiction of the 1999 population and housing census*. Berne: Swiss National Centre of Competence in Research (NCCR) North-South/University of Berne.

Epprecht, M., & Robinson, T. P. (Eds.). (2007). *Agricultural atlas of Vietnam. A depiction of the 2001 rural agriculture and fisheries census*. Rome: Pro-Poor Livestock Policy Initiative (PPLPI) of the United Nation's Food and Agriculture Organization (FAO); Hanoi: General Statistics Office (GSO), Government of Vietnam.

Fang, L., de Vlas, S. J., Liang, S., Looman, C. W. N., Gong, P., Xu, B., et al. (2008). Environmental factors contributing to the spread of H5N1 avian influenza in mainland China. *PLoS ONE, 3*(5), e2268.

Fforde, A., & Vylder, S. D. (1996). *From plan to market—The economic transition in Vietnam*. Boulder: Westview Press.

Fielding, R., Lam, W. W. T., Ho, E. Y. Y., Lam, T. H., Hedley, A. J., & Leung, M. (2005). Avian influenza risk perception, Hong Kong. *Emerging Infectious Diseases, 11*(5), 677–682.

Figuié, M., & Fournier, T. (2008). Avian influenza in Vietnam: Chicken-hearted consumers? *Risk Analysis, 28*(2), 441–451.

Fox, J. (1992). The problem of scale in community resource management. *Environmental Management, 16*(3), 289–297.

Frewer, L. J., Miles, S., & Marsh, R. (2002). The media and genetically modified foods: Evidence in support of social amplification of risk. *Risk Analysis, 22*(4), 701–711.

Friedmann, J. (2005). *China's urban transition*. Minneapolis: University of Minnesota Press.

Giddens, A. (1992). *Consequences of modernity*. Oxford: Polity Press.

Gilbert, M., Chaitaweesub, P., Parakamawongsa, T., Premashthira, S., Tiensin, T., Kalpravidh, W., et al. (2006). Free-grazing ducks and highly pathogenic avian influenza, Thailand. *Emerging Infectious Diseases, 12*(2), 227–234.

Gilbert, M., Xiao, X. M., Chaitaweesub, P., Kalpravidh, W., Premashthira, S., Bole, S., et al. (2007). Avian influenza, domestic ducks and rice agriculture in Thailand. *Agriculture Ecosystems & Environment, 119*(3–4), 409–415.

Gilbert, M. X., Xiangming, X., Pfeiffer, D. U., Epprecht, M., Boles, S., Czarnecki, C., et al. (2008a). Mapping H5N1 highly pathogenic avian influenza risk in Southeast Asia. *Proceedings of the National Academy of Sciences, 105*, 4769–4774.

Gilbert, M., Xiangming, X., Pfeiffer, D. U., Epprecht, M., Boles, S., Czarnecki, C., et al. (2008b). Mapping H5N1 highly pathogenic avian influenza risk in Southeast Asia. *PNAS, 105*, 4769–4774.

Gilbert, M., Xiangming, X., Wint, W., & Slingenbergh, J. (2014). Livestock production dynamics, bird migration cycles, and the emergence of highly pathogenic avian influenza in East and Southeast Asia. In R. Saueborn (Ed.), *Global environmental change and infectious diseases: Impacts and adaptations*. Berlin: Springer.

Grace, J. B., & Bollen, K. A. (2005). Interpreting the results from multiple regression and structural equation models. *Bulletin Ecological Society of America, 86*, 283–295.

Graham, J., Gurian, P., Corella-Barud, V., & Avitia-Diaz, R. (2004). Periurbanization and in-home environmental health risks: The side effects of planned and unplanned growth. *International Journal of Hygiene and Environmental Health, 207*, 447–454.

Grainger, A. (1995). The forest transition: An alternative approach. *Area, 27*(3), 242–251.

Grossman, G. M., & Krueger, A. B. (1995). Economic growth and the environment. *Quarterly Journal of Economics, 110*(2), 353–377.

Hall, D. (2004). Smallholders and the spread of capitalism in rural Southeast Asia. *Asia Pacific Viewpoint, 45*(3), 401–414.

Henning, J., Pfeiffer, D. U., & Vu, L. T. (2009). Risk factors and characteristics of H5N1 Highly Pathogenic Avian Influenza (HPAI) post-vaccination outbreaks. *Veterinary Research, 40*(3), 15.

Hofstede, G. (1984). *Culture's consequences*. Newbury Park: Sage.

Holldren, J. P., & Smith, K. R. (2000). Energy, the environment, and health. In J. Goldemberg, J. W. Baker, S. Ba-N'Daw, H. Khatib, & A. Popescu (Eds.), *World energy assessment*. New York: U.N. Development Programme, U.N. Dep. Economics and Social Affairs, World Energy Council.

Holling, C. S. (1973). Resilience and stability of ecological systems. *Annual Review of Ecology and Systematics, 4*, 1–23.

Holling, C. S. (2003). Foreword: The backloop to sustainability. In F. Berkes, J. Colding, & C. Folke (Eds.), *Navigating social-ecological systems: Building resilience for complexity and change* (p. xv). Cambridge: Cambridge University Press.

Horwitz, P., & Wilcox, B. (2005). Parasites, ecosystems and sustainability: An ecological and complex systems perspective. *International Journal for Parasitology, 35*, 725–732.

Kapan, D. D., Bennett, S. N., Ellis, B. N., Fox, J., Lewis, N. D., Spencer, J. H., Saksena, S., & Wilcox, B. A. (2006). Avian influenza (H5n1) and the evolutionary and social ecology of infectious disease emergence. *EcoHealth, 3*(3), 187–194.

Kasperson, J. X., Kasperson, R. E., & Pidgeon, N. (2003). The social amplification of risk: Assessing fifteen years of research and theory. In N. Pidgeon, R. E. Kasperson, & P. Slovic (Eds.), *The social amplification of risk* (pp. 13–46). London: Cambridge University Press.

Laughlin, D., & Abella, S. (2007). Abiotic and biotic factors explain independent gradients of plant community composition in ponderosa pine forests. *Ecological Modeling, 205*, 231–240.

Levin, S. A. (1999). *Fragile dominion*. Reading: Perseus Books.

Mallin, M. A., & Cahoon, L. B. (2003). Industrialized animal production? A major source of nutrient and microbial pollution to aquatic ecosystems. *Population and Environment, 24*, 369–385.

Mather, A. S. (1992). The forest transition. *Area, 24*(4), 367–379.

Mather, A. S. (2007). Recent Asian forest transitions in relation to forest transition theory. *International Forestry Review, 9*(1), 491–502.

McDade, T. W., & Adair, L. S. (2001). Defining the 'urban' in urbanization and health a factor analysis approach. *Social Science and Medicine, 53*(1), 55.

McGranahan, G., Jacobi, P., Songsore, J., Surjadi, C., & Kjellen, M. (2000). *Citizens at risk: From urban sanitation to sustainable cities*. London: Earthscan.

Montgomery, M. R., Stren, R., Cohen, B., & Reed, H. E. (Eds.). (2004). *Cities transformed: Demographic change and its implications in the developing world*. London: Earthscan.

Morse, S. (2005). Factors in the emergence of infectious diseases. *Emerging Infectious Diseases, 1*(1), 7–15.

O'Brien, K. (1993). Improving survey questionnaires through focus groups. In D. Morgan (Ed.), *Successful focus groups: Advancing the state of the art* (pp. 105–117). Newbury Park: Sage.

O'Neill, R. V. (2001). Is it time to bury the ecosystem concept? With full military honors of course! *Ecology Law Quarterly, 82*(12), 3275–3284.

Oliveira, C. G., Lacerda, H. G., Martins, D. R. M., Barbosa, J. D. A., Monteiro, G. R., Queiroz, J. W., et al. (2004). Changing epidemiology of American cutaneous leishmaniasis (ACL) in Brazil: A disease of the urban–rural interface. *Acta Tropica, 90*(2), 155–162.

Otte, J., Pfeiffer, D. U., Tiensin, T., & Price, L. (2006). Evidence-based policy for controlling HPAI in poultry: Bio-security revisited. In E. Silbergeld (Ed.), *Pro-poor livestock policy initiative research report*. Baltimore: John Hopkins Bloomberg School of Public Health.

Park, H., Russell, C., & Lee, J. (2007). National culture and environmental sustainability: A cross-national analysis. *Journal of Economics and Finance, 31*(1), 104–121.

Pfeiffer, D. U., Minh, P. Q., Martin, V., Epprecht, M., & Otte, M. J. (2007). An analysis of the spatial and temporal patterns of highly pathogenic avian influenza occurrence in Vietnam using national surveillance data. *Veterinary Journal, 174*(2), 302–309.

Phillips, J. D. (1999). Methodology, scale, and the field of dreams. *Annals of the Association of American Geographers, 89*, 754–760.

Pidgeon, N., Kasperson, R., & Slovic, P. (Eds.). (2003). *The social amplification of risk*. Cambridge: Cambridge University Press.

Pope, C., & Mays, N. (1995). Researching the parts other methods cannot reach: An introduction to qualitative methods in health and health services research. *British Medical Journal, 311*, 42–45.

Rigg, J. (2005). Land, farming, livelihoods, and poverty: Rethinking the links in the rural south. *World Development, 34*(1), 180–202.

Rigg, J., & Vandergeest, P. (2012). *Revisiting rural places*. Honolulu: University of Hawaii Press.

Rudel, T. K. (1998). Is there a forest transition? Deforestation, reforestation, and development. *Rural Sociology, 63*(4), 533–552.

Shipley, B. (2000). *Cause and correlation in biology: A user's guide to path analysis, structural equations, and causal inference*. Cambridge: Cambridge University Press.

Slovic, P. (2000). *The perception of risk*. London: Earthscan.

Smith, K. R. (1990). The risk transition. *International Environmental Affairs, 2*, 227–251.

Smith, K. R. (1997). Development, health, and the environmental risk transition. In B. S. L. G. Shahi, A. Binger, T. Kjellstrom, & R. Lawrence (Eds.), *International perspectives in environment, development, and health* (pp. 51–62). New York: Springer.

Smith, K. R., & Akbar, S. (2003). Health-damaging air pollution: A matter of scale. In G. McGranahan & F. Murray (Eds.), *Health and air pollution in rapidly developing countries* (pp. 1–20). London: Earthscan.

Smith, K. R., & Ezzati, M. (2005). How environmental health risks change with development: The epidemiologic and environmental risk transitions revisited. *Annual Review of Environmental Resources, 30*, 291–333.

Spencer, J. H. (2007). Innovative systems to create peri-urban infrastructure: Assessment of a local partnership to provide water to the poor in Vietnam. *International Development Planning Review, 29*(1), 1–22.

Spencer, J. H. (2013). The urban health transition hypothesis: Empirical evidence of an avian influenza kuznets curve in Vietnam? *Journal of Urban Health, 90*(2), 343–357.

Vlahov, D., & Galea, S. (2002). Urbanization, urbanicity, and health. *Journal of Urban Health-Bulletin of the New York Academy of Medicine, 79*(4), S1–S12.

Walsh, S. J., Evans, T. P., Welsh, W. F., Entwisle, B., & Rindfuss, R. R. (1999). Scale dependent relationships between population and environment in Northeast Thailand. *Photogrammetric Engineering and Remote Sensing, 65*, 97–105.

Wilcox, B., & Colwell, R. (2005). Emerging and reemerging infectious diseases: Biocomplexity as an interdisciplinary paradigm. *EcoHealth, 2*, 1–14.

Wilcox, B., & Gubler, D. (2005). Disease ecology and the global emergence of zoonotic pathogens. *Environmental Health and Preventive Medicine, 10*(September), 263–272.

Woolhouse, M. E. J., & Gowtage-Sequeria, S. (2005). Host range and emerging and reemerging pathogens. *Emerging Infectious Diseases, 11*(12), 1842–1847.

World Health Organization. (2012). Cumulative number of confirmed human cases for avian influenza A(H5N1) reported to WHO, 2003–2012. WHO/GIP, data in HQ as of 07 June 2012. www.who.int/influenza/human_animal_interface/EN_GIP_20120607CumulativeNumberH5N1cases.pdf

Xiao, X. M., Boles, S., Liu, J., Zhuang, D., Frolking, S., Lia, C., et al. (2005). Mapping paddy rice agriculture in southern China using multi-temporal MODIS images. *Remote Sensing of the Environment, 95*, 480–492.

Xiao, X. M., Boles, S., Frolking, S., Li, C., Babu, J. Y., Salas, W., et al. (2006). Mapping paddy rice agriculture in South and Southeast Asia using multi-temporal MODIS images. *Remote Sensing of the Environment, 100*, 95–113.

Zhang, Q., Wang, J., Peng, X., Gong, P., & Shi, P. (2002). Urban built-up land change detection with road density and spectral information from multi-temporal Landsat TM data. *International Journal of Remote Sensing, 23*(15), 3057–3078.

Chapter 6
Studying Power with the Social-Ecological System Framework

Graham Epstein, Abigail Bennett, Rebecca Gruby, Leslie Acton, and Mateja Nenadovic

6.1 Introduction

There is no concept that has captivated philosophers, historians, geographers, and political scientists, quite like *power*. Scholars have long posed theoretical questions concerning the existence, origins, and manifestations of power without settling on anything resembling consensus (Machiavelli [1532] 1988; Hobbes [1651] 2010). Normative questions regarding who should rule, under what conditions, and for what purposes have similarly been mired in centuries of debate that offer perspectives and insights, but no clear answers (Wilson 1887; Waldo 1948; Ostrom 2008).

Differential treatments of power also lie at the heart of a long-standing divide among social scientific traditions in the study of social-ecological systems (SESs). Power is central in the interdisciplinary field of political ecology, where it is understood as a core driver of social-ecological outcomes (Lebel et al. 2010). In contrast, the "Bloomington School" of new institutionalists (grounded in the work of Vincent and Elinor Ostrom et al.) deliberately moved away from the focus on power that dominated twentieth-century political science—a focus they felt to be "extreme and limiting" (Aligica and Boettke 2009, p. 30). Instead, they directed their attention to

G. Epstein (✉)
School of Public Affairs and Department of Political Science,
The Vincent and Elinor Ostrom Workshop in Political Theory
and Policy Analysis, Indiana University, 513 N. Park Ave,
Bloomington, IN 47408-3895, USA
e-mail: gepstein@indiana.edu

A. Bennett • L. Acton • M. Nenadovic
Nicholas School of the Environment, Duke Marine Lab, Duke University,
135 Duke Marine Lab Road, Beaufort, NC 28516, USA

R. Gruby
Department of Human Dimensions of Natural Resources, Colorado State University,
1480 Campus Delivery, Fort Collins, CO 80523-1480, USA

M.J. Manfredo et al. (eds.), *Understanding Society and Natural Resources*,
DOI 10.1007/978-94-017-8959-2_6, © The Author(s) 2014

institutions and how they affect the prospects for self-organized governance of common-pool resources (Ostrom 1990).

Institutions refer to the formal and informal rules, norms, and shared strategies (or conventions) that structure human interactions at all levels of social organization (Ostrom 2005). They are linguistic statements that specify what actions must, must not, or may be taken given certain conditions, and, as such, they may exist in written form, in the minds of individuals, or both (Crawford and Ostrom 1995). New institutionalists focus on how groups can create credible commitments to limit individual selfishness and obtain greater benefits for the collective (Dietz et al. 2002). When groups are able to communicate and develop trust, they are sometimes able to extricate themselves from predicted tragedies by forming institutions that prescribe cooperative behavior (Ostrom et al. 1994). This approach tends to assume that the outcomes of collective action benefit the group as a whole and that members of a group share a common understanding of desired outcomes. These assumptions give this work an air of equality and symmetry that often overshadows the importance of power and distributional inequalities. As a result, the new institutionalist view that social-ecological sustainability is primarily a function of implementing the 'right kinds' of institutions is often seen as overly optimistic and simplistic (Agrawal 2003; Clement 2010).

In recent years, new institutional theories and frameworks—inclusive of common-pool resource (CPR) theory, the institutional analysis and development (IAD) framework and social-ecological system (SES) framework—have faced increasing criticism for failing to adequately attend to issues of power, politics, and inequality and how they affect environmental governance processes (Agrawal 2013). For example, Mosse (1997, p. 470) has argued that "historically-specific structures of power, rather than simply calculated pay-offs (or traditional wisdom) underlie the norms and conventions of collective resource use, and account for the occurrence and persistence of local institutions of resource use." Agrawal (2003) has similarly suggested that commons research does not adequately attend to intragroup politics, power, and resistance. He argues that the relationship between power and rights to access and use natural resources should complement the narrow focus of new institutionalist scholars on internal institutions and rules.

This chapter takes these critiques as a point of departure to begin to develop a systematic, interdisciplinary approach to integrate power with institutional studies of SESs. Our main goal is to assess whether diverse concepts of power can be explored and analyzed with the SES framework and whether such an endeavor is potentially fruitful. To this end, we structure our study in four stages. First, we provide an overview of the SES framework, which aims to enhance cross-disciplinary theory-building by providing "the most general set of variables [or attributes] that should be used to analyze all types of [SES] settings" (Ostrom 2005, p. 28). Second, we outline a process for operationalizing various concepts of power through this framework. Third, we illustrate how this process may be used to test a hypothesis—in this case, that power affects SES outcomes. In this third stage, we review how some new institutionalists have thought about, defined, and studied power and then classify these definitions using existing attributes in the SES framework. We then identify operational indicators of these attributes using data from a collaborative forest governance database—International Forestry Resources and Institutions

(IFRI)—and use these to conduct an illustrative quantitative analysis of the relationship between power and the combined social-ecological outcome. Although we use quantitative data analysis techniques in our study, qualitative, quantitative, and mixed-methods approaches all stand to make distinct and complementary contributions to understand the role of power in resource governance. Fourth, we reflect on this analysis and its conclusions to consider the extent to which the SES framework can be used to integrate power within institutional approaches to studying SESs.

This chapter contributes four main arguments relevant to scholars interested in bridging power-centered and institution-centered approaches. First, power is, and always has been, part of new institutionalist thinking, although the term power is rarely invoked explicitly. Second, if the SES framework is to provide a metatheoretical structure for interdisciplinary, systematic, and diagnostic studies of sustainability as it intends, then this structure must be able to account for power. Third, the SES framework can be used to integrate power-centered approaches with institutional analysis, at least with regards to institutional forms of power. Lastly, there remains a need to consider more diverse conceptions of power across the social sciences and to determine whether broader integration is possible, and what if any implications this has for the SES framework and the study of sustainability.

6.2 Incorporating Power Within The SES Framework

The SES framework is a particularly noteworthy addition to the set of frameworks, theories, and models used for the study of sustainability (Ostrom 2007, 2009). However, the SES framework, like its predecessor the IAD framework, appears mostly silent on questions of power with the notable absence of terms such as "power" or "politics." Perhaps the primary challenge in incorporating power into an analysis using the SES framework is grappling with the many competing and overlapping conceptualizations of power that exist across social scientific disciplines. While the range of conceptualizations of power may at first seem overwhelming, and reviewing them in detail is indeed beyond the scope of this chapter, it is nonetheless helpful to delineate some broad categories. For example, one branch of political ecology emphasizes the primacy of materialist conceptions of power, drawing on ideas rooted in the scholarship of Marx. The focus here is on differing control over and access to natural resources and the influence of material conditions on social and ecological outcomes. As Robbins (2004) asserts, "no explanation of environmental change is complete, therefore, without serious attention to who profits from changes in control over resources, and without exploring who takes what from whom," (Robbins 2004, p. 52). Other social theorists are more concerned with discursive forms of power, or those ways of talking about, representing, and generating knowledge about the world that influence human-environment relations. Discourses can both create and limit the realm of possibility for how humans may think, act, and behave with regards to the natural world. Post-structural approaches to power, such as Foucault's, conceive of a discourse that includes not just the way actors talk about and represent nature and nature's governance, but also the everyday institutions and activities that shape actors' perceptions of themselves, their desires, and their relationships with the world around them.

In this chapter, we choose to focus on operationalizing and measuring institutional conceptions of power, which are distinct from yet always interrelated with materialist and discourse approaches. We chose to focus on institutional conceptions of power for two reasons. First, the fact that power is an integral aspect of institutions is almost always underemphasized in the current literature. Second, due to the SES framework's disciplinary proximity to institutionalism, testing institutional forms of power is a reasonable first step. Only after showing that institutional forms of power can be taken into account by the SES framework might we move forward in conceptualizing how an analysis of materialist, discursive, or post-structural accounts of power might be applied within the framework. As we will explain, though power is not explicitly included in the SES framework, several key potential indicators of institutional power, such as the operational rules governing the system, *are* included. These attributes can be employed to ask questions concerning how different levels of access and control over resources are shaped by institutional characteristics of the system and how, in turn, these relationships may influence social-ecological outcomes. Before turning to this question, we briefly describe the SES framework, and then consider how it may be used to study the effects of power on sustainability (for a more comprehensive description, we refer readers to Ostrom 2007; Basurto and Ostrom 2009; Ostrom and Cox 2010).

6.3   Overview of the SES Framework

The SES framework explicitly aims to bridge disciplinary and methodological boundaries while facilitating the synthesis of disparate studies by providing a common classificatory framework containing potentially important SES attributes and relationships. Derived from the IAD framework, the SES framework retains the action situation (Fig. 6.1), a general game-theoretic model of interdependent choice, and carries with it much of the intellectual history of the Bloomington School (Kiser and Ostrom 1982; Ostrom et al. 1994; Crawford and Ostrom 1995; Ostrom 2005). In general, outcomes are understood to be the aggregate result of individual interactions and decisions in action situations structured by attributes of four core components: resources (RU), resource systems (RS), governance systems (GS), and actors (A). Although this simple model is thought to encompass and explain diverse outcomes in SESs, analytical complexity emerges from the wide range of attributes that collectively define each component and their interactions within action situations. The most recent elaboration of the SES framework (Epstein et al. 2013) includes more than 30 potentially influential attributes pertaining to the 4 core components of SESs (Table 6.1). Since the SES framework is structured as a multi-tiered classificatory system, each of these attributes can be further unpacked into types and subtypes such that the full suite of potentially relevant conditions is effectively unknown (Ostrom and Cox 2010).

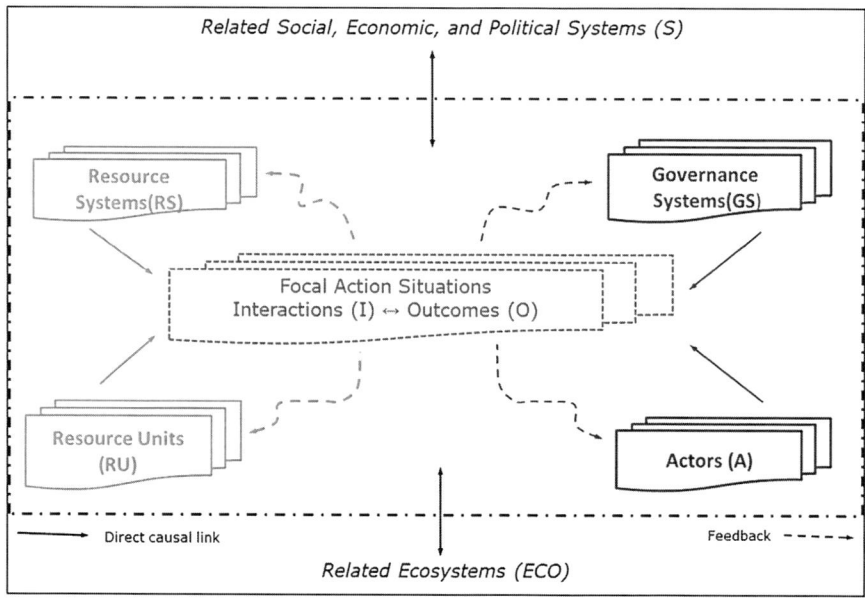

Fig. 6.1 Analytical structure of the social-ecological system framework (*Source*: Based on McGinnis and Ostrom forthcoming)

To date, the SES framework has been used to study a wide range of systems, including forests, fisheries, irrigation systems, and nature-based tourism (Fleischman et al. 2010; Blanco 2011; Gutierrez et al. 2011; Basurto and Nenadovic 2012; Cinner et al. 2012; Basurto et al. 2013). In adopting a common framework, these studies may advance knowledge more rapidly by generating observations on a common set of attributes that can be readily compared or integrated for large-n analysis. Alternatively, individual case studies may be used to add diagnostic pieces to the overall puzzle of sustainability (Basurto and Ostrom 2009).

The structure of the framework is somewhat flexible, allowing for the integration of additional concepts and attributes to improve the study of SESs. Although we know of no studies of power that have derived from engagement with the SES framework, we do not rush to add attributes here. Given the wide range of conceptualizations of power from different fields and strands of literature, adding a single attribute, "power," would likely create considerable confusion regarding how such an attribute could be operationalized or measured, working contrary to the goal of providing a common classificatory system for SES research. Instead, close examination of various conceptualizations of power reveals that many indicators thereof are already included among the existing attributes of the framework. Thus, our analysis focuses on the extent to which the existing attributes of the SES framework, whether individually or in combination, can be used to operationalize and measure power. We then apply these measures to conduct an illustrative analysis of the

Table 6.1 The social-ecological system framework

Social, economic, and political settings (S)	
S1 – Economic development. S2 – Demographic trends. S3 – Political stability	
S4 – Other governance systems. S5 – Markets. S6 – Media organizations. S7 – Technology	
Resource Systems (RS)	Governance Systems (GS)
RS1 – Sector (e.g., water, forests, pasture)	GS1 – Government organizations
RS2 – Clarity of system boundaries	GS2 – Nongovernment organizations
RS3 – Size of resource system	GS3 – Network structure
RS4 – Human-constructed facilities	GS4 – Property-rights systems
RS5 – Productivity of system	GS5 – Operational-choice rules
RS6 – Equilibrium properties	GS6 – Collective-choice rules
RS7 – Predictability of system dynamics	GS7 – Constitutional-choice rules
RS8 – Storage characteristics	GS8 – Monitoring and sanctioning rules
RS9 – Location	
Resource Units (RU)	Actors (A)
RU1 – Resource unit mobility	A1 – Number of relevant actors
RU2 – Growth or replacement rate	A2 – Socioeconomic attributes
RU3 – Interaction among resource units	A3 – History or past experiences
RU4 – Economic value	A4 – Location
RU5 – Number of units	A5 – Leadership/entrepreneurship
RU6 – Distinctive characteristics	A6 – Norms (trust-reciprocity)/social capital
RU7 – Spatial and temporal distribution	A7 – Knowledge of SES/mental models
	A8 – Importance of resource (dependence)
	A9 – Technologies available
Action situations: Interactions (I) → Outcomes (O)	
Activities and processes	Outcome criteria
I1 – Harvesting	O1 – Social performance measures (e.g., efficiency, equity, accountability, sustainability)
I2 – Information sharing	O2 – Ecological performance measures (e.g., overharvested, resilience, biodiversity, sustainability)
I3 – Deliberation processes	O3 – Externalities to other SESs
I4 – Conflicts	
I5 – Investment activities	
I6 – Lobbying activities	
I7 – Self-organizing activities	
I8 – Networking activities	
I9 – Monitoring activities	
I10 – Evaluative activities	
Related Ecosystems (ECO)	
ECO1 – Climate patterns. ECO2 – Pollution patterns. ECO3 – Flows into and out of focal SES	

Left margin: Ecological Rules (*ER*) ER1 – Physical rules. ER2 – Chemical rules. ER3 – Biological rules

Source: Adapted from McGinnis and Ostrom (forthcoming)

general hypothesis that "power matters" with regard to SES outcomes. Our application is demonstrative in the sense that its primary purpose is to illustrate how such an analysis may be conducted; the results then are not intended to be interpreted in any conclusive sense.

6.4   Operationalizing Research on the Role of Power in Social-Ecological Systems

Building on Adock and Collier (2001), this section explicates a four-step process for operationalizing studies of power using the SES framework (Fig. 6.2). This process is designed to help quantitative and qualitative researchers avoid or at least be more aware of threats to validity that emerge in the transition from theory to measurement and on to evaluation or causal inference. While these insights are not exclusively relevant to the current endeavor, they are worth highlighting here given the historic lack of attention to these important issues in the study of SESs.

The first step is to explicitly adopt particular definitions or theories of power relevant to an SES puzzle. The critique that "power matters" is an authoritative comment regarding a relationship between a condition and an outcome. However, it is also quite vague given the diverse ways in which power has been defined. Although a few studies of SESs have attempted to bring power-centered and institution-centered theories into constructive dialogue (e.g., Clement 2010; Gruby and Basurto 2013), these initial efforts reflect a small subset of the diverse ways in which social scientists have thought about, defined, and studied power. Without explicit agreement on what power is, it seems unlikely that any one test of a theory that "power matters" can produce the types of evidence required to support or reject such a general hypothesis. The challenge then for scholars seeking to bridge these two approaches is to answer which of the many conceptualizations of power matter and under which conditions.

The second step in this process is to either classify the chosen definition in terms of one or more attributes of the SES framework, or add attributes that appear to be missing. In cases where definitions directly map onto attributes, this process is straightforward; in others (i.e., definitions of power); the classification process typically involves a number of assumptions that must be made explicit. For example,

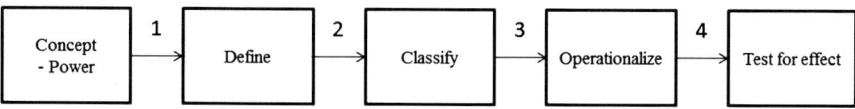

Fig. 6.2 Steps in testing the effects of power with the social-ecological system framework. *Source*: Elaborated from Adcock and Collier (2001)

Clement (2010) attempts to explain variations between policy intentions and outcomes by "politicizing" the IAD framework and adding two classes of attributes, namely "discourse" and "political-economic context." While she develops a convincing argument that "power matters" and illustrates its effect through a qualitative case study, her addition of "discourse" to the IAD framework reflects only one of many possible classifications of this concept. In fact, one of the core goals of the SES framework is to systematically organize concepts and their definitions such that results are driven by empirical relationships rather than competing definitions or measures (Ostrom 2007).

Upon classification of a definition, the third step is to choose how to operationalize or measure that attribute for empirical analysis. This can be as simple as establishing the presence or absence of some attribute, or involve more complex multivariate measures or qualitative descriptions. Finally, the fourth step is to analyze the effects of measured attributes on the outcomes of interest. Qualitative researchers might analyze these effects by using process tracing to bring together multiple pieces of evidence in order to systematically evaluate the claims of competing hypotheses (George and Bennett 2005; Collier 2011). Quantitative researchers may examine data by using some form of significance test and statistical model.

6.5   Analyzing Power Within The SES Framework

The preceding discussion highlights the complexity inherent in testing a hypothesis that "power matters." It demonstrates the importance of being deliberate and explicit about the necessary and potentially value-laden choices concerning definitions, classifications, and measurement, not to mention those imposed by the choice of inferential methods. In this section, we seek to illustrate how the SES framework may be used to organize a rigorous, broad research agenda on the effects of power by proceeding through each step of the process just outlined. Our analysis is divided into three main subsections, each of which proceeds through all four steps for specified institutional conceptualizations of power. More specifically, within each subsection, we discuss (1) the distinct institutional definition(s) of power that we are seeking to test, (2) the author(s) associated with that definition, and (3) how that definition may be classified within the SES framework. Finally, we use the IFRI database to (4) operationalize the attribute(s) and test whether there is a statistically significant relationship between each measure of power and a social-ecological outcome.

The IFRI database is perhaps the single most influential and contemporary source of information with which commons scholars develop and test hypotheses concerning the interactions of people, the environment, and institutions in small-scale SESs. The database is composed of a variety of continuous, categorical, and descriptive variables—including a wide range of attributes present in the SES framework—that are collected using a consistent case-study approach (Wertime et al. 2007). The database enables multiple-methods research, although in recent years, as the number of case studies have increased to include more than 400 forests and 600 user groups, IFRI scholars have increasingly turned to large-n quantitative studies that have historically been absent in the commons literature (Andersson and Agrawal 2011; Chhatre and Agrawal 2008, 2009; Persha et al. 2011; Coleman 2009; Coleman and Fleischman 2012). The IFRI database was chosen for this analytical exercise due to its rigor and its reso-nance with the SES framework. A comparable database for the study of large scale systems is the International Regimes database which asks questions con-cerning the formation, boundaries, and processes of international regimes in response to a wide variety of social, economic and ecological problems (Breitmeier et al. 2006; Young and Zürn 2006). Although certainly useful for a quantitative study of power, the international regimes database, because of its emphasis on international-scale processes, is less suited to respond to specific critiques from political ecology that tend to emphasize the effects of power on individuals and communities.

The sample used in this study was constructed in the following way. First, we selected the user group as the unit of analysis. Next, we dropped cases in the fol-lowing order: (1) repeat observations of a user group, (2) groups found in the United States, and (3) those with missing data on any of the dependent variables. The omission of the US cases is common, as they differ substantially from the other countries in terms of economic development and the ways in which forest resources are used. Finally, we randomly dropped duplicate observations of user groups that use multiple forests, as well as forests containing multiple user groups, in order to generate a sample including a maximum of one observation per forest and user group. The dependent variable measures social-ecological benefits and is constructed by summing two multifactor indexes that measure social and ecological benefits, respectively. Social or livelihood benefits are mea-sured by performing a factor analysis similar to that of Chhatre and Agrawal (2009), based on the contributions of a forest to the fuelwood, fodder, and timber needs of a group. Ecological benefits, on the other hand, are measured by per-forming a factor analysis on the polychoric correlation matrix of (1) a forester's perception of vegetation density, (2) a forester's assessment of species diversity, and (3) user group perceptions of the condition of the forest. These attributes were similarly used in Andersson and Agrawal (2011), although they simply averaged these figures.

The results are compiled in Table 6.2, which records the one-way relationship between a particular measure of power and the combined social-ecological benefits (the dependent variable). In most cases we report differences in means between groups that possess and lack power. However, polyserial and pairwise correlations are used for Ostrom's and North's definitions given that they are measured using continuous and ordinal indicators, respectively. We generally predict a positive rela-tionship between the power of a group and the dependent variable.

Table 6.2 A preliminary assessment of the effects of institutional power on social-ecological benefits derived from forests

Classification	Operationalization	Effect
Commons (1924) and Riker (1980): Institutional control		
The power of a group depends upon their rights and responsibilities with regards to the use and management of forest resources		
GS5 Operational rules	Perceived fairness of operational rules (1=Fair; 0=Unfair)	+0.443**
GS4 Property-rights system	Owner(s) of forest is a member of the user group (1=Yes; 0=No)	−0.334
GS6 Collective choice rules	User group is responsible for rulemaking (1=Yes; 0=No)	−0.029
GS8 Monitoring and sanctioning	User group monitors use of forest commons (1=Seasonally, Year round; 0=Occasionally, Never)	+0.450**
Ostrom (2005): Extent of control and value of opportunity		
The power of a group depends upon the level of control over collective choice situations and the economic value of resources		
GS6 Collective choice rules and RU4 Economic value	User group is responsible for rulemaking (1=Yes; 0=No)*	0.095
	The commercial value of forest commons (0=Low; 4=High)	
Lukes (2005): First face of power		
A group lacks power when they participate in collective choice processes, but policies are not congruent with their subjective interests		
GS5 Operational rules and GS6 Collective choice rules	User group participates in rulemaking and does not perceive the rules as fair (1=Yes; 0=No)	−0.102
Lukes (2005): Second face of power		
A group lacks power when rules are not congruent with their subjective interests and they do not participate in collective choice processes		
GS5 Operational rules and GS6 Collective choice rules	User group is not responsible for rulemaking and does not perceive the rules as fair (1=Yes; 0=No)	−0.434**
North (1990): Path dependence and bargaining power		
The power of group covaries with the age of a group or organization		
A3 History of use	The age of the forest user group (years)	0.159***

***$p < 0.01$; **$p < 0.05$; *$p < 0.10$

6.6   Institutional Power

The conceptualizations of power discussed in this section, while distinct from each other, share a common assumption that institutions—again: rules, norms, and shared strategies—carry within their particular form and structure the ability to influence societal outcomes. During the first half of the twentieth century, an important theoretical innovation of "old" institutionalism was to highlight the distinction between institutional and non-institutional aspects of social phenomena. Institutional factors are those that the policy process can directly influence through laws, rules, and regulations, while non-institutional factors can only be influenced indirectly by means of the particular institutions that are created (e.g., economic or demographic conditions) (Ostrom 1976). Embedded within this distinction is an opportunity to study the influence of the former on the latter, or of the ability of institutions to make a difference on a range of existing conditions.

For example, John R. Commons, a particularly prominent "old" institutionalist, proposed an institutional theory of markets that sought to explain the ways in which economic power resulting from the accumulation of wealth, uneven access to resources, and/or monopolies on the means of production could be mitigated by what he referred to as *working rules*. These working rules are scripts that tell individuals what they may or may not, must or must not do as they transact with others (Commons 1931). Through an emphasis on working rules, Commons (1924, p. 6) examined the "principles of collective control of transactions through associations and governments, placing limits on selfishness, that are more recently included in economic theory" to build a foundation for understanding how the social injustices of laissez-faire capitalism might be mitigated. For Commons, institutions were embodiments of power and thus carried with them the possibility of rectifying what he saw as the problems of the day. Since then, multiple institutional theories have been proposed that each point to the central role that institutions play in allowing individuals and groups to make and adhere to choices in complex environments. Nonetheless, Commons' typically positive view of institutional power has receded as scholars such as Riker (1980) lament the democratic implications of an institutional theory of political processes. The most notable of these implications is that political outcomes are not only the result of the will or "tastes" of the people but also of the institutions that are used to make decisions and the political skills or "artistry" of those who seek to manipulate agendas and exploit opportunities for their own ends. In other words, institutional power may be used and manipulated by individuals or groups in pursuit of their own interests, and thus can serve as the source of, as well as the solution to, social problems.

The Bloomington School of new institutionalists does not often explicitly include power as a distinct or circumscribed concept in its frameworks and analyses. Nonetheless, for scholars interested in questions of institutional power, the Bloomington School has adopted a broad definition of institutions (Crawford and Ostrom 1995; Ostrom 2005) that can be interpreted as a potential carrier of power. First, institutions are said to include *de jure* institutions (rules-in-form or rules on paper) and *de facto* institutions (rules-in-use), as well as social norms and shared strategies. All of which are nested into layers of institutions that determine how institutions at other levels may be changed; these layers are differentiated into operational-, collective-, and constitutional-choice situations. In the context of resource governance, *operational-level* institutions govern how resources are accessed and used. Such rules may state, for example, that only members of a given community—not outsiders—may access a forest, as with community forest concessions in Mexico (Alcorn and Toledo 2000; Bray et al. 2006). Other rules might determine what types of resources may be extracted, during what seasons, and using which harvesting tools. *Collective-choice* institutions, in turn, provide a framework for how—and by whom—operational-level rules are created and modified. A collective-choice institution might state that a majority of forest users are required to approve a rule change or, alternatively, that a local leader has the power to unilaterally create or alter operational rules. Operational-level and collective-choice-level rules are almost always nested in at least one more institutional level—*a constitutional level*—that sets the constraints within which collective-choice rules are determined. The US Constitution is an example of a set of constitutional-level institutions that determine the procedures through which, and the bounds within which, other rule-making procedures are themselves modified (Ostrom 2008). For an institutionalist concerned with power and its effect on individuals and groups, the implication is that one must explore not only the effects of operational rules but also the formal and informal institutions that affect how operational rules are chosen, as well as the configurations of actors that hold power to initiate and manipulate these processes.

We begin the analysis with the most basic definition of institutional power, which suggests that any and all institutions have the capacity to privilege some groups, at the expense of others (Riker 1980; Immergut 1998; Pierson 2000). Thus all institutions, including operational rules (GS5), property-rights systems (GS4), collective-choice (GS6) and constitutional rules (GS7), as well as monitoring and sanctioning rules (GS8) merit consideration as unique classifications of institutional power. The IFRI database is replete with such details, which allow us to measure group-level subjective perceptions of operational rules, to determine whether users participate in collective-choice and monitoring processes, and to establish whether any member(s) of a group possess property rights over the forest commons. The assumptions tested by these measures are that the institutional power of a group is higher when (1) operational rules reflect subjective interests of the group of resource users, (2) group members participate in the rule-making process, (3) group members monitor conformance to rules, and (4) group members hold enforceable property rights. Broadly speaking the results indicate a positive relationship between social-ecological benefits and groups that possess power in the form of favorable operational rules (GS5) and participate in monitoring and sanctioning processes (GS8). Power as characterized by participation in collective-choice processes (GS6) and property-rights systems (GS4) did not have a significant relations with the dependent variable, although this is possibly the result of the bivariate analysis that fails to control for additional sources of heterogeneity (Table 6.2).

6.7 Elinor Ostrom's Definition of Power

Although the Bloomington School is often criticized for the general absence of power in related studies, Ostrom (2005) offers a clear and concise definition of power in her seminal work on the IAD framework, *Understanding Institutional Diversity*. According to Ostrom:

> the "power" of an individual in a situation is the value of the opportunity (the range in the outcomes afforded by the situation) times the extent of control. Thus, an individual can have a small degree of power, even though the individual has absolute control if the amount of opportunity in a situation is small. The amount of power may also be small when the opportunity is large, but the individual has only a small degree of control. (2005, p. 50)

This definition has several implications for power and how it can be studied. First, it suggests that a value that corresponds to power can be assigned to each actor, and does not necessarily imply a zero sum situation. Second, power also varies with the expected benefits and costs of a situation, such that the power of actors holding a small amount of control over a valuable opportunity may be equivalent or greater than that of an actor holding a large degree of control over a less valuable opportunity. For example, an individual vote among many on a very important and potentially rewarding issue may offer more power than unilateral control over a situation with a less valuable outcome. Finally, power can be measured and said to exist as a "power to" do something regardless of whether an actor chooses to make use of it.

Ostrom's (2005) definition of power could be operationalized at any of the institutional levels (i.e., operational, collective choice, constitutional), although we chose to focus on the collective-choice level. Collective-choice rules are often seen as particularly important sources of power because they allow participants to modify the rules that govern operational situations from which flow the majority of instrumental benefits and costs. For instance, when forest users operating under a set of operational rules are confronted by a new disturbance or threat, such as external poachers (Fleischman et al. 2010), participation in collective-choice processes allows them to rapidly adjust those rules to changing conditions. We measured power as the product of a binary measurement of participation in decision-making processes regarding operational rules (GS5) and an ordinal measure of the commercial value of the forest (RU4). Thus, the power of a group is highest when they participate in collective-choice processes and the commercial value of the forest is high, while it is lowest when they do not participate in collective-choice processes and the commercial value of the forest is low. Participation is just one of many potential measures of the concept of control that is indicated in Ostrom's definition of power and may not carry a strong correspondence with control over decision-making processes. The results suggest that Ostrom's definition of power has a positive but insignificant relationship with the combined social-ecological benefit measure used in this study.

6.8   Steven Lukes's Three Faces of Power

A particularly prominent treatise on power that draws upon institutionalist thinking is Steven Lukes's (2005) *Power: A Radical View*, initially published in 1974. Lukes defines power in terms of the realized ability of one group to affect the other in a way that is contrary to their interests. A clear distinction from Ostrom's (2005) definition of power is that, for Lukes, power exists only when it is exercised and only in situations where one of those groups possesses "power over" the other. He is also particularly attentive to multiple manifestations or "faces" of power. Lukes views these "faces" of power as three distinct processes that individuals or groups use to exercise their power over others, two of which are clearly linked to institutional processes.

The first face of power is by far the easiest to identify and study, as it relies upon the observation of overt conflict between two or more groups participating in some political environment (Lukes 2005). When decisions are ultimately made that favor one group at the expense of the other, power is said to exist. As an example, a group possessing a 50 % plus one majority in a two-party legislature using majority rule could be said to hold "power over" the other group, assuming there are differences in subjective interests. While Lukes acknowledges the general validity of this view, he also points to its inadequacy for explaining situations where power is exercised by limiting the participation of some groups. This is the second face of power, wherein groups with identifiable interests or grievances are prevented from even representing their interests in political processes by virtue of the overt and covert actions of some other group. For example, one group may exercise power over another by preventing the first group from voting, or by constructing institutional barriers that increase the costs of participation in political or administrative decision-making processes (Yackee and Yackee 2006; Obar and Schejter 2010), thereby producing policies that favor the subjective interests of the dominant group. Lukes also offers a third face of power centered on the manipulation of the subjective interests of a group as described below:

> Is it not the most supreme and insidious exercise of power to prevent people, to whatever degree, from having grievances by shaping their perceptions, cognitions and preferences in such a way that they accept their role in the existing order of things, either because they can see or imagine no alternative to it, or because they see it as natural and unchangeable or because they value it as divinely ordained and beneficial? (2005, p. 28)

This conception may or may not be institutional, depending upon the ways in which one conceptualizes the relationship, if any, between institutions and "perceptions, cognitions, and preferences." While the neoclassical economic model of the individual assumes that preferences are stable, recent advances from various fields provide strong evidence that preferences and perceptions are influenced by cultural experience and participation over time in particular institutional environments (Agrawal 2005; Henrich et al. 2006). Alternatively, institutions—as prescriptive, linguistic constructs—can, in some sense, themselves be considered a type of belief. According to Crawford and Ostrom (1995), a shared strategy is a linguistic statement consisting of actions to be taken by individuals defined by some attribute(s) under certain conditions. As an example, they offer a situation where an individual who initiates a call that is disconnected will call back. This simple strategy or social convention addresses a simple coordination problem where either both parties wait for the other to call, or perhaps try simultaneously and receive busy signals by generating shared expectations. Although beliefs about others' actions are certainly representative of the type of shared strategy envisioned by Crawford and Ostrom, it is less clear that the same could be said for beliefs about policies or rules that lack a social dimension.

Lukes's (2005) three faces of power provide three different conceptualizations, two of which we are able to operationalize in this study. Lukes's third face of power was not operationalized due to the difficulty of analyzing outcomes based on belief systems and an inability to distinguish between subjective and objective interests using the data stored in the IFRI database. Lukes's first face of power focuses on subjective perceptions of policy outcomes or operational rules (GS5) between two or more groups that participate in collective choice venues (GS6). This was operationalized by distinguishing between groups that participate in rule-making processes that fail to produce rules that align with their subjective interests and all other groups, with the assumption that the former lacks, or is subject to, the power of another group. The results indicate that the first face of power has little impact on the combined social-ecological outcome.

The second face of power refers to groups whose subjective interests are not met by operational rules (GS5), but who also do not participate in collective-choice processes (GS6). This is operationalized by distinguishing between groups where operational rules are perceived to be unfair and do not participate in rulemaking, and all other combinations. As opposed to the first face of power, Lukes's second face has a statistically significant and negative relationship with the combined social-ecological outcome. There is, however, an important caveat to this claim. Lukes clearly situates his definitions of power in relative terms, where one group is able to affect another in a way that is contrary to the second group's interests. This analysis assumes the existence of that other group, be it the state or another user group, ignoring the possibility that the lack of participation and dissatisfaction with rules is a result of other factors, most notably intragroup processes or a collective failure to self-organize.

6.9   Douglass North and the Institutional Matrix

While both Ostrom's and Lukes's new institutionalist conceptualizations of power can be easily abstracted from any particular situation, some institutional political economists embed conceptualizations of power within historically contingent contexts that are difficult to account for in quantitative approaches. For example, Douglass North (1990), the economic historian, asked why the economies of some countries performed better than others, and why those countries that fared worse did not simply adopt institutions that enhanced performance. The answer, according to North, is a set of institutions that resists change via a variety of structural and active processes. The most commonly cited process is increasing returns, wherein institutions generate a positive feedback process that favors movement in the same direction of prior decisions by virtue of some combination of benefit flows and increasing exit costs (Pierson 2000; Arthur 1989). Power enters the discussion of increasing returns when institutions privilege some members of a group or society with a greater share of the benefits and greater institutional control that enhances their ability to bargain in collective-choice settings. Pierson (2000) draws upon the community power debate and Lukes (2005) to discuss how, over time, increasing returns processes may transform power from an overt expression of wills to a latent and then hidden conflict as the institutional matrix reinforces itself:

> Increasing returns processes can transform a situation of relatively balanced conflict, in which one set of actors must openly impose its preferences on another set ("the first face of power"), into one in which power relations become so uneven that anticipated reactions ("the second face of power") and ideological manipulation ("the third face") make open political conflict unnecessary. Thus, positive feedback over time simultaneously increases power asymmetries and renders power relations less visible. (Pierson 2000, p. 259)

These path-dependent processes suggest that the greatest indicator of power may not be found in individual institutions or their simple interactions but rather in the continuity of a particular form of organization to manage transactions or resolve a policy problem. Pierson (1996), for instance, discusses how, once the welfare state has been established in democracies, it tends to persist because it generates a set of incentives that make change particularly costly for politicians. North (1990), in a more negative light, suggests that the lack of economic development in some countries is the result of inefficient forms of economic organization that persist because those that have invested in that form of organization generate increasing returns and greater bargaining power to ensure its continuity. In any case, both point to the time dimensions or historicity of institutions as an important indicator of their power.

Path-dependent forms of power bound up in institutional matrices are perhaps the most abstract of our definitions of power. We classify path-dependent power in terms of a user group's history of use (A3), although we recognize that this attribute relates to several potentially influential dimensions of use. Nonetheless, we assume that the longer a group has existed in a recognizable form using a particular resource, the more likely they will have built up a set of institutions, or an institutional matrix that creates power for the group against other groups and the state. In contrast, a user group that lacks power is unlikely to be able to maintain a recognizable form, and would instead be characterized by the formation and decay of different groups in the same geographical area. This follows North's argument that the persistence of organizations (i.e., formal and informal groups) in a given environment is tied to their bargaining power. Thus, power can be measured indirectly by considering the length of time that a group has been organized in a recognizable form. This classification was fairly easy to operate with the IFRI database, which allows us to measure the approximate age of the user group that participates in rule-making processes. The results show that this path-dependent form of institutional power is associated with positive social-ecological outcomes (Table 6.2).

6.10   Discussion

The measures of power used in this illustration reflect a pragmatic attempt to operationalize a study of how institutional forms of power relate to social-ecological benefits. Some of our findings appear straightforward. For instance, our results indicate that groups with power, as exercised through operational rules and monitoring and sanctioning processes, are associated with better social-ecological outcomes. In addition, we find that Lukes's (2005) second face of power is associated with a particularly large and negative social-ecological outcome. That is, groups that are dissatisfied with operational rules but are unable to enter collective-choice situations and modify those rules, are less likely to develop long-term sustainable patterns of use. These results are not entirely surprising given that they correspond to Ostrom's (1990) design principles and continue to receive support from various sources (Chhatre and Agrawal 2008; Coleman 2009; Cox et al. 2010). Notwithstanding the patterns of association found in this study, important questions remain as to whether power is accurately captured by the classifications and measures that were used, and the extent to which the evidence presented provides a basis for causal inference. This section engages in self-critique to examine some limitations of the considered approach to analysis.

   Although some definitions of power, particularly those that refer to specific institutions, were readily classified, it is far from certain that their operationalization accurately reflects the power of a group. For instance, with regards to institutional control, groups are assumed to hold greater power if operational rules are perceived to be fair, if they participate in rulemaking, or if they own the forest commons. There is, however, considerable room for debate as to whether a group could be said to be powerful if it possesses one of these attributes but not others. In addition, North's (1990) view of bargaining power as an output of early choices that generate a process of increasing returns and path dependence is equally problematic. According to our results using North's conceptualization of power, groups that manage to persist in a recognizable form over an extended period of time are more likely to be associated with positive social-ecological outcomes. The proposed explanation is that groups develop a matrix of supportive institutions that set them on a distinct historical trajectory (North 1990; Pierson 2003). However, while a group can be seen as an informal organization whose survival depends upon its ability to generate a continuous stream of benefits to its members, measures of its age may be prone to suffer from idiosyncratic measurement error or measure concepts completely unrelated to power, or even its inverse. India's caste system, for instance, has for centuries been used to define groups; however, it systematically assigns power to some of these groups while withholding it from others. In other words, a group may persist precisely because it finds itself on the less powerful side of socially, politically, or institutionally entrenched power inequities.

Finally, as definitions become more specific to involve interactions among attributes and measures, important questions concerning the level of measurement must be considered. For instance, Ostrom's (2005) definition involving the extent of control (presumably varying between 0 and 1) and value of opportunity (presumably a continuous variable) was operationalized using a binary and ordered variable, respectively. The way in which Ostrom's definition of power was operationalized reflects the availability of data, but also draws attention to the potentially confounding role of measurement of the dependent and independent variables.

Even if one accepts the general validity and assumptions related to definitions, classifications, and measurements offered in this study, there are several reasons why one might still reject the evidence provided. To begin with, most causal inference in the positivist paradigm rests upon the general validity of three attributes of an analysis: (1) association between a cause and an outcome, (2) isolation of potential causes from other attributes of the environment, and (3) the direction of effects (Bollen 1989). Association is generally the least controversial, and in this study were measured using standard methods such as difference of means and correlations. Isolation and direction are typically more problematic, as they ask the researcher to separate causes from all other attributes that may bias estimates and establish whether the "cause" is in fact responsible for producing the "effect." Randomization, matching, or quasi-experiments are often considered the best means with which to isolate factors (Holland 1986; Shadish et al. 2002; Rubin 2005), although in some cases structural models and even linear regression may be sufficient for pseudo-isolation (Pearl 2012). Establishing the direction of causal effects from cross-sectional observational data is even more problematic. Temporal priority (i.e., a gap between the observation of a cause and the outcome), or direct manipulation in an experimental environment, are usually sufficient to infer the direction of a causal effect (Brady 2008). However, in the absence of either, most directional claims using observational data rest upon logical and theoretical understandings of the phenomena. In this case, the analysis neither seeks to isolate power from other influences, nor does the cross-sectional data allow us to infer whether a measure of power precedes the outcome, or instead whether the outcome is actually a cause of power.

A final critique of the illustrative analysis is that many of the same measures already appear in the literature on the commons (Chhatre and Agrawal 2008, 2009; Coleman 2009) and have been merely recast in terms of power. Institutional control, for instance, is typically studied in isolation from its normative power-laden implications. This highlights several points that merit additional discussion. First, power is, and always has been, a feature of the commons literature, which should be self-evident from many of the design principles (Ostrom 1990). Minimal rights to organize, participation in collective-choice processes, and the accountability of monitors all concern different types of institutional power held by a group of resource users. That is far from being a power-neutral approach to the study of social-ecological phenomena; power is as an integral part of commons theory and the SES framework.

However, the critiques are not entirely without merit. Compared to some disciplines, such as political ecology, new institutionalism often softens or hides the normative implications of power imbalances behind a veil of game-theoretic terminology and a pragmatic emphasis on designing institutions that produce beneficial societal outcomes, however those may be defined. In other words, accompanying the shift in language is a sense that something meaningful is lost. Thus, bringing together multiple disciplines to study power within the SES framework compels researchers to engage explicitly with challenging and inevitable tradeoffs between critical and pragmatic approaches. Moreover, the emphasis on groups, broadly labeled resource users, likely overlooks a wide range of power relations within groups, most notably differential power between elite and non-elite members (Vedeld 2000; Iversen et al. 2006; Mwangi 2006). Finally, it is clear that, while power exists implicitly in the SES framework, it is not given a prominent position; and if trends continue, the range of theorizing and studying power in the institutionalist tradition will remain overly narrow.

6.11   Conclusions: An Interdisciplinary Agenda for the Study of Power in SESs

This chapter has illustrated that the SES framework holds great potential for social science integration, and may serve as a bridge between power-centered approaches and institution-centered approaches to the study of social-ecological systems. It further demonstrates that the SES framework is equipped with a wide range of attributes that can be used to study to several definitions or theories of power. Although the analysis presents empirical results with associated significance, the study *does not* provide definitive answers to the questions of whether any individual type of power matters, or which of the many alternatives best captures the concept of power. Instead, our primary goal was to assess whether asking such questions with the SES framework is possible and whether such an endeavor is potentially fruitful. We believe that the answer to both questions is yes, but that there remains considerable work to be done with regards to other theories of power, measurement, and evaluation before the framework could be said to facilitate such an endeavor.

The four methodological steps that we applied in this study provide guidance that other researchers can use to integrate other theories of power within the SES framework. Rather than simply assuming that power exists in some objectively observable way, researchers must attend to the ways in which (1) the values of existing SES attributes differentially affect different actors and groups and (2) different actors and groups contest and reshape the value of SES attributes. For example, instead of asking *what type of operational rules produce better social-ecological outcomes*, we asked *how the perceived fairness of operational rules influences outcomes*. We reoriented questions about the form of collective-choice institutions to ask whether groups have the power to control the outputs of institutional decision-making processes, and what effect this has on social-ecological outcomes. If we use the SES framework to conceptualize a social-ecological system made up of the four key subsystems and the variable subsets within them, then to wrestle with the role of power within such a system is to examine the shadow that those variables cast on the material, institutional, and discursive attributes of a varied set of actors.

Questions of power must be investigated in a space of inquiry that is once-removed from the social-ecological system; it does not consist of the subsystems and variables within those subsystems but rather the heterogeneous effects of those variables on different groups, as well as the process through which heterogeneous actors contest those variables. In studying the effects of power, we are not posing questions about the direct relation between the variables and outcomes but about the effects that the differentiated meanings and implications of those variables for different key actors have on social-ecological outcomes. This is why we make the claim, at least regarding institutional conceptualizations of power that "power" or "politics" need not appear as attributes, themselves, within the SES framework. Rather, as we suggest, institutional conceptualizations of power are realized in the relationships between existing attributes and their implications for a specified group of actors.

Similarly, future research about the relationships between power and sustainability need not, necessarily, add new power-related attributes the framework. Rather, the process of research design and collection should carefully attend to the connections between indicators of existing attributes, their implications for particular groups and their relationship with social and ecological outcomes. Implicit in an approach that locates power not as a single, discrete attribute but as relationships between one or more attributes and a group of actors, is a claim about the ontology of power itself. Specifically, it suggests that power is a composite theoretical construct made up of attributes and relationships. This claim is further supported by the existence of a wide range of distinct conceptualizations of power from across disciplines. Thus, to engage in a cross-disciplinary study of power in the context of SESs requires us to deconstruct the vague and variegated concept, *power*, and specify its component parts and the relationships among them. The SES framework is well suited for this task.

The general approach adopted in this chapter to study institutional forms of power may be used to advance the study of other conceptualizations. Many materialist approaches from political ecology, for example, suggest that power exists as a result of unequal access and control over wealth, natural resources, or the means of production. An initial glance at the SES would suggest that many of the attributes, including the economic value of the resource, socioeconomic characteristics of the resource users, resource users' dependence on the resource, and property rights regimes, may be put to use to develop appropriate measures of materialist conceptions of power. Moreover it seems likely that the framework could similarly structure studies of discursive conceptualizations of power in terms of communicated knowledge, norms, and mental models that shape individuals' beliefs and behavior. Indeed, some attributes of the SES, such as knowledge of the SES/mental models as well as social norms, may provide an opportunity to better understand what, if any, differences exist between knowledge and discourse, and how they are transmitted across groups.

Ultimately, however, whether the SES is fully equipped in its current form to facilitate research on the role of power across all disciplines will require further theoretical, conceptual, and empirical work that is beyond the scope of this chapter. Nonetheless, this general strategy, which focuses on identifying existing SES attributes and the relationships among groups with respect to those attributes, encourages researchers to embrace interdisciplinary approaches to power, while thinking rigourously about how to move through the research process from conceptualization to operationalization and measurement. The SES framework was designed precisely to facilitate such interdisciplinary work and to provide a foundation upon which multiple disciplinary approaches to research may find, if not agreement, then mutual intelligibility.

Finally, further issues arise as researchers move from measurement to analysis of whether particular conceptualizations of power matter. This analysis presumed to evaluate the relationship between power and social-ecological benefits by positing a single causal step from the indicator to the outcome. However, many scholars view power in terms of a complex web of self-reinforcing historical processes, institutions, and resources that collectively privilege some groups over others (Pierson 2000; Benjaminsen et al. 2009). Furthermore, studying some individual indicator of power in isolation from others may fundamentally conflict with the ways in which power operates to either sustain or degrade social-ecological systems. This reflects a growing debate in the social sciences concerning the ways in which attributes or variables are understood to affect social phenomena. The classic approach that corresponds to multivariate quantitative methods is to assume that variables have a conditionally independent and additive effect on a dependent variable (Freedman 1999). In contrast, many qualitative methodologists view outcomes in terms of a unique confluence of slow- and fast-moving causes that interact in complex ways to produce often unexpected results (Pierson 2003). More recently, a third perspective has emerged that seeks to strike a balance between these two extremes and suggests that outcomes depend upon the state of combinations of attributes that collectively define a case (Ragin 2000; Basurto and Ostrom 2009). We suggest that the SES framework offers scholars engaging diverse theoretical and methodological approaches an opportunity to structure their debates in systematic and coherent terms.

The SES framework is a bold and ambitious tool meant to serve a diverse audience of interdisciplinary scholars, many of whom focus explicitly on questions of power and inequality. It is unfortunate that the framework has yet to take greater strides in this direction, forcing scholars to develop ad hoc solutions, or, more likely, to choose alternative, more disciplinary-focused analytical tools. In perpetuating the shift toward a positive theory of environmental governance, the SES framework neglects the important normative question as to why we should care in the first place. The fields of environmental governance in particular and public policy in general exist to confront the problems of society and promote "human dignity" (Lasswell 1951). Power is an integral part of human affairs, and we believe that power ought to be given greater attention within institutional studies of SESs. However, such an endeavor must seek to explicate the positive and normative implications of diverse forms of power that characterize "alternatives futures" (Ostrom 2008).

References

Adcock, R., & Collier, D. (2001). Measurement validity: A shared standard for qualitative and quantitative research. *American Political Science Review, 95*(3), 529–546.

Agrawal, A. (2003). Sustainable governance of common-pool resources: Context, methods, and politics. *Annual Review of Anthropology, 32,* 243–262.

Agrawal, A. (2005). *Environmentality: Technologies of government and the making of subjects.* Durham: Duke University Press.

Agrawal, A. (2013). Studying the commons, governing common-pool resource outcomes: Some concluding thoughts. *Environmental Science & Policy.* http://www.sciencedirect.com/science/article/pii/S1462901113001615. Accessed 22 Oct 2013.

Alcorn, J. B., & Toledo, V. M. (2000). Resilient resource management in Mexico's forest ecosystems: The contribution of property rights. In F. Berkes, C. Folke, & J. Colding (Eds.), *Linking social and ecological systems: Management practices and social mechanisms for building resilience* (pp. 216–249). Cambridge: Cambridge University Press.

Aligica, P. D., & Boettke, P. J. (2009). *Challenging institutional analysis and development: The Bloomington school.* New York: Routledge.

Andersson, K., & Agrawal, A. (2011). Inequalities, institutions, and forest commons. *Global Environmental Change, 21*(3), 866–875.

Arthur, W. B. (1989). Competing technologies, increasing returns, and lock-in by historical events. *The Economic Journal, 99*(394), 116–131.

Basurto, X., & Nenadovic, M. (2012). A systematic approach to studying fisheries governance. *Global Policy, 3*(2), 222–230.

Basurto, X., & Ostrom, E. (2009). Beyond the tragedy of the commons. *Economia delle fonti di energia e dell'ambiente, 52*(1), 35–60.

Basurto, X., Gelcich, S., & Ostrom, E. (2013). The social-ecological system framework as a knowledge classificatory system for benthic small-scale fisheries. *Global Environmental Change., 23*(6), 1366–1380.

Benjaminsen, T. A., Maganga, F. P., & Abdallah, J. M. (2009). The Kilosa killings: Political ecology of a farmer–herder conflict in Tanzania. *Development and Change, 40*(3), 423–445.

Blanco, E. (2011). A social-ecological approach to voluntary environmental initiatives: The case of nature-based tourism. *Policy Sciences, 44*(1), 35–52.

Bollen, K. A. (1989). *Structural equations with latent variables.* New York: Wiley.

Brady, H. E. (2008). Causation and explanation in social science. In J. M. Box-Steffensmeier, H. E. Brady, & D. Collier (Eds.), *The Oxford handbook of political methodology* (pp. 217–270). New York: Oxford University Press.

Bray, D. B., Antinori, C., & Torres-Rojo, J. M. (2006). The Mexican model of community forest management: The role of agrarian policy, forest policy and entrepreneurial organization. *Forest Policy and Economics, 8*(4), 470–484.

Breitmeier, H., Young, O. R., & Zürn, M. (2006). *Analyzing international environmental regimes.* Cambridge: MIT Press.

Chhatre, A., & Agrawal, A. (2008). Forest commons and local enforcement. *Proceedings of the National Academy of Sciences of the United States of America, 105*(36), 13286–13291.

Chhatre, A., & Agrawal, A. (2009). Trade-offs and synergies between carbon storage and livelihood benefits from forest commons. *Proceedings of the National Academy of Sciences, 106*(42), 17667–17670.

Cinner, J. E., McClanahan, T. R., MacNeil, M. A., Graham, N. A. J., Daw, T. M., Mukminin, A., Feary, D. A., Rabearisoa, A. L., Wamukota, A., Jiddawi, N., Campbell, S. J., Baird, A. H.,

Januchowski-Hartley, F. A., Hamed, S., Lahari, R., Morove, T., & Kuange, J. (2012). Comanagement of coral reef social-ecological systems. *Proceedings of the National Academy of Sciences, 109*(14), 5219–5222.

Clement, F. (2010). Analysing decentralised natural resource governance: Proposition for a 'politicised' institutional analysis and development framework. *Policy Sciences, 43*(2), 129–156.

Coleman, E. A. (2009). Institutional factors affecting biophysical outcomes in forest management. *Journal of Policy Analysis and Management, 28*(1), 122–146.

Coleman, E. A., & Fleischman, F. D. (2012). Comparing forest decentralization and local institutional change in Bolivia, Kenya, Mexico, and Uganda. *World Development, 40*(4), 836–849.

Collier, D. (2011). Understanding process tracing. *PS: Political Science & Politics, 44*(4), 823–830.

Commons, J. R. (1924). *Legal foundations of capitalism*. New Brunswick: Transaction Publishers.

Commons, J. R. (1931). Institutional economics. *American Economic Review, 21*, 648–657.

Cox, M., Arnold, G., & Villamayor Tomas, S. (2010). A review of design principles for community-based natural resource management. *Ecology and Society, 15*(4), 38.

Crawford, S. E. S., & Ostrom, E. (1995). A grammar of institutions. *American Political Science Review, 89*(3), 582–600.

Dietz, T., Dolšak, N., Ostrom, E., & Stern, P. C. (2002). The drama of the commons. In T. Dietz, N. Dolšak, E. Ostrom, P. C. Stern, S. Stovich, & E. U. Weber (Eds.), *The drama of the commons* (pp. 3–35). Washington, DC: National Academies Press.

Epstein, G., Vogt, J. M., Mincey, S. K., Cox, M., & Fischer, B. (2013). Missing ecology: Integrating ecological perspectives with the social-ecological system framework. *International Journal of the Commons, 7*(2), 432–453.

Fleischman, F., Boenning, K., Garcia-Lopez, G., Mincey, S., Schmitt-Harsh, M., Daedlow, K., López, M. C., Basurto, X., Fischer, B. C., & Ostrom, E. (2010). Disturbance, response, and persistence in self-organized forested communities: Analysis of robustness and resilience in five communities in southern Indiana. *Ecology and Society, 15*(4), 9.

Freedman, D. (1999). From association to causation: Some remarks on the history of statistics. *Statistical Science, 14*(3), 243–258.

George, A. L., & Bennett, A. (2005). *Case studies and theory development in the social sciences*. Cambridge, MA: MIT Press.

Gruby, R. L., & Basurto, X. (2013). Multi-level governance for large marine commons: politics and polycentricity in Palau's protected area network. *Environmental Science & Policy, 33*, 260–272.

Gutierrez, N. L., Hilborn, R., & Defeo, O. (2011). Leadership, social capital and incentives promote successful fisheries. *Nature, 470*(7334), 386–389.

Henrich, J., McElreath, R., Barr, A., Ensminger, J., Barrett, C., Bolyanatz, A., Cardenas, J. C., Gurven, M., Gwako, E., Henrich, N., Lesorogol, C., Marlowe, F., Tracer, D., & Ziker, J. (2006). Costly punishment across human societies. *Science, 312*(5781), 1767–1770.

Hobbes, T. ([1651] 2010). *Leviathan: Or the matter, forme, and power of a common-wealth ecclesiasticall and civill* (I. Shapiro, Ed.). New Haven: Yale University Press.

Holland, P. W. (1986). Statistics and causal inference. *Journal of the American Statistical Association, 81*(396), 945–960.

Immergut, E. M. (1998). The theoretical core of the new institutionalism. *Politics and Society, 26*, 5–34.

Iversen, V., Chhetry, B., Francis, P., Gurung, M., Kafle, G., Pain, A., & Seeley, J. (2006). High value forests, hidden economies and elite capture: Evidence from forest user groups in Nepal's Terai. *Ecological Economics, 58*(1), 93–107.

Kiser, L. L., & Ostrom, E. (1982). The three worlds of action: A metatheoretical synthesis of institutional approaches. In E. Ostrom (Ed.), *Strategies of political inquiry* (pp. 179–222). Beverly Hills: Sage. (Reprinted from *Polycentric games and institutions: Readings from the workshop in political theory and policy analysis*, pp. 56–88, by M. McGinnis, Ed., 2000, Ann Arbor: University of Michigan Press)

Lasswell, H. (Ed.). (1951). *The policy orientation*. Stanford: Stanford University Press.

Lebel, L., Mungkung, R., Gheewala, S. H., & Lebel, P. (2010). Innovation cycles, niches and sustainability in the shrimp aquaculture industry in Thailand. *Environmental Science & Policy, 13*(4), 291–302.

Lukes, S. (2005). *Power: A radical view* (2nd ed.). Basingstoke: Palgrave Macmillan.

Machiavelli, N. ([1532] 1988). *The prince* (Q. Skinner & R. Price, Eds.). Cambridge: Cambridge University Press.

McGinnis, M., & Ostrom, E. (forthcoming). SES framework: Initial changes and continuing challenges. *Ecology and Society.*

Mosse, D. (1997). The symbolic making of a common property resource: history, ecology and locality in a tank-irrigated landscape in South India. *Development and Change, 28*, 467–504.

Mwangi, E. (2006). The footprints of history: Path dependence in the transformation of property rights in Kenya's Maasailand. *Journal of Institutional Economics, 2*(2), 157–180.

North, D. C. (1990). *Institutions, institutional change and economic performance.* Cambridge: Cambridge University Press.

Obar, J. A., & Schejter, A. M. (2010). Inclusion or illusion? an analysis of the FCC's public hearings on media ownership 2006–2007. *Journal of Broadcasting & Electronic Media, 54*(2), 212–227.

Ostrom, V. (1976). John R. Commons's foundations for policy analysis. *Journal of Economic Issues, 10*(4), 839–857.

Ostrom, E. (1990). *Governing the commons: The evolution of institutions for collective action.* New York: Cambridge University Press.

Ostrom, E. (2005). *Understanding institutional diversity.* Princeton: Princeton University Press.

Ostrom, E. (2007). A diagnostic approach for going beyond panaceas. *Proceedings of the National Academy of Sciences of the United States of America, 104*(39), 15181–15187.

Ostrom, V. (2008). *The intellectual crisis in American public administration* (3rd ed.). Tuscaloosa: University of Alabama Press.

Ostrom, E. (2009). A general framework for analyzing sustainability of social-ecological systems. *Science, 325*(5939), 419–422.

Ostrom, E., & Cox, M. (2010). Moving beyond panaceas: A multi-tiered diagnostic approach for social-ecological analysis. *Environmental Conservation, 37*(4), 451–463.

Ostrom, E., Gardner, R., & Walker, J. (1994). *Rules, games, and common-pool resources.* Ann Arbor: University of Michigan Press.

Pearl, J. (2012). The causal foundations of structural equation modeling. In R. H. Hoyle (Ed.), *Handbook of structural equation modeling* (pp. 68–91). New York: Guilford Press.

Persha, L., Agrawal, A., & Chhatre, A. (2011). Social and ecological synergy: Local rulemaking, forest livelihoods, and biodiversity conservation. *Science, 331*(6024), 1606–1608.

Pierson, P. (1996). The new politics of the welfare state. *World Politics, 48*, 143–179.

Pierson, P. (2000). Increasing returns, path dependence, and the study of politics. *American Political Science Review, 94*(2), 251–267.

Pierson, P. (2003). Big, slow-moving, and.. invisible. In J. Mahoney & D. Rueschemeyer (Eds.), *Comparative historical analysis in the social sciences* (pp. 177–207). Cambridge: Cambridge University Press.

Ragin, C. C. (2000). *Fuzzy-set social science.* Chicago: University of Chicago Press.

Riker, W. H. (1980). Implications from the disequilibrium of majority rule for the study of institutions. *American Political Science Review, 74*(2), 432–446.

Robbins, P. (2004). *Political ecology: A critical introduction.* Malden: Blackwell.

Rubin, D. B. (2005). Causal inference using potential outcomes. *Journal of the American Statistical Association, 100*(469), 322–331.

Shadish, W. R., Cook, T. D., & Campbell, D. T. (2002). *Experimental and quasi-experimental designs for generalized causal inference.* Stamford: Wadsworth Cengage learning.

Vedeld, T. (2000). Village politics: Heterogeneity, leadership and collective action. *Journal of Development Studies, 36*(5), 105–134.

Waldo, D. (1948). *The administrative state: A study of the political theory of American public administration.* New Brunswick: Transaction Publishers.

Wertime, M., Ostrom, E., Gibson, C., & Lehoucq, F. (2007). *International Forestry Resources and Institutions (IFRI) research program: Field manual, version 13*. Bloomington: Indiana University, Center for the Study of Institutions, Population and Environmental Change (CIPEC).

Wilson, W. (1887). The study of administration. *Political Science Quarterly, 2*(2), 197–222.

Yackee, J. W., & Yackee, S. W. (2006). A bias towards business? Assessing interest group influence on the U.S. bureaucracy. *Journal of Politics, 68*(1), 128–139.

Young, O. R., & Zürn, M. (2006). The international regimes database: Designing and using a sophisticated tool for institutional analysis. *Global Environmental Politics, 6*(3), 121–143.

Chapter 7
Considerations in Representing Human Individuals in Social-Ecological Models

Michael J. Manfredo, Tara L. Teel, Michael C. Gavin, and David Fulton

7.1 Purpose

The most troubling problems in conservation – deforestation, land degradation, biodiversity loss, and climate change – are difficult to isolate and examine as independent phenomena. Increasingly, the view from science casts these as outcomes from complex interactions within and between human society and its biophysical context. Reductionist science is poorly suited for representing such complexity, and that has given rise to multidisciplinary, multi-level systems approaches. This increase in multidisciplinary approaches has created a transformative wave of change as the existing institutions of conservation science absorb, adapt, and give way to innovations that can advance such approaches.

In this chapter, we examine how conservation science that focuses on the human individual – particularly the tradition of social science research that has emerged under the flag of "human dimensions of natural resources" – might fit within a systems approach. Our examination of this topic has a dual purpose: to suggest the implications for (1) how ecosystem sciences can integrate the human individual into dynamic, multi-level models, and (2) how human dimensions research can envision the individual and direct new research initiatives in a broader social-ecological context.

M.J. Manfredo (✉) • T.L. Teel • M.C. Gavin
Department of Human Dimensions of Natural Resources, Colorado State University,
1480 Campus Delivery, Fort Collins, CO 80523-1480, USA
e-mail: michael.manfredo@colostate.edu

D. Fulton
Department of Fisheries, Wildlife and Conservation Biology, University of Minnesota,
1980 Folwell Ave #200, St Paul, MN 55108, USA

M.J. Manfredo et al. (eds.), *Understanding Society and Natural Resources*,
DOI 10.1007/978-94-017-8959-2_7, © The Author(s) 2014

7.2 Impetus for Change Emanating from Ecological Sciences

Historically, biological traditions have set the direction of natural resources research and that is also the case in the drive toward multi-scale, multi-level and multi-disciplinary approaches. More specifically, this new direction in natural resources research is borne from the shift toward systems science in ecology. C. S. Holling (1998) described the ecosystems approach by contrasting two cultures in the ecological sciences. The traditional approach was reductionist, narrow and targeted, experimentally focused, concerned with Type I error, hypothesis testing and standard statistics. In this culture, the environment is viewed as largely fixed and at a single scale, and causation is considered single and separable. The other fast-emerging culture was seen as broad, exploratory, multi-disciplinary and integrative, and it is focused on multiple lines of converging evidence. It uses non-standard statistics and is concerned with Type II error. It takes a systems view of the environment, describing the environment's dynamic qualities as self-organizing with multiple interactions, and operating at multiple scales.

This systems approach views nature as complex, dynamic, and adaptive. The system reveals both chaos and order, has continuous and discontinuous elements, and is marked by abrupt change (Holling et al. 2002). Hierarchy is central to this conceptualization. Phases of change are proposed to occur within multi-scale, multi-level structures that are nested within a broader hierarchy. The structures move at separate speeds and are multi-directional in their effects. Each level experiences its own change cycle, but slower and larger scales set conditions for faster, smaller ones, whereas the latter are the sites of variation that can generate functional shifts at higher scales. Systems are seen as having varying degrees of resilience – a reference to their ability to retain crucial functions during episodes of change. Adger et al. (2005) suggest that "the concept of resilience is a profound shift in traditional perspectives, which attempt to control changes in systems that are assumed to be stable, to a more realistic viewpoint aimed at sustaining and enhancing the capacity of social-ecological systems to adapt to uncertainty and surprise" (p. 1036).

7.3 A Need for Greater Inclusion of the Individual in Ecosystem Models

Humans were largely absent from the early ecosystem models (e.g., Noss 1990), then were added as macro, driving forces that cause change in biological systems (e.g., Forester and Machlis 1996). But quickly, attention was given to integrating the social component in describing "social-ecological systems" (SES). Broad questions for the social aspects of resilience ask about human response and adaptation, how reorganization follows collapse or sudden dramatic change, and how social learning accumulates (Gunderson and Holling 2002). Political scientists, economists, geographers and anthropologists (Abel and Stepp 2003; Collins et al. 2011; Kok and Veldkamp 2011), working at the group or institutional level, have been quicker to respond to this trend than were those focused on individuals in the social psychology tradition. Abel and Stepp (2003), for example, called for a new "ecosystems ecology in anthropology", a discipline that has a long history of an ecological approach to cultural change. However, increasingly, the importance of including individuals in the internal dynamics of SES models has been recognized (Redman et al. 2004; Collins et al. 2011). Inclusion of the individual addresses an important weakness evident in SES modeling: representing the capacity of humans to make choices that affect the system. Davidson (2010) suggests that to be fully inclusive of a social component, concepts of resilience will need to account for the fact that, unlike other organisms in an ecological system, humans have the ability to postpone ecological effects, have unequal agency (due to power differentials) and the ability to imagine, anticipate, and invoke collective action.

While research on the individual predominates the published literature in the human dimensions of natural resources (HDNR) area (Manfredo et al. 2004), very little of this work fits readily into the SES paradigm. With few exceptions, most HDNR research is borne from the information processing paradigm made popular in psychology in the 1970s. This paradigm viewed psychological attributes as isolated, static, and enduring and did not account for the influence of factors such as culture or context on cognition (Gardner 1987). More recently, there have been explicit attempts to include individuals in SES models that have focused on finding simple social variables for use in techniques such as agent-based modeling (Janssen and Ostrom 2006; Buizer et al. 2011). Researchers have also proposed concepts or frameworks such as mental models (Jones et al. 2011), consumer preferences (Baumgärtner et al. 2011), and people-environment transactions (Stokols et al. 2013) as ways of representing individuals in SES.

Independent of these efforts, areas such as Cognitive Ecology, Evolutionary Psychology, Social-Ecological Psychology and Cross-Cultural Psychology have led the drive toward more complex, dynamic, adaptive multi-level approaches that might offer guidance in bringing individuals into SES models. Drawing from these areas, we make suggestions about three basic questions in developing approaches for bringing the individual into SES models: (1) is human thought conceptualized as a dynamic and adaptive process, (2) is the individual placed in a multi-level context, and (3) is human thought seen as mutually constructed with the social and natural environment. We address each of these questions throughout the remainder of this chapter.

7.4   Human Thought as Dynamic and Adaptive

Growing threats such as climate change and desertification have prompted concerns about the human ability to anticipate, adapt, and alter behavior to avoid undesirable results. Certainly, history shows that humans are remarkably adaptive. This adaptive success is attributed to our ability to create culture, accumulate knowledge and transmit that across generations. It is the cognitive processes of the human mind that create outcomes such as technological advances, social collaboration, and institutional invention that generate effective adaptive success. The view from evolutionary psychology would propose that our cognitive systems and fundamental cognitions such as values are critical mechanisms for adapting to our social and ecological surroundings. Even a simple depiction of the dynamic and adaptive nature of thought processes, as we present here, can have meaningful implications for SES modeling.

7.4.1   Dual Adaptive Systems in Humans

Theory suggests there are two separate systems that drive human thought and behavior (Evans and Stanovich 2013). System one evolved early in humans, and responses stemming from it tend to be automatic, fast and intuitive, and non-conscious. System two evolved more recently, and response is slow, requiring working memory and the deliberation of existing knowledge. The first system is developed gradually over time through the process of associative learning, repeat experience, and trial and error. This system accumulates an individual's learning into quick response mechanisms. Given the cumulative process, a single incident is unlikely to have a big effect on the responses of System one; further, the more incidents accumulated, the less likely change is to occur. As a consequence, a significant amount of foundational associative learning occurs in one's early years of life. System one gives a person instantaneous response to a constantly-changing surrounding. For a given motivational or goal state, this system shapes subconscious perception of the environment and its opportunities and dangers. It drives the automatic course of action in a fluid and "online" manner.

System two is based on semantic or symbols-based learning, storage of information, recall and deliberation in new and novel ways. System two is used episodically as needed when a response situation rises to a conscious level. Information that is drawn from the environment and processed with information from memory to anticipate consequences of response can be stored for retrieval at a later time. System two processing is considered *slow in the sense that it requires considerable cognitive effort and time to reach a conclusion.* While it is slow in that sense, it is *fast in its ability to change and adapt to new situations.* For example, where a new incident is unlikely to affect System one processing, new information can readily change an attitudinal and behavioral response borne from System two.

System one is obviously not independent of System two. The foundational aspects of perception, assumptions, and evaluation shape our awareness, under-standing, and acceptance of new information. Yet, the two systems are believed to sometimes act in conflict with one another in difficult decision choices (e.g., when a person carefully analyzes pros and cons and has a gut feeling different from the result of that critical analysis). Evidence for the dual systems approach, and the conflict between systems that may arise, is supported by studies that show different areas of the brain are active when different systems are engaged (Goel 2008).

How can the dual systems view be useful in SES? These two systems paint a complex picture of human ability to adapt. On the one hand, System one facilitates continuity and predictability. It is the essence of cultural transmission by which customs, practices and meanings are carried through generations. System two prepares humans for abrupt and sudden changes in their surroundings. It allows people to quickly (relative to the effect of information on System one) assimilate information, weigh it against information stored in memory and develop a response to maximize positive outcomes.

A few examples reveal ways that the dual systems model brings perspectives to conservation problems. Recent articles have indicated that in order to attain social and ecological sustainability, human values and subsequent behavior must change (Burns 2012; Ehrlich and Kennedy 2005; Karp 1996; Vlek and Steg 2007). In some cases, initiatives have been undertaken to change human values. These initiatives will likely face an inordinate challenge, particularly with attempts at traditional rational appeal. As proposed by Kitayama and Uskul (2011), values are not entities but the "water we swim in". They are learned through slow processes of associative learning and reinforced by more explicit, System two learning. Values arise gradually through the continued repetition of cultural practices, stories and myths, beliefs and meanings. They are not merely learned, but are "embrained", or integrated within the mental processes controlled by the brain, and evidence is emerging that suggests there is a genetic basis for such culturally-delineated patterns.

Recent findings exemplify how the long-term durability of values might affect conservation. Manfredo et al. (2013) provide data indicating that Americans' wildlife value orientations can be traced to their ancestry, or country of origin, with shifts in thought patterns occurring slowly as states in the U.S. become more modernized. This view casts doubt on the ability to engineer an effect on cultural values and transfers attention to the key question of how values adapt (and at what rate) to a rapidly changing world. That is, when there is significant interruption in surrounding life circumstances, such as warfare, massive environmental change or migration, how do values affect the adaptation process?

In another example, Weber (2006, 2010) explains that the slow acceptance of climate change may be related to the general lack of personal experience (climate events) that would inform the associative system of risk and produce negative evaluations or feelings such as fear that motivate action. In other words, System one's intuitive influence contrasts with the deliberative process of System two on climate change response. Simply providing more facts will not change the situation. Research by Kahan et al. (2011) offers support for this explanation. Scientific literacy, according to their findings, had minimal influence on perceptions of climate change in America, whereas cultural value effects were strong and guided people's assessment of the credibility of climate change information (suggesting a strong System one influence).

A final example comes from the study of traditional ecological knowledge (TEK), which encompasses empirical knowledge of natural resources, resource management systems, social institutions that guide management, and worldviews that provide meaning to the role of humans in ecosystems (Berkes 1999). TEK is far more than factual information learned cognitively (System two). It is also a cumulative knowledge system transmitted across generations through associative learning and System one processes. An increasing body of evidence demonstrates that TEK is adapted by each generation and supplemented as new information becomes available through individual and group experiences with resource management (Berkes et al. 2000). In other words, TEK can be thought of as a cultural system and not merely as a body of cognitive information.

We conclude this section by proposing that whether or not the dual systems approach is applied, theoretical approaches and problem statements in SES research should emphasize the dynamic, adaptive nature of human thought. In doing so, research should explore alternative methods to the conventional interview or survey response methodology. This change will undoubtedly be challenging given that research sponsors often request traditional survey methods, HDNR researchers are trained primarily in these methods, and the availability of alternatives is somewhat limited. However, alternatives do exist, and recent examples illustrate the potential of experimental approaches (e.g., game theory), cognitive ability and styles tests, implicit attitude tests, longitudinal studies, and physiological and brain imaging measures.

7.5   The Individual in a Multi-level Context

A systems approach to understanding the individual in SES views "the brain, body and world in coupled motion" (Hutchins 2010, p. 709). Such an approach requires that the researcher reach across scales and influences in explaining human behavior. It requires the adoption of a broad, inclusive meta-theory but also implies non-traditional types of statistical methods (e.g., hierarchical linear modeling, social network analysis, agent-based modeling) and methodological concerns (e.g., the ecological fallacy). Hutchins (2010) has noted that applying systems approaches to understanding real-life behavior in psychology has been challenging. This is in no small part due to the complexity of humans and the near endless permutations of levels of effects. For those working in conservation, the hierarchy applied in research will depend on the way the problem is defined. Here we briefly overview three broad hierarchical categories that will be useful to consider: within the individual, individual-group, and institutional and structural factors.

7.5.1   Hierarchies Within the Individual

Humans are driven by a variety of interrelated processes, and each has a separate literature and breadth of theories including, for example, theories of needs and motivation, perception, cognition and evaluation, affect and emotion, and learning and memory. We understand just a fraction of how these processes operate together, are formed and adapt people to their social-ecological surroundings. Hierarchical approaches would work toward bridging understanding of the interdependence of these processes and their impact on human judgments, decisions and behavior.

There is little current research in HDNR that takes a hierarchical approach, with one exception – the value-attitude-behavior hierarchy (VAB). VAB research brings together, in hierarchical form, the guiding influence of the more slowly-formed cognitive processes of values (System one) and the more rapid evaluative processes of attitudes (System two). There are a variety of examples of the VAB approach being used in the natural resources arena (e.g., Fulton et al. 1996; Vaske and Donnelly 1999; Hrubes et al. 2001; Oreg and Katz-Gerro 2006; Milfont et al. 2010). An important area for future research will likely explore the contexts in which behavioral response breaks from System one thought patterns (like values in the climate change example above) and information processing takes precedence in defining individual choice.

While VAB and other goal hierarchy approaches became popular in the 1980s, more recent advances reach across various sub-disciplines of psychology, joining self-report cognitive measures with genetic, biological, and physiological measures. The latter have been particularly useful in explaining dual systems models and also supporting evolutionary explanations of human behavior (Goel 2008). As an example of these advancements, Chiao and Blizinsky (2013) provided evidence of the linkage among social-ecological conditions (mental health and disease prevalence), human genetics (serotonin gene transporter), and cultural value types (individualist-collectivist). With data spanning across 29 nations, their research revealed that the Short allele of the 5-HTTLPR was more prevalent in collectivist cultures as an adaptive response to higher levels of disease prevalence. Collectivist cultures and associated customs not only support preventative behavior for disease spread, but they serve to provide social support that mediates the negative emotion and fear avoidance behavior associated with the Short allele 5-HTTLPR.

In another example, Greene et al. (2004) found evidence that personal moral decisions stimulate areas of the brain associated with emotion and social cognition (more primitive responses, available before language) while impersonal moral decisions are related to areas of the brain associated with in-depth processing. Considering its potential to inform future directions in HDNR research, to what extent might this finding be applied to understanding human-wildlife relationships, an area of focus within HDNR? More specifically, as an illustration, to what extent might these different decision paths be associated with mutualism versus domination wildlife value orientations identified in the literature (see Manfredo et al. 2009; Teel and Manfredo 2010)? Moral decisions would typically be those involving humans, but a key difference between those with a mutualism versus domination orientation is that the former views wildlife as family or companions, deserving of rights like humans, while the latter "de-personizes" wildlife. It would be reasonable to pursue the explanation that differences between the value orientation types on judgments about wildlife treatment are rooted in the two different cognitive systems examined by Greene et al. (2004).

7.5.2 The Individual-Group Hierarchy

Humans are driven strongly by social affiliation motives. That tendency has spawned high levels of cooperation and altruism among humans. As Fehr and Fischbacher (2004) note, "human societies represent a spectacular outlier with respect to all other animal species because they are based on large-scale cooperation among genetically unrelated individuals" (p. 185). This tendency is seen to be, in part, an evolutionary response to being prey species for many predators in formative times (Hart and Sussman 2005), and it also posed adaptive advantages for evolving, competitive cultural groups (Boyd and Richerson 2009).

The human need to attach to groups is widely evident. People define themselves through a hierarchy of identities – e.g., friendship groups, sports team groups, chat groups, professions and professional associations, governance groups (local, state, nation), being human, etc. The group itself can be considered an emergent property in systems terms, with characteristics and dynamics beyond the mere aggregation of individuals. The process of attaining group membership is elemental in forming our social world. The processes and effects of group membership are explained through Social Identity (SIT) and Self-Categorization (SCE) theories which Spears (2011) claims are "possibly as close as we come in contemporary social psychology to a grand theory" (p. 208). A social group exists when people share a definition of who they are, how they relate and how they are different from those not in the group. People have many different group identities, and these identities vary in how important they might be to a person and how accessible they are in a given situation. Once a social identity is accessed in a given situation, it is important because it shapes one's social perception of a situation, appropriate social conduct and one's own self-definition. This is a dynamic process, however, because as the situation changes, so might the salience of the identity. For example, a representative of the coal industry and a representative of the environmental community, while in a public debate about global warming, may take on highly adversarial and oppositional identities; yet, in other contexts they may define themselves as members of the same group (e.g., mothers, alumni of the same university, Americans, fans of a given team, etc.). The salience of a given identity is seen as dependent on the situational context as well as a person's commitment to the identity. The more salient and committed a person is to an identity, the more likely that person will act out and seek the identity role (Stryker and Serpe 1982).

The process of self-categorization into a group offers explanation of how social identity affects the thoughts and behavior of the individual. As a member of the group, and contingent on one's commitment and emotional attachment to the group, one will learn or infer the appropriate norms, attributes, and attitudes associated with that group. Norms, which are beliefs about how one ought to behave or think, are a critical aspect of group maintenance. Norms represent ideal or prototypical thoughts or behaviors that unify and ensure compliance and agreement within the group (Turner 1991). As a group member, the person adopts those norms or attitudes in situations where their group identity is salient. Theorists have coined this as "depersonalizing" – a process whereby a person acts in accordance with group norms and perceives oneself as representing the group, not as an individual.

The use of SIT and SCE in understanding environmental topics has been relatively neglected (Twigger-Ross et al. 2003), and yet the available applications offer promise in considering how these frameworks may be integrated into HDNR studies on individual-group dynamics in the future. For example, Bonaiuto et al. (1996) showed that perceptions of beach pollution among samples of youth in six separate resort towns were associated with the person's attachment to either the town or the nation (Great Britain). SIT was borne from an interest in understanding conflict among groups of people, and Stoll-Kleemann (2004) illustrates such a use. The theory was applied in this case to address intergroup conflict among farmers, conservationists and forest managers in biodiversity management in Germany's protected areas. In another application, Carpenter and Cardenas (2011) explored the relevance of social identity in common pool resource experimentation. They found that when students were placed in an experimental context where they knew they were engaged in a cross-cultural "game" situation (i.e., their national social identity was made salient), participants were more likely to represent the country group proto-type position: U.S. students were more likely to emphasize conservation strategies in allocating forest resources, while Columbian students tended to emphasize resource extraction.

In a multi-level application, Burton and Wilson (2006) provided an analysis of the shift in Western farming regimes from productivist (focused on maximizing food production) to post-productivist (focused on consumption and sustainability) and multi-functional (both productivist and post-productivist, separated spatially and temporally) regimes. While structural variables (i.e., policy, political economy) would suggest the transition is occurring, the identities that farmers reported did not suggest such a transition at the individual level (i.e., individual farmers sustained an emphasis on production).

SIT and SCE have also been applied in understanding prejudice, conformity, crowding behavior, organizational behavior, leadership deviance and group cohesiveness, all of which are important topics in HDNR (Hogg 2006). There is strong potential for a dynamic model of individual-group involvement in researching conservation topics. The research methodology might involve more intense in-person assessments (e.g., groups, situational salience, group norms and attitudes) across actors and times. Such assessments could serve as the basis for examining the relative influence of group versus independent action on key policy and behavioral outcomes and how groups, individuals, and the environment interact and change in that process.

7.5.3 Institutional and Structural Factors

How do human psychological processes interact with various elements of context such as modes of economy, technological capabilities, power differentials, demographic trends, political structures and ecological conditions? While these questions have received attention since the earliest efforts of the social sciences, they have become mainstream to psychology in only the past couple of decades (Triandis 2007). Here we review just a few of the categories that have received attention in the literature.

7.5.3.1 Economic Development

Economic development has historically been a central focus of social science theory. It is particularly strong as a result of the influence of Marx who had a broad and lasting influence across the social sciences, especially in sociology and political science. A contemporary articulation of his modernization theory is found in Post-materialism theory, as introduced by Inglehart (1997), who focuses on modern-day (post World War II) global cultural change. Inglehart's theory proposes that increases in modernization (wealth, education, and urbanization) reduce the prevalence of subsistence needs among citizens of a nation. As subsistence needs decline, needs for affiliation and self-esteem become more prominent, and as needs change, so do people's values. Materialist values, emphasizing a concern for basic physical and economic security, are replaced by self-expressive (also referred to as post-materialist) values, and in the realm of religion, traditional values are replaced by secular values. This shift in values is proposed to have important implications for many areas of social life (Pippa and Inglehart 2004; Inglehart and Welzel 2005). For example, when materialist values pre-dominate, individuals are more willing to subordinate their own preference for the greater good, but with self-expressive values, individuals tend to pursue their own preference and an active voice in government. This, Inglehart contends, produces a trend toward participatory decision-making and away from hierarchical authority.

Post-materialism theory has been used to address environmental topics in two areas. The first was to provide an explanation for the growth of environmentalism. Inglehart initially proposed that the rise of pro-environmental attitudes was associated with the shift toward post-materialist values: as countries became modernized and less concerned about meeting basic needs, citizens would become more concerned about the environment. Findings from his own data, however, showed a preponderance of pro-environmental attitudes in developing countries (Brechin and Kempton 1994), prompting Inglehart (1995) to suggest that environmentalism will arise: (1) in situations where there are "objective conditions" of increased environmental degradation, but also (2) due to "subjective conditions" of cultural shift, in countries of increasing modernization. This issue continues to be explored in a multi-level context with an interest in understanding the cause of both in-country differences and intra-individual differences in explaining the growth of environmentalism. Findings generally support the chain of events in which an increase in economic well-being brings about a shift toward post-materialist values, and that shift in turn yields a rise in pro-environmental attitudes (Gelissen 2007; Franzen and Meyer 2010; Haller and Hadler 2008).

A second conservation-related application of Post-materialism theory is found in research by Manfredo et al. (2009). These researchers proposed that the changing context of social life has led to a shift from domination to mutualism value orientations toward wildlife in the U.S. While domination prioritizes human well-being over wildlife and promotes treatment of wildlife in utilitarian terms, mutualism views wildlife as capable of relationships of trust with humans and is defined by a desire for companionship with wildlife. This theoretical perspective argued that the reduced reliance on wildlife for material goods, the human tendency toward anthropomorphizing and a growing need for affiliation in post-materialist society have fueled the trend toward mutualism wildlife value orientations. This shift, in turn, has had an important impact on people's relationships with wildlife and their attitudes toward wildlife policy-related issues. A multi-level study in 19 western U.S. states revealed a strong contextual effect of modernization variables – i.e., individual differences in wildlife value orientation scoring could be explained by state-level influences of urbanization, income, and education. Higher levels of these state-level predictors were also associated with higher percentages of mutualists in a state. Moreover, those with a mutualism versus domination orientation were less likely to favor traditional wildlife management techniques (e.g., lethal control) and participate in recreational hunting, revealing the connection between wildlife value orientations and wildlife-related attitudes and behaviors (Manfredo et al. 2009).

7.5.3.2 Governance Systems

Over the past two decades, a growing body of work has centered around questions of governance in SES research, recognizing the importance of institutional mechanisms in influencing system dynamics and resilience (Gerlak 2013; Anderies et al. 2004; Walker et al. 2004). Broadly, governance can be defined in this context as creating the conditions for collective action and ordered rule as well as the set of formal and informal rules that constitute the social system's institutions (Walker et al. 2006). More specifically, environmental governance, defined as the "set of regulatory processes, mechanisms and organizations through which political actors influence environmental actions and outcomes" (Lemos and Agrawal 2006, p. 298), has received increasing attention in the literature.

An area ripe for research considering the role of the individual is reflected in the move toward new approaches to environmental governance, including adaptive governance or adaptive co-management which relies on collaborative networks that connect individuals, organizations, and institutions at multiple levels for managing ecosystems (Dietz et al. 2003; Folke et al. 2003; Olsson et al. 2004). These approaches build upon the extensive tradition of research on community-based governance of common-pool resources advanced by Ostrom (1990, 1997, 2007a) and others (e.g., see Agrawal 2002; Schlager 2004) and are reflective of the rise in more participatory, decentralized forms of decision-making, or collaborative governance (Ansell and Gash 2007; Rogers and Weber 2010).

Using a series of case studies from Sweden and Canada, Olsson et al. (2004) point out how individual actors and their characteristics, including leadership and trust-building capabilities as well as cultural values and local ecological knowledge, are often a critical component of the self-organizing process that defines adaptive co-management systems. They go on to demonstrate the potential of such systems for building resilience by enhancing community capacity to deal with uncertainty and change. The individual characteristics they identified have also been described as important elements of social capital, a term widely used across the social sciences to represent the collective capacity of individuals in social networks (Walker et al. 2006; Ostrom and Ahn 2003; Pretty 2003). According to Ostrom and Ahn (2009), social capital can be viewed as "an attribute of individuals and of their relationships that enhance their ability to solve collective-action problems" (p. 20).

In addition to social capital and related elements such as leadership, a host of other factors designed to connect individual and group-level characteristics to governance regimes were identified in a recent SES framework advanced by Ostrom (2007b). Under the category of resource "user" variables, Ostrom included, for example, individuals' knowledge or mental models of the SES, their dependence on the resource and history of use, and socioeconomic characteristics. Recognizing concerns over recommended panaceas or blueprint approaches to the governance of complex social-ecological problems, the framework was intended to serve as a diagnostic tool for analysis by detailing an array of variables posited by prior research and theory (including a review by Agrawal [2001]) to impact patterns of SES interactions and outcomes. While Ostrom cautioned that not all variables would be relevant in every study, demanding an assessment of which variables and at what levels would be relevant in terms of their potential impact on human behavior and SES outcomes, the framework was proposed by the author as "a step toward building a strong interdisciplinary science of complex, multilevel systems" that would facilitate future research to match governance strategies to particular problems in the SES context (Ostrom 2007b, p. 15181).

Emerging work in this area shows the potential for establishing stronger linkages between environmental governance and the psychological characteristics of individuals within a given social structure. For example, Newig and Fritsch (2009) conducted a meta-analysis of 47 case studies to assess the effectiveness of participatory, multi-level forms of governance. While polycentric governance (consisting of multiple centers of decision-making authority) was correlated with environmental outcomes, the environmental preferences of individuals (averaged across all participants) had by far the strongest effect. In another example, Paciotti et al. (2005) examined the linkage between collaborative personality styles and the adoption of social justice institutions in comparing the Sukuma and Pimbwe ethnic groups in Tanzania. Game theory methods showed a much stronger sharing tendency in the Sukuma compared to the Pimbwe. This finding was true in situations of within-group and between ethnic group games. Sharing and trust served as a foundation for the Sukuma institutional form of justice called Sungusungu. The Pimbwe's attempt to adopt this form of governance was simply not successful due to their lack of cooperative style. The researchers suggested that the alignment of personality characteristics and governance styles serves to adapt social groups to their surroundings.

Arguably, as indicated by these examples, psychological research could contribute to further addressing certain individual-level variables and interactions that Ostrom (2007b) and others have identified as well as aid in expanding the list of individual characteristics (including measures of psychological constructs such as values, attitudes, etc.) worth considering in the governance, and broader SES context. In addition, inclusion of governance considerations in HDNR research would expand understanding of the broader institutional and structural factors that can influence individual thought and behavior.

7.5.3.3   Geographic Regions

Geographic variation on psychological attributes has long been an interest in the social sciences (Allik and McCrae 2004; Hofstede and McCrae 2004; Rentfrow et al. 2008). There has been a strong focus on national differences, but differences have also been identified at regional levels (Nisbett 1993; Kitayama et al. 2006; Rentfrow et al. 2008). Attributes found to vary have included personality characteristics, cultural values, and emotional expression. As an illustration, Rentfrow et al. (2008) revealed how the "Big 5" personality characteristics varied considerably at the state level across the U.S. Further, state-level personality correlated strongly with other social quality variables such as crime rate, social involvement, and religiosity (all increased with levels of extraversion).

While the geographic differences found in such studies are interesting, the real value of their findings is the provocation to explain, or theorize, why these differences exist. For example, Rentfrow et al. (2008) proposed that the explanation for geographic variation in their study was tied to "founder migration" – non-random groups of people with distinct attributes and perhaps genetic make-up settling in an area. The characteristics in a region are perpetuated through selective migrations (people moving to the area) and social and environmental influences. Similarly, Nisbett (1993; Nisbett and Cohen 1996) suggested geographic differences in white male violence could be attributed to cultural factors associated with historical patterns of economy and migration. Kitayama et al. (2006) also relied on a similar cultural mechanism to explain regional differences when comparing "frontier" settlement areas with other areas across nations, but Kitayam and Uskul (2011) went on to argue that cross-cultural psychological and behavioral differences are the result of complex interactions among different systems that make up human individuals (genes, brains, minds) and collectives (social networks, cultures, broader environments). As these examples reveal, understanding geographic differences in psychological attributes will require holistic theorizing and multi-level approaches that recognize the complexity of the human context. Recent advances integrating culture in psychological research, discussed in more detail below, offer promise in informing new directions in this area (Oishi et al. 2009; Kitayama and Cohen 2007).

7.5.3.4   Cultural Groups

Cross-cultural psychology emerged in the 1960s and 1970s, and as recently as 1998 Segall et al. (1998) proclaimed "psychology in general has long ignored 'culture' as a source of influence on human behavior and still takes little account of theories or data from other than Euro-American cultures" (p. 1102). Yet, cross-cultural psychology has given fresh insight into the pursuit of identifying human universals, considered one of the primary goals of psychology (Jahoda and Krewer 1997). Interestingly, the growing body of cross-cultural research has found that many of the central theories of psychology, which originated from research in North America, have not generalized well to other cultures (Norenzayan and Heine 2005).

But the contrasting goal of psychology, understanding diversity, has been well served by cross-cultural psychology, and the logic for such diversity coincides with the main goal of an SES approach – determining the adaptive interrelationships of humans and their environments. As Triandis (2007) states:

> [people]…determine ways of organizing information, symbols, evaluations, and patterns of behavior; intellectual, moral, and aesthetic standards; knowledge, religion, and social patterns…systems of government, systems of making war; and expectations and ideas about correct behavior that are more or less effective [functional] in adapting to their ecosystem. (p. 64)

Further, the methods and theory of cross-cultural psychology can be key to informing SES approaches to conservation issues that include individuals and account for their varied cultural backgrounds.

One area of research in cross-cultural psychology that has received enduring attention is the diversity of cultural values and understanding why they exist (e.g., Hofstede 2001; Schwartz 2006). This research has been guided by practical concerns of improving intergroup relations, increasing success in global markets, and international diplomacy. A frequent focus in this area has been on the documented difference between the collectivist values of Southeast Asian cultures and the individualistic values of cultures of Europe and North America (Triandis 1995). Two recent studies illustrate the social-ecological nature of the explanation for this difference, highlighting the relevance of cross-cultural psychology in an SES context as well as the potential for this sub-discipline to inform future directions in HDNR research aimed at understanding the broader influences on individual thought/behavior. Kitayama et al. (2010) proposed a production-adoption model of cultural change which they used to explain the strong individualistic and independence "ethos" in the U.S. They argued that novel values and practices arise within a social group to cope with major adaptive challenges for biological, economic, and/or political survival, whereas adoption of existing practices is motivated by a desire to achieve prestige and higher status within one's community. In the U.S., values and practices associated with independence and self-reliance emerged as settlers moved West and had to adapt to sparsely-populated, harsh environmental conditions. Given the economic success of Westward expansion, residents in the Eastern U.S. imitated these values in achieving higher social status and prestige.

In another recent example, Gelfand et al. (2011) proposed exploring the collective-independent difference through a cultural systems model of "tightness-looseness" that links ecological threats, social processes (norms and tolerances of deviant behavior), socio-political institutions, and psychological processes. Overall, based on empirical findings with data from 33 nations, they suggested that as ecological and human threats increase, the need for strong norms and punishment of deviant behavior also increases because these mechanisms facilitate social coordination in response to the threats. Such coordination enhances the chance of survival. Given that institutions are a reflection of norms and tolerances, societies with strong norms ("tight" nations) would have more restricted press, more laws, criminal justice systems with higher monitoring and severe punishment, and stronger religion, while "loose" nations would be the opposite. Also, in tight cultures, there is a higher degree of structure and constraint in day-to-day situations as well as individual-level psychological adaptations like self-regulation and self-monitoring.

7.6 Mutually Constructed Nature of Human Thought and the Social and Natural Environment

Implicit in this question is the assumption that the structure and organization of individual thought serves to adapt humans to their social-ecological surroundings. This assumption has been an emphasis since the origins of psychology and, more recently, a particular focus of evolutionary psychology. Schwartz (2006) proposed, for example, that value orientations serve to guide people in a cultural group in how to maintain the individual-group relationship, how to act to preserve the social fabric, and how to manage relationships with the natural world. As another illustration, Fredrickson and Branigan (2001) argued that while negative emotions have served to support basic survival, positive emotions are believed to have fostered exploration, expansion and pioneering among humans. A final example is research by Uskul et al. (2008) who showed different ecological niches occupied by humans affect economic activities, which, in turn, produce different cognitive styles that help adapt human activities to the niche.

What is generally missing in this literature is the feedback effect that human adaptation has on the environment, an essential aspect of SES modeling. The criticism that ecology has not looked at the reciprocal effects of humans and the environment can be applied equally to the social sciences. This is one of the critical challenges for the future recognized by Oishi and Graham (2010) who introduce "socio-ecological psychology", which would examine "how mind and behavior are shaped in part by their natural and social habitats and *how natural and social habitats are in turn shaped partly by mind and behavior* [emphasis added]" (p. 356).

A better understanding of reciprocal effects will be difficult to obtain without research taking on an expanded time frame that might be achieved by: (1) the integration of ethnographic and historical perspectives with traditional social psychological approaches (e.g., Haggerty and Travis 2006), and/or (2) the increased use of longitudinal research (e.g., Boone and Galvin, Chap. 9, this volume). A classic example of the former is Rappaport's (1968) *Pigs for the Ancestors: Ritual in the Ecology of a New Guinea People.* Rappaport based the book on his ethnographic work with the Tsembaga, a group of Maring speakers living in the highlands of Australian New Guinea (now Papua New Guinea). He presented a systems approach that proposed beliefs about religion and the resultant rituals served as the regulatory mechanism creating a homeostasis among the Tsembaga, other human groups, and the environment. The rituals served to maintain biotic communities, limit warfare among groups, provide a basis for establishing allies, and distribute protein (from a ritual involving the widespread slaughter of pig populations) throughout the local population at the time of greatest need. While Rappaport's work was criticized on a number of counts, the simplicity and elegance of his account led to it becoming a classic. It provided a compelling story of humans in a social-ecological system with ideology serving a central, adaptive role.

At present, there is little emphasis on the dynamic interplay of human thought and the social-ecological context. Far more abundant when it comes to research on the social aspects of SES is literature that: (1) is normative, with an emphasis on ways to increase collaborative approaches to governance; (2) includes individual-level variables that give token representations of human influences in a system; and (3) consists of broad-based conceptual and structural models that depict broad categories of individual-level variables and feedbacks, with sparse articulation of specific effects. More uncommon, but emerging, are approaches that predict how people will behave when given new information or a particular set of circumstances, which in turn creates a myriad of social, policy and ecological outcomes (see Boone and Gavin, Chap. 9, this volume; Fischer et al. 2013). Approaches to SES that adequately represent the mutual construction of society, individuals, and the environment are arguably one of the most important goals for the future.

7.7 Conclusion

The ultimate purpose of an SES approach is to inform questions about human resilience and adaptation in the face of environmental change. Humans' remarkable success in adaptation to date is linked to cognitive abilities of innovation, social learning, and combining different sources of information into new understandings of the world (Cosmides and Tooby 2002; Boyd et al. 2011). Is there some way that we can understand and direct that innovation toward effective mitigation? The emergence and adoption of social innovation is a topic of new and growing interest among conservation researchers, particularly in response to climate change (Nicholls and Murdock 2012; Rodima-Taylor et al. 2012). Yet, it is also a topic that has received considerable attention in organizational sciences over the past four decades, where it is generally believed that innovation is necessary for long-term organizational success (Hage 1999; Willis and Mastrofski 2011). Meta-analyses in this area suggest that in team situations, innovation is related to process variables such as support for innovation, vision, task orientation, and external communication (Hülsheger et al. 2009). In another analysis of innovation in work situations, Hammond et al. (2011) found a complex mix of factors produced innovation, including individual factors, characteristics of the job, and environmental factors. Other literature reviews have found inconsistent and inconclusive results among the many empirical studies (Wolfe 1994; Anderson et al. 2004). Anderson et al. (2004) concluded that: (1) future research should look at innovation processes as cyclical, longitudinal, and iterative; and (2) context and a multi-level approach (individual-group-organization-culture) are critical for exploring this topic. Interestingly, this proposal converges on the conclusion that broadly-generalizable panaceas for complex social-ecological problems are simply not forthcoming (Ostrom 2007b).

It leads us to conclude that an understanding of adaptation and innovation in SES should attend to the considerations we raise here: the dynamic aspects of human thought, the importance of individuals' involvement with and attachment to groups, and the influence of a broad array of social and ecological contextual variables.

We began this chapter by suggesting that a more complete inclusion of individuals in SES models has implications for both ecosystem science and HDNR researchers. The role of humans in the conceptual approaches of ecosystem science has moved through phases of increasing integration over the past three decades. Initially, humans were viewed as external to ecosystems; then humans were seen as drivers of impacts to ecosystems. More recently, humans have been cast as active agents that impact and respond to ecosystems that are in constant shift. We are just beginning to move toward a fully integrative view that humans are participants in a co-constructed, co-evolving, dynamic system. The complexity of social systems is in need of more attention in SES models which will remain poorly specified until there is a representation of the multi-level context of human individuals. We support the view that individuals occupy a unique and central role here – they are the primary unit of evolutionary succession; and causal processes, both up and down scales, must circulate through them (Schank 2001). In other words, change at other social levels aggregated upward, such as cultural evolution, institutional change, technological advances, innovation, etc., all must occur in the minds and actions of individuals.

Ecosystem science sees the system as hierarchies nested within broader hierarchies, each operating at different speeds and cycles of change. For those in HDNR, we propose that such an approach works well for examining individuals in their social-ecological context. We propose a view of the psychological attributes of the individual as dynamic, in a multi-level context, and mutually constructed with society and environment.

We conclude by reinforcing the importance of understanding the role of human individuals in the complex social-ecological interactions that produce daunting global environmental challenges such as climate change, land degradation, and loss of biodiversity. The impacts of humans on ecosystems are registered one behavior and one individual at a time. But each behavior exists in a somewhat patterned tapestry of behavioral choices across many individuals, across time and space. A better understanding of human behavior in its broader tapestry is important if our science is to effectively inform decisions that influence resilience to growing environmental stress.

References

Abel, T., & Stepp, J. R. (2003). A new ecosystems ecology for anthropology. *Conservation Ecology, 7*(3), 12.

Adger, W. N., Hughes, T. P., Folke, C., Carpenter, S. R., & Rockström, J. (2005). Social-ecological resilience to coastal disasters. *Science, 309*, 1036–1039.

Agrawal, A. (2001). Common property institutions and sustainable governance of resources. *World Development, 29*, 1649–1672.

Agrawal, A. (2002). Common resources and institutional sustainability. In E. Ostrom, T. Dietz, N. Dolsak, P. C. Stern, S. Sonich, & E. U. Weber (Eds.), *The drama of the commons* (pp. 41–86). Washington, DC: National Academy Press.

Allik, J., & McCrae, R. R. (2004). Toward a geography of personality traits: Patterns of profiles across 36 cultures. *Journal of Cross-Cultural Psychology, 35*, 13–28.

Anderies, J. M., Janssen, M. A., & Ostrom, E. (2004). A framework to analyze the robustness of social-ecological systems from an institutional perspective. *Ecology and Society, 9*(1), 8.

Anderson, N., De Dreu, C. K. W., & Nijstad, B. A. (2004). The routinization of innovation research: a constructively critical review of the state-of-the-science. *Journal of Organizational Behavior, 25*, 147–173.

Ansell, C., & Gash, A. (2007). Collaborative governance in theory and practice. *Journal of Public Administration Research and Theory, 18*(4), 543–571.

Baumgärtner, S., Derissen, S., Quaas, M. F., & Strunz, S. (2011). Consumer preferences determine resilience of ecological-economic systems. *Ecology and Society, 16*(4), 9.

Berkes, F. (1999). *Sacred ecology: Traditional ecological knowledge and resource management.* Philadelphia: Taylor & Francis.

Berkes, F., Colding, J., & Folke, C. (2000). Rediscovery of traditional ecological knowledge as adaptive management. *Ecological Applications, 10*, 1251–1262.

Bonaiuto, M., Breakwell, G. M., & Cano, I. (1996). Identity processes and environmental threat: The effects of nationalism and local identity upon perception of beach pollution. *Journal of Community and Applied Social Psychology, 6*, 157–175.

Boyd, R., & Richerson, P. J. (2009). Culture and the evolution of human cooperation. *Philosophical Transactions of the Royal Society B: Biological Sciences, 364*, 3281–3288.

Boyd, R., Richerson, P. J., & Henrich, J. (2011). The cultural niche: Why social learning is essential for human adaptation. *Proceedings of the National Academy of Sciences, 108*, 10918–10925.

Brechin, S., & Kempton, W. (1994). Global environmentalism: A challenge to the postmaterialism thesis? *Social Science Quarterly, 75*, 245–269.

Buizer, M., Arts, B., & Kok, K. (2011). Governance, scale, and the environment: The importance of recognizing knowledge claims in transdisciplinary arenas. *Ecology and Society, 16*, 1.

Burns, T. R. (2012). The sustainability revolution: A societal paradigm shift. *Sustainability, 4*, 1118–1134.

Burton, R. J. F., & Wilson, G. A. (2006). Injecting social psychology theory into conceptualisations of agricultural agency: Towards a post-productivist farmer self-identity? *Journal of Rural Studies, 22*(1), 95–115.

Carpenter, J., & Cardenas, J. C. (2011). An inter-cultural examination of cooperation in the commons. *Journal of Conflict Resolution, 55*(4), 632–651.

Chiao, J. Y., & Blizinsky, K. D. (2013). Population disparities in mental health: Insights from cultural neuroscience. *American Journal of Public Health.* doi:10.2105/AJPH.2013.301440.

Collins, S., et al. (2011). An integrated conceptual framework for long-term social-ecological research. *Frontiers in Ecology and the Environment, 9*(6), 351–357.

Cosmides, L., & Tooby, J. (2002). Unraveling the enigma of human intelligence: Evolutionary psychology and the multimodular mind. In R. J. Sternberg & J. C. Kaufman (Eds.), *The evolution of intelligence* (pp. 145–198). Mahwah: Lawrence Erlbaum Associates.

Davidson, D. J. (2010). The applicability of the concept of resilience to social systems: Some sources of optimism and nagging doubts. *Society and Natural Resources, 23*, 1135–1149.

Dietz, T., Ostrom, E., & Stern, P. (2003). The struggle to govern the commons. *Science, 302*, 1907–1912.

Ehrlich, P. R., & Kennedy, D. (2005). Millennium assessment of human behavior. *Science, 309*(22), 562–563.

Evans, J. S. B. T., & Stanovich, K. E. (2013). Dual-process theories of higher cognition: Advancing the debate. *Perspectives of Psychological Science, 8*(3), 223–241.

Fehr, E., & Fischbacher, U. (2004). Social norms and human cooperation. *Trends in Cognitive Sciences, 8*(4), 185–190.

Fischer, A. P., Korejwa, A., Kock, J., Spies, T., Olsen, C., White, E., & Jacobs, D. (2013). Using the forest, people, fire agent-based social network model to investigate interactions in social-ecological systems. *Practicing Anthropology, 35*(1), 8–13.

Folke, C., Colding, J., & Berkes, F. (2003). Synthesis: Building resilience and adaptive capacity in social-ecological systems. In F. Berkes, J. Colding, & C. Folke (Eds.), *Navigating social-ecological systems: Building resilience for complexity and change* (pp. 352–387). Cambridge: Cambridge University Press.

Forester, D. J., & Machlis, G. E. (1996). Modeling human factors that affect the loss of biodiversity. *Conservation Biology, 10*(4), 1253–1263.

Franzen, A., & Meyer, R. (2010). Environmental attitudes in cross-national perspective: A multilevel analysis of the ISSP 1993 and 2000. *European Sociological Review, 26*(2), 219–234.

Fredrickson, B. L., & Branigan, C. (2001). Positive emotions. In T. J. Mayne & G. A. Bonanno (Eds.), *Emotions: Current issues and future directions* (pp. 123–151). New York: Guilford Press.

Fulton, D. C., Manfredo, M. J., & Lipscomb, J. (1996). Wildlife value orientations: A conceptual and measurement approach. *Human Dimensions of Wildlife, 1*(2), 24–47.

Gardner, H. (1987). *The mind's new science: A history of the cognitive revolution*. New York: Basic Books.

Gelfand, M. J., et al. (2011). Differences between tight and loose cultures: A 33-nation study. *Science, 332*, 1100–1104.

Gelissen, J. (2007). Explaining popular support for environmental protection: A multilevel analysis of 50 nations. *Environment and Behavior, 39*, 392–415.

Gerlak, A. K. (2013). Policy interactions in human-landscape systems. *Environmental Management*. doi:10.1007/s00267-013-0068-y.

Goel, V. (2008). Anatomy of deductive reasoning. *Trends in Cognitive Sciences, 11*, 435–441.

Greene, J. D., Nystrom, L. E., Engell, A. D., Darley, J. M., & Cohen, J. D. (2004). The neural bases of cognitive conflict and control in moral judgment. *Neuron, 44*(2), 389–400.

Gunderson, L. H., & Holling, C. S. (Eds.). (2002). *Panarchy: Understanding transformations in human and natural systems*. Washington, DC: Island Press.

Hage, J. T. (1999). Organizational innovation and organizational change. *Annual Review of Sociology, 25*, 597–622.

Haggerty, J. H., & Travis, W. R. (2006). Out of administrative control: Absentee owners, resident elk and the shifting nature of wildlife management in southwestern Montana. *Geoforum, 37*, 816–830.

Haller, M., & Hadler, M. (2008). Dispositions to act in favor of the environment: Fatalism and readiness to make sacrifices in a cross-national perspective. *Sociological Forum, 23*(2), 281–311.

Hammond, M. H., Neff, N. L., Farr, J. L., Schwall, A. R., & Zhao, X. (2011). Predictors of individual-level innovation at work: A meta-analysis. *Psychology of Aesthetics, Creativity, and the Arts, 5*, 90–104.

Hart, D., & Sussman, R. W. (2005). *Man the hunted: Primates, predators, and human evolution*. Boulder: Westview Press.

Hofstede, G. (2001). *Culture's consequences: Comparing values, behaviors, institutions and organizations across nations*. Thousand Oaks: Sage.

Hofstede, G., & McCrae, R. R. (2004). Personality and culture revisited: Linking traits and dimensions of culture. *Cross-Cultural Research, 38*, 52–88.

Hogg, M. A. (2006). Social identity theory. In P. J. Burke (Ed.), *Contemporary social psychological theories* (pp. 111–136). Palo Alto: Stanford University Press.

Holling, C. S. (1998). Two cultures of ecology. *Conservation Ecology, 2*(2), 4.

Holling, C. S., Gunderson, L. H., & Ludwig, D. (2002). In quest of a theory of adaptive change. In L. H. Gunderson & C. S. Holling (Eds.), *Panarchy: Understanding transformations in human and natural systems* (pp. 3–22). Washington, DC: Island Press.

Hrubes, D., Ajzen, I., & Daigle, J. (2001). Predicting hunting intentions and behavior: An application of the theory of planned behavior. *Leisure Sciences, 23*, 165–178.

Hülsheger, U. R., Anderson, N., & Salgado, J. F. (2009). Team-level predictors of innovation at work: A comprehensive meta-analysis spanning three decades of research. *Journal of Applied Psychology, 94*(5), 1128–1145.

Hutchins, E. (2010). Cognitive ecology. *Topics in Cognitive Science, 2*(4), 705–715.

Inglehart, R. (1995). Public support for environmental protection: Objective problems and subjective values in 43 societies. *PS: Political Science and Politics, 28*(1), 57–72.

Inglehart, R. (1997). *Modernization and postmodernization.* Princeton: Princeton University Press.

Inglehart, R., & Welzel, C. (2005). *Modernization, cultural change and democracy: The human development sequence.* Cambridge: Cambridge University Press.

Jahoda, G., & Krewer, B. (1997). History of cross-cultural and cultural psychology. In J. W. Berry, Y. H. Poortinga, & J. Pandey (Eds.), *Handbook of cross-cultural psychology: Theory and method* (Vol. 1, pp. 1–42). Boston: Allyn & Bacon.

Janssen, M. A., & Ostrom, E. (2006). Empirically based, agent-based models. *Ecology and Society, 11*(2), 37.

Jones, N. A., Ross, H., Lynam, T., Perez, P., & Leitch, A. (2011). Mental models: An interdisciplinary synthesis of theory and methods. *Ecology and Society, 16*(1), 46.

Kahan, D. M., Wittlin, M., Peters, E., Slovic, P., Ouellette, L. L., Braman, D., & Mandel, G. N. (2011). *The tragedy of the risk-perception commons: Culture conflict, rationality conflict and climate change* (Temple University Legal Studies Research Paper No. 2011-26; Cultural Cognition Project Working Paper No. 89; Yale Law & Economics Research Paper No. 435; Yale Law School, Public Law Working Paper No. 230). http://ssrn.com/abstract=1871503. Accessed 30 Aug 2013.

Karp, D. A. (1996). Values and their effect on pro-environmental behavior. *Environment and Behavior, 28*(1), 111–133.

Kitayama, S., & Cohen, D. (2007). *Handbook of cultural psychology.* New York: Guilford.

Kitayama, S., & Uskul, A. K. (2011). Culture, mind, and the brain: Current evidence and future directions. *Annual Review of Psychology, 62*, 419–449.

Kitayama, S., Ishii, K., Imada, T., Takemura, K., & Ramaswamy, J. (2006). Voluntary settlement and the spirit of independence: Evidence from Japan's "northern frontier". *Journal of Personality and Social Psychology, 91*(3), 369–384.

Kitayama, S., Conway, L. G., III, Pietromonaco, P. R., Park, H., & Plaut, V. C. (2010). Ethos of independence across regions in the United States: The production–adoption model of cultural change. *American Psychologist, 65*(6), 559–574.

Kok, K., & Veldkamp, T. (2011). Scale and governance: Conceptual considerations and practical implications. *Ecology and Society, 16*(2), 23.

Lemos, M. C., & Agrawal, A. (2006). Environmental governance. *Annual Review of Environment and Resources, 31*, 297–325.

Manfredo, M. J., Teel, T. L., & Bright, A. D. (2004). Application of the concepts of values and attitudes in human dimensions of natural resources research. In M. J. Manfredo, J. J. Vaske, B. L. Bruyere, D. R. Field, & P. J. Brown (Eds.), *Society and natural resources: A summary of knowledge* (pp. 271–282). Jefferson: Modern Litho.

Manfredo, M. J., Teel, T. L., & Henry, K. L. (2009). Linking society and environment: A multi-level model of shifting wildlife value orientations in the western U.S. *Social Science Quarterly, 90*(2), 407–427.

Manfredo, M. J., Dietsch, A., & Teel, T. L. (2013). *A dynamic view of environmental values: Tracing the trajectory of value orientations toward wildlife.* Unpublished draft manuscript, Department of Human Dimensions of Natural Resources, Colorado State University, Fort Collins.

Milfont, T. L., Duckitt, J., & Wagner, C. (2010). A cross-cultural test of the value-attitude-behavior hierarchy. *Journal of Applied Social Psychology, 40*(11), 2791–2813.

Newig, J., & Fritsch, O. (2009). Environmental governance: Participatory, multi-level – and effective? *Environmental Policy and Governance, 19*(3), 197–214.

Nicholls, A., & Murdock, A. (Eds.). (2012). *Social innovation: Blurring boundaries to reconfigure markets.* New York: Palgrave Macmillan.

Nisbett, R. E. (1993). Violence and U.S. regional culture. *American Psychologist, 48*, 441–449.

Nisbett, R. E., & Cohen, D. (1996). *Culture of honor: The psychology of violence in the south.* Boulder: Westview.

Norenzayan, A., & Heine, S. J. (2005). Psychological universals: What are they and how can we know? *Psychological Bulletin, 131*(5), 763–784.

Noss, R. F. (1990). Indicators for monitoring biodiversity: A hierarchical approach. *Conservation Biology, 4*(4), 355–364.

Oishi, S., & Graham, J. (2010). Social ecology : Lost and found in psychological science. *Perspectives on Psychological Science, 5*(4), 356–377.

Oishi, S., Keseber, S., & Snyder, B. H. (2009). Sociology: A lost connection in social psychology. *Personality and Social Psychology Review, 13*, 334–353.

Olsson, P., Folke, C., & Berkes, F. (2004). Adaptive comanagement for building resilience in social-ecological systems. *Environmental Management, 34*(1), 75–90.

Oreg, S., & Katz-Gerro, T. (2006). Predicting proenvironmental behavior cross-nationally: Values, the theory of planned behavior, and value-belief-norm theory. *Environment and Behavior, 38*, 462–483.

Ostrom, E. (1990). *Governing the commons: The evolution of institutions for collective action.* New York: Cambridge University Press.

Ostrom, E. (1997). Institutional rational choice: An assessment of the institutional analysis and development framework. In P. A. Sabatier (Ed.), *Theories of the policy process* (pp. 21–64). Boulder: Westview Press.

Ostrom, E. (2007a). A general framework for analyzing sustainability of social-ecological systems. *Science, 325*, 419–422.

Ostrom, E. (2007b). A diagnostic approach for going beyond panaceas. *Proceedings of the National Academy of Sciences, 104*(39), 15181–15187.

Ostrom, E., & Ahn, T. K. (Eds.). (2003). *Foundations of social capital.* Cheltenham: Edward Elgar.

Ostrom, E., & Ahn, T. K. (2009). The meaning of social capital and its link to collective action. In G. T. Svendsen & G. L. H. Svendsen (Eds.), *The handbook of social capital: The troika of sociology, political science and economics* (pp. 17–35). Northampton: Edward Elgar.

Paciotti, B., Hadley, C., Holmes, C., & Mulder, M. B. (2005). Grass-roots justice in Tanzania. *American Scientist, 93*, 58–65.

Pippa, N., & Inglehart, R. (2004). *Sacred and secular: Religion and politics worldwide.* Cambridge: Cambridge University Press.

Pretty, J. (2003). Social capital and the collective management of resources. *Science, 302*, 1912–1914.

Rappaport, R. A. (1968). *Pigs for the ancestors: Ritual in the ecology of a New Guinea people.* New Haven: Yale University Press.

Redman, C. L., Grove, J. M., & Kuby, L. H. (2004). Integrating social science into the long-term ecological research (LTER) network: Social dimensions of ecological change and ecological dimensions of social change. *Ecosystems, 7*(2), 161–171.

Rentfrow, P. J., Gosling, S. D., & Potter, J. (2008). A theory of the emergence, persistence, and expression of geographic variation in psychological characteristics. *Perspectives on Psychological Science, 3*(5), 339–369.

Rodima-Taylor, D., Olwig, M. F., & Chhetn, N. (2012). Adaption as innovation, innovation as adaptation: An institutional approach to climate change. *Applied Geography, 33*, 107–111.

Rogers, E., & Weber, E. (2010). Thinking harder about outcomes for collaborative governance arrangements. *American Review of Public Administration, 40*(5), 546–567.

Schank, J. C. (2001). Beyond reductionism: Refocusing on the individual with individual-based modeling: Computer-simulated surrogates in modeling. *Complexity, 6*(3), 33–40.

Schlager, E. (2004). Common-pool resource theory. In R. F. Durant, D. J. Fiorino, & R. O'Leary (Eds.), *Environmental governance reconsidered: Challenges, choices, and opportunities* (pp. 145–176). Cambridge: MIT Press.

Schwartz, S. H. (2006). A theory of cultural value orientations: Explication and applications. *Comparative Sociology, 5*, 136–182.

Segall, M. H., Lonner, W. J., & Berry, J. W. (1998). Cross-cultural psychology as a scholarly discipline: On the flowering of culture in behavioral research. *American Psychologist, 53*(10), 1102–1110.

Spears, R. (2011). Group identities: The social identity perspective. In S. J. Schwartz, K. Luyckx, & V. L. Vignoles (Eds.), *Handbook of identity theory and research* (Vol. 1, pp. 201–224). New York: Springer.

Stokols, D., Lejano, R. P., & Hipp, J. (2013). Enhancing the resilience of human-environment systems: A social-ecological perspective. *Ecology and Society, 18*(1), 7.

Stoll-Kleemann, S. (2004). The social-psychological dimension of biodiversity conservation. In S. S. Light (Ed.), *The role of biodiversity conservation in the transition to rural sustainability* (pp. 147–159). Amsterdam: Ios Press.

Stryker, S., & Serpe, R. T. (1982). Commitment, identity salience, and role behavior: Theory and research example. In W. Ickes & E. S. Knowles (Eds.), *Personality, roles, and social behavior* (pp. 199–218). New York: Springer.

Teel, T. L., & Manfredo, M. J. (2010). Understanding the diversity of public interests in wildlife conservation. *Conservation Biology, 24*(1), 128–139.

Triandis, H. C. (1995). *Individualism and collectivism*. Boulder: Westview.

Triandis, H. C. (2007). Culture and psychology: A history of the study of their relationship. In S. Kitayama & D. Cohen (Eds.), *Handbook of cultural psychology* (pp. 59–76). New York: Guilford Press.

Turner, J. C. (1991). *Social influence: Mapping social psychology series*. Belmont: Thompson Brooks/Cole Publishing.

Twigger-Ross, C., Bonaiuto, M., & Breakwell, G. (2003). Identity theories and environmental psychology. In M. Bonnes, T. Lee, & M. Bonaiuto (Eds.), *Psychological theories for environmental issues* (pp. 203–233). Burlington: Ashgate.

Uskul, A. K., Kitayama, S., & Nisbett, R. E. (2008). Ecocultural basis of cognition: Farmers and fishermen are more holistic than herders. *Proceedings of the National Academy of Sciences, 105*(25), 8552–8556.

Vaske, J. J., & Donnelly, M. P. (1999). A value-attitude-behavior model predicting wildland preservation voting intentions. *Society and Natural Resources, 12*, 523–537.

Vlek, C., & Steg, L. (2007). Human behavior and environmental sustainability: Problems, driving forces, and research topics. *Journal of Social Issues, 63*(1), 1–19.

Walker, B., Holling, C. S., Carpenter, S. R., & Kinzig, A. (2004). Resilience, adaptability and transformability in social-ecological systems. *Ecology and Society, 9*(2), 5.

Walker, B., Gunderson, L., Kinzig, A., Folke, C., Carpenter, S., & Schultz, L. (2006). A handful of heuristics and some propositions for understanding resilience in social-ecological systems. *Ecology and Society, 11*(1), 13–27.

Weber, E. U. (2006). Experience-based and description-based perceptions of long-term risk: Why global warming does not scare us (yet). *Climatic Change, 77*, 103–120.

Weber, E. U. (2010). What shapes perceptions of climate change? *Wiley Interdisciplinary Reviews: Climate Change, 1*(3), 332–342.

Willis, J. J., & Mastrofski, S. D. (2011). Innovations in policing: Meanings, structures, and processes. *Annual Review of Law and Social Science, 7*, 309–334.

Wolfe, R. (1994). Organizational innovation: Review, critique, and suggested research directions. *Journal of Management Studies, 31*, 405–431.

Part III
Methodological Advances for Facilitating Social Science Integration

Chapter 8
The Representation of Human-Environment Interactions in Land Change Research and Modelling

Peter H. Verburg

8.1 Introduction: Land Change and Spatial Models

Land change is the result of multiple human-environment interactions operating across different scales. Land change research needs to account for processes ranging from global trade of food and energy to the local management of land resources at farm and landscape level. Land change has a pronounced impact on the local and global environment. Land change may cause degradation of the living environment through soil degradation or changes in the aesthetic qualities of the landscape. At the same time, land change may lead to aggregate impacts on larger spatial and temporal scales, examples include the impacts on global climate and food security. Such impacts affect human well-being and often feedback on land use practices and decision making by adapting to the changing environmental and socio-economic context. Human-environment interactions in the land system are, therefore, connected across scales with multiple feedbacks, leading to so-called 'teleconnections' or 'telecoupling' in the earth system. The same process may cause different trajectories of land change in different world regions: globalization of food production can cause deforestation in tropical regions while marginal agricultural landscapes in other regions are abandoned. The local environmental and socio-economic context determines how the same global changes lead to different trajectories of land change in different parts of the world.

Land change occurs at the interface of human and environmental systems and is crucial in understanding both the causes and consequences of global environmental change. Local land change decisions are often made by individual land owners.

P.H. Verburg (✉)
Institute for Environmental Studies, VU University Amsterdam,
De Boelelaan 1087, 1081HV Amsterdam, The Netherlands
e-mail: peter.verburg@vu.nl

M.J. Manfredo et al. (eds.), *Understanding Society and Natural Resources*,
DOI 10.1007/978-94-017-8959-2_8, © The Author(s) 2014

In some cases land owners decide on land practices of small agricultural plots in terms of farming practices as well as having the opportunity to sell their land or buy adjacent plots. In other cases land owners have authority to make land use decisions over large areas of land managed by multiple individual farmers. When ownership is linked to the state or community, decision making on land resources is either the outcome of a political process (e.g. in assigning concessions for deforestation) or a result of communal decision making. Irrespective of the land ownership, land change decisions are steered by both the preferences of the land owner and managers and the way in which the decision process is influenced by the environmental conditions, commodity markets, socio-economic context and other driving factors. The spatial and temporal diversity of the actors of land change, the environment and the socio-economic and cultural context lead to a wide array of different land change trajectories with processes operating across multiple spatial scales. Such diversity expresses itself in a diverse mosaic of land use within the landscape and in the development of widely diverging trajectories of landscape change worldwide.

Effective management of land resources and the transition towards sustainable natural resource management can only be achieved based on a thorough understanding of the complex interactions and feedbacks in the land system. Land science has developed a wide portfolio of methods to investigate land system change, ranging from local case studies aimed at understanding the land change decisions leading to land change to global scale integrated assessment models that evaluate the impacts of land change on the earth system functioning. One of the major challenges of the land change community is to reconcile the different methodological approaches at different scales and make complementary use of the different types of knowledge generated. Computer simulation models play an important role in land science. Models provide a platform for formalized synthesis of the knowledge on the functioning of the land system, allow hypothesis testing and allow the exploration of alternative development trajectories and intervention options. This chapter will review how human-environment interactions are conceptualized in land science and land change models in particular. The chapter will explicitly address how social science knowledge is integrated in land change models and discuss a research agenda for further improving the representation of human agency in land change models.

8.2   The Representation of Human-Environment Interactions in Land Change Models

Conceptual models of human-environment interactions in land science

A theory of land system change should conceptualize the relationships between the driving and conditioning forces and land use change; including the relationships among the driving forces and human behavior and organization underlying these relationships. Existing disciplinary theories can help to analyze aspects of land change in specific situations and under well-defined assumptions. However, the paradigms and theories applied by the different disciplines are often difficult to integrate and their specific foci do not easily combine into an integrated understanding of land change. So far researchers have not yet succeeded in defining an all-compassing theory of land change and it can be questioned if the formulation of such theory is within reach. The lack of such overarching theory hampers the design of (conceptual) models to represent the human-environment interactions underlying land change.

Theories from multiple disciplines, such as economics, geography, ecology and anthropology, contribute to the explanation of land change. Often, these theories are related to specific land conversion processes or sectors, e.g. Boserupian theory concerning the effects of population on land use intensity (Boserup 1965; Turner and Fischer-Kowalski 2010; Turner and Ali 1996), neo-Thünen theory about moving frontiers and urban markets (Walker 2004; Walker and Solecki 2004) and the theories of Fujita and Krugman about urban development (Fujita et al. 1999a, b) as notable examples. Most theories cannot adequately explain the complexity of land use decision making underlying the observed land changes. Assumed agent behaviors in the common rational choice paradigm are very restricted and a variety of alternative decision making models are available (Meyfroidt 2012). Rational choice theory may reasonably explain land use decisions under the bid-rent paradigm. However, in reality individuals may rather seek to minimize risks or take them, as the case may be (Rabin 1998). Poorly defined property rights are not conducive to the competitive bidding process that leads to the equilibrium rent profile, which is most frequently underlying urban and agricultural models (Parker and Filatova 2008). In a recent review of the representation of decision making in land change research, Meyfroidt (2012) concludes that in land change science the cognitive aspects of decision making are underrepresented. His overview of alternative decision making models is synthesized by the notion that (i) land use choices result from multiple decision-making processes and rely on various motives, influenced by social norms, emotions, beliefs, and values toward the environment; (ii) social–ecological feedbacks are mediated by the environmental cognitions, that is, the perception, interpretation, evaluation of environmental change, and decision-making; (iii) human agents actively re-evaluate their beliefs, values, and functioning to adapt to unexpected environmental changes (Meyfroidt 2012).

The different, alternative, representations of decision making in land change and land change models are discussed by Hersperger et al. (2010) who describe 4 conceptual models that (often implicitly) underlie much land change model representations. Figure 8.1 summarizes the three most important models identified by Hersperger. We have added a fourth model that explicitly addresses the socio-ecological feedbacks and re-evaluation of decision making upon environmental change.

The first model looks for a direct relation between driving factors and land change, e.g. between population and agricultural intensity or between road building and deforestation. The identification of the underlying driving factors of land change has been a popular research topic and many papers have, for specific case studies, revealed the locally most important drivers of land change. Decision making that moderates the relation between driving factors and land change is often implicit and not analyzed explicitly. The relations between driving factors and land change can

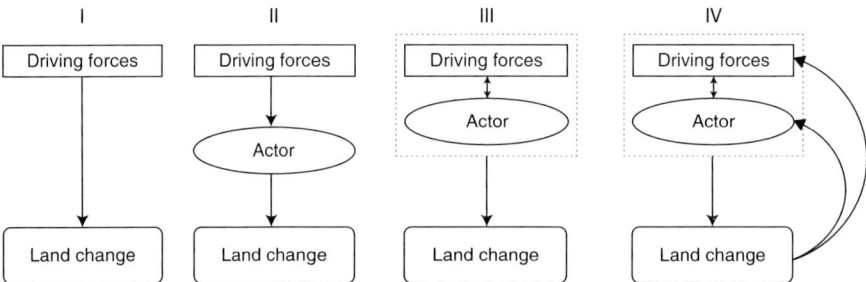

Fig. 8.1 Conceptual models for the representation of the relation between driving factors and land change (Modified after Hersperger et al. 2010)

be established by empirical analysis using observed land change data and statistical techniques, either based on spatial data or household interviews (Bürgi et al. 2004; Verburg et al. 2004a; Walsh et al. 1999). When using spatial data, statistical models are estimated that relate locations of observed land change (as dependent factor) to the spatial distribution of the driving factors (as independent factors). For example, locations of urbanization may be associated with locations of improved accessibility, resulting in a statistical model that relates accessibility to urbanization.

The second model represents the chain from driving factors to actor to land change. Although the actor has an explicit role in this sequence, the decision making of the actors itself may not be studied in detail and uniform decision making structures may be assumed. In addition, the driving factors are assumed to be independent of the actors. Examples of the application of this conceptual model include many economic land change models in which all actors are assumed to behave according to an uniform rational choice model (Happe et al. 2006). In such models the actors are supposed to make decisions based on land rent. Land rent is then explained as a function of driving factors, e.g. soil suitability and transportation costs.

The third conceptual model explicitly addresses the decision making process and accounts for the fact that the same driving factor may lead to a different land change outcome depending on variations in the decision making process. Examples include many social science studies in which variations in decision making between groups of the population are studied. As an example, Overmars et al. (2007) identified that in a case study in the Philippines, different ethnic groups have different land use decision strategies based on cultural tradition and knowledge. In many agent-based land change models a typology of agents is made in which the different groups are represented by different decision making rules towards land change (Valbuena et al. 2008). In the model of Valbuena et al. (2010a) hobby farmers are distinguished from commercial farmers as the decision making of both groups is governed by different objectives and motivations.

The fourth conceptual model, which we have added in addition to the models of Hersperger et al., represents an explicit feedback from land change to the actor and the driving factors. These feedbacks cause an impact of land change on the driving factors of land change, or invoke changes in the decision making strategy as result of actor learning, adaptation and perception in response to the experienced land change. Feedbacks between land change and decision making are not always straightforward and direct. Often the feedback operates across different spatial or temporal scales. Local land changes add up to impacts on the global climate system, in turn leading to local impacts in vulnerable regions in terms of changes in cropping conditions or increased flood risks to which people adapt their decision. The importance of such feedbacks was stressed by van Noordwijk et al. (2011) and Meyfroidt (2012). Unfortunately, only a small number of examples of the study of such feedbacks are available in the land science literature, mostly due to the difficulty of observing and quantifying such feedback mechanisms (Claessens et al. 2009; Verburg 2006).

8.2.1 Different Perspectives and Research Approaches

To obtain a full understanding of the causes and consequences of land change a complementary use has to be made of different research approaches. These can be classified as the narrative, the empirical and the modeling approaches (Lambin et al. 2003). The results of the narrative and empirical approach are often used as input to the modeling approach that aims at formalizing the identified relations in a structured framework.

The narrative approach seeks depth of understanding through historical detail and interpretation. It tells the land change story, providing an empirical and interpretative baseline by which to assess the validity and accuracy of the other visions. It is especially beneficial in identifying stochastic and random events that significantly affect land change but might be missed in approaches employing less expansive time horizons or temporal sampling procedures (Briassoulis 2000). The narrative approach is mostly valid at the level of individual actors and one of the challenges of the approach is to link it with the features of land change that occur at more aggregate levels of analysis. This has given rise to efforts to better link 'people and pixels' through georeferencing narrative research and efforts to link the narrative approach to empirical approaches using geographical data (Liverman and Cuesta 2008; Rindfuss et al. 2003; Rindfuss and Stern 1998). By linking household data to the spatial units of land managed by those households, it becomes possible to relate household characteristics to the actual land management applied in the field.

The empirical approach builds on the narrative approach but takes a more quantitative perspective by identifying significant relations and pattern in the collected data while testing hypothesis that are either based on the narrative research approach or through deductive reasoning (Pfaff and Sanchez-Azofeifa 2004). Such empirical analysis can take place at various levels of spatial and temporal aggregation, ranging from the analysis of household survey data (Overmars and Verburg 2005) or the analysis of spatial units, i.e. pixels or polygons, organized in geographic data layers (Chomitz and Gray 1996; Veldkamp et al. 2001) to the analysis of time series of country-level statistics (Rudel et al. 2009). A major drawback of the empirical quantification of relations between land use and its supposed drivers is the induced uncertainty with respect to the causality of the supposed relations. The danger lies in leaping directly from the exploratory stage, or even from statistical tests based on descriptive models, to conclusions about causes (James and McCulloch 1990). Besides, most causal explanations are valid at the scale of study, mostly the individual actor of land change, and therefore subject to upscaling problems. This asks for validation of the causality of empirically derived relations. A combination of the narrative perspective with the empirical perspective can help to test the validity of the empirical relations. An example of such a combined approach is a study of Overmars in the Philippines (Overmars and Verburg 2005). Overmars used an approach that evaluates the results of statistical models based on geographic data by a household-level analysis of decision making.

The modeling approach uses theoretical, assumed or empirical relations to construct a model that allows the exploration of land change dynamics across historic (observed) or future time periods. Models especially allow the analysis of 'what-if' questions through acting as an artificial laboratory for conducting controlled experiments which are very difficult to establish in the real world. Similarly to the empirical perspective, land change models are aimed at a wide variation of different spatial and temporal scales. Local agent-based models mostly represent individual actors within a community or small region (Matthews et al. 2007) while spatial models often are applied at the regional level, simulating the changes in land use of land units or pixels. Land use is also an explicit part of larger scale models operating at the global level, ranging from global equilibrium models of the world economy (Hertel et al. 2010) to integrated assessment models of global environmental change (Thomson et al. 2010). The following section will describe the way in which human-environment interactions are addressed in land change models in more detail.

From the above it is clear that both the different research approaches and the different spatial scales of analysis are able to provide complementary insights. However, the linking of the approaches across the different scales may not be straightforward. Coleman (1990) developed a framework that describes the interaction between micro and macro levels for social systems. The same framework can also be applied to land change models. Land change assessments made at the regional level, using remote sensing and geographic data, are often explained by specifying a micro-level mechanism. Figure 8.2, based on the work of Coleman (1990), depicts the relations between the macro and micro levels. Macro-level analyses (pathway A) of land use are normally based on empirical techniques, e.g. the analysis of spatial patterns of land use derived from remote sensing. Pathway B explains the underlying processes from which the different land use patterns have emerged, e.g. the individual decisions in response to the (changing) socio-economic and physical context. Aggregated, these individual decisions lead to changes in land use pattern that can be analyzed in the more macro-scale analysis. This aggregation may not be straightforward due to non-linear relationships causing the 'ecological fallacy' or 'modifiable area unit problem' (Easterling 1997; Marceau and Hay 1999).

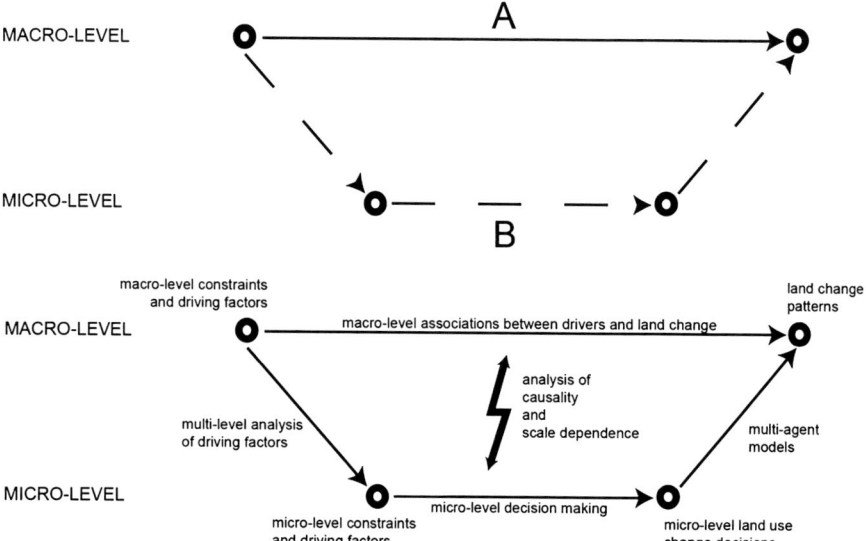

Fig. 8.2 Illustration of the relations between macro and micro-level analysis of land change (Based on Coleman 1990)

These terms relate to the bias that is introduced when non-linear relations at individual level are applied to aggregate data. Also, interactions between agents, e.g. leading to collective behavior, as well as the role of institutions and other 'collective' agents lead to aggregate results that deviate from the sum of individual decisions (Gibson et al. 2000; Liu et al. 2007). Tools have been developed to analyze the role of processes across multiple scales, e.g. multi-level statistics (Neumann et al. 2011; Overmars and Verburg 2006; Pan and Bilsborrow 2005) and agent-based models, that model the emergence of patterns from individual decision making (Parker et al. 2008). Still, the importance of scalar dynamics in analyzing human-environment interactions is still frequently overlooked.

8.2.2 Using Social Science Case-Studies to Help Parameterize Land Change Models

The disconnection between the different research perspectives, and the disciplinary communities involved in the different approaches, causes land change models to neglect the knowledge gained by the narrative and empirical perspectives. A specific approach to bridge the different research approaches in land change research and generalize local findings across larger regions has been the use of meta-analysis of case studies. Meta-analysis is a form of systematic review aimed at the statistical evaluation of a large number of case studies and can provide the empirical base for designing simulation models. Meta-analysis is especially useful if new (and possibly more structured) data collection is not feasible due to the large time and financial investments required. Such systematic review of studies is useful in land science since globally valid explanations of what factors drive land use change remain largely incomplete (Rudel 2008). Common understanding of the causes of land change is dominated by simplifications that, in turn, underlie many land change models. Within case studies of land change, based on either the narrative or empirical research approach, a wealth of in-depth information on decision making in human-environment interactions is available. Meta-analysis can help to identify common-alities across these case studies and identify which factors (variables) cause different cases to behave differently. Case studies on land change often contain information on the proximate causes of land change and their underlying driving factors and provide insight in the decision making processes leading to changes in land use and management. The main approach to systematic review of the knowledge in case studies in the field of land science has been the synthesis of proximate causes and driving factors for specific land change processes resulting in a listing of the globally most frequently mentioned drivers of land change. Examples of such systematic review or meta-analysis are available for deforestation (Geist and Lambin 2002; Rudel 2005), desertification (Geist and Lambin 2004), agricultural intensification (Keys and McConnell 2005) and shifting cultivation (van Vliet et al. 2012). These meta-analysis support the conclusion that the simple answers found in population growth, poverty and infrastructure rarely provide an adequate understanding of land change. Rather, individual and social responses follow from changing economic conditions, mediated by institutional factors. Opportunities and constraints for new land uses are created by markets and policies, increasingly influenced by global factors (Lambin et al. 2001). A weakness of the existing meta-analysis in land use is that it is mainly tended towards understanding the broad, macro-scale social forces that affect nature-society relationships and less attention is given to the role of the space-time context in determining these relationships, i.e. mostly the human-environment system is investigated following the first conceptual model in Fig. 8.1. At the same time, the case studies included tend to be biased towards the most interesting regions with dramatic land changes.

For a more limited set of case studies Rindfuss et al. (2007) tried to more specifi-cally identify the important factors explaining differences in land change processes between frontier regions. However, as case studies are often made by different teams and with different objectives, the quantitative comparison of such cases turned out to be more troublesome; indicating the need for more clearly document-ing common sets of case study findings and harmonizing case study methods in order to be able to contextualize case study findings. Such harmonization will ensure that case study results can more easily be contextualized, allowing the use of the findings in land change models.

8.2.3 Representation of Human-Environment Interactions in Land Change Models

A wide variety of land change models have been developed over the past two decades that have been reviewed numerous times (Agarwal et al. 2001; Priess and Schaldach 2008; Verburg et al. 2004b) based on different criteria. In this section we do not aim to provide an exhaustive review of these models, but rather will address the variation in ways that human-environment interactions are conceptualized in the different models. In contrast to a classification or representation based on the specific modeling technique used, e.g. cellular automata or agent-based modeling, the methods employed to represent human-environment interactions may be classified on a scale from deductive, theory-led approaches to fully empirical, inductive approaches to modeling. Overmars et al. (2007) provide such a scale from deductive to inductive reasoning and conclude that many of the existing models are neither fully deductive or inductive. But, still large differences exist in the role of theory and empirical data in conceptualizing the model. Especially the way in which decision making on land change is represented differs. In some approaches an almost completely deductive approach is taken by assuming rational agents that optimize income and tailor land change decisions towards that goal. Some of these models operate at the level of individual decision makers, e.g. farmers (Piorr et al. 2009) while others operate at the level of large world regions in which decision making is conceptualized for an aggregate (representative) agent (Havlík et al. 2011; Souty et al. 2012; Van Meijl et al. 2006). It may be questioned under what conditions the same behavioral assumptions are valid for both individual and highly aggregate agents. On the other end of the spectrum models that employ machine learning methods to relate land change to its determinants are found. Many machine learning techniques do not provide insight into the estimated relations and it is only the observed data that determine the relations employed in the model to simulate future land changes. Many other models fall somewhere in between these extremes. So called 'factor-led induction' (Overmars et al. 2007) employs theory to identify the factors driving land change decisions while the actual relations between these factors and land change are established using empirical estimation of statistical coefficients using observed data (Chomitz and Gray 1996; Nelson and Hellerstein 1997). Such a theory-based approach is important to explore for several reasons. It structures the model around the critical human-environment relationships identified within the theory, and focuses attention on the data required to explore those relationships. Similarly, many agent-based models of land change employ a range of empirical techniques to make a typology of different decision making types and parameterize the decision making rules in the model based on household survey results (Robinson et al. 2007; Smajgl et al. 2011; Valbuena et al. 2008). The latter group of models is of specific interest to the study of human-environment interactions. Multi-agent models simulate decision making by individual agents of land use change, explicitly addressing interactions among individuals. The explicit attention for interactions between agents makes it possible for this type of model to simulate emergent properties of systems. These are properties at the macro scale that are not predictable from observing the micro units in isolation. If the decision rules of the agents are set such that they sufficiently look like human decision making they can simulate behavior at the meso-level of social organization, i.e. the behavior of heterogeneous groups of actors. Multi-agent-based models of land change are particularly well suited to representing complex spatial interactions under heterogeneous conditions (Bousquet and Le Page 2004; Parker et al. 2003). Multi-agent systems are able to formalize decision-forming behavior of individual stakeholders, either based on theory (Happe et al. 2006), or based on observations and statistical analysis (Bousquet et al. 2001; Robinson et al. 2007; Valbuena et al. 2010a). In the initial years of application of agent-based models to land change, most multi-agent models focused on either hypothetical or simplified representations of the real world to explore interactions between agents and between agents and the environment. Especially the parameterization of agent behavior in models for real case-studies turned out to be very complex. However, more recently a larger number of applications of agent-based models to real case studies worldwide have been published, showing the potential of the approach to explore the land change dynamics in local to regional level case studies (Le et al. 2012; Robinson et al. 2012; Valbuena et al. 2010b). At larger spatial scales, ranging from the region to the global level the principles of agent-based modeling have not yet been applied in simulation models, leaving most models at that level with highly simplified representations of human-environment interactions (Rounsevell and Arneth 2011). The possibilities for either upscaling or outscaling agent-based models have been described by Rounsevell and colleagues (Rounsevell et al. 2012), but have not yet been applied in operational models.

8.3 Land Change Models as a Platform for Social Science Integration

The review and discussion in the previous sections has illustrated the importance of the social sciences for studying land change processes. Often, the social sciences have taken the narrative or empirical approach for studying land change. The modeling perspective is often dominated by natural scientists and in many models the social drivers of land change are underrepresented. This underrepresentation can, to some extent, be attributed to the lack of spatial data representing the social drivers. For the physical factors such data are often better available, e.g. soil maps and climate data. At the same time, the poor representation of social science in land change models is due to the difficulty to generalize social science findings outside the context of a specific case study, and the lack of an overarching theory of land change that includes the social dimensions. Still, there are several advances and prospects that allow land change models to act as a platform for social science integration in natural resource studies.

It is not likely that the complete richness of human-environment interactions leading to land change will easily and completely be described by one single, all compassing theory that can inform the design of land change models. Different existing theories describe specific land change processes and are valid under specific conditions or at a specific scale; together the different theories help explain part of the total variation in human-environment interactions leading to land change. The combination and integration of narrative research with empirical investigations will help to better define the conditions under which certain land change processes occur and when theories and conceptual models are valid. Such understanding will help to define under what conditions land change models based on these conceptual relations can adequately capture the system dynamics. The complementary use of narrative, empirical and model-based explorations requires the interdisciplinary collaboration and exchange of insights across the different research perspectives and disciplines. Land change models may be designed based on the narrative and theoretical understanding of the human-environment interactions in a particular context. At the same time, social science perspectives may be formalized by representing them in simulation models, enabling to test the implications for system dynamics.

Besides interdisciplinary collaboration it is also required to broaden the perspective of the individual disciplinary approaches. The most effective way to reap the benefits of more deductive work is not to rigidly 'go deductive' and stay there. Such a 'process-led approach' may blind the analyst to alternative processes at work (Overmars et al. 2007). Rather, the message should be that researchers will profit most from developing a consciousness of the whole spectrum between the inductive and deductive extremes, and an awareness of the advantages of the variation in research routines, and then seeking the most fertile sequences and interactions between inductive and deductive work. Ultimately, this will contribute to theory development in the field of land change while at the same time helping the development of modeling tools to explore the dynamics in land systems and possible responses to policy interventions.

The lack of social science integration in models of land change is exemplified by the, often, very simplistic representation of human-environment interactions in operational land change models which does not do justice to the complexity of decision making. Especially at larger spatial and temporal scales models assume in most cases profit optimizing strategies at the level of either spatial units or for highly aggregate representative agents. In contrast, at local scales much advancement has been made in the representation of human behavior and decision making in agent-based models of land change. Ignoring spatial and temporal variation in decision making and responses to environmental change leads to inaccuracies in global assessment outcomes and difficulties in using these models to design place-based natural resource management and adaptation and mitigation strategies. The upscaling and/or outscaling of agent-based models of land change is restricted by the lack of empirical data to support the parameterization of the human-environment interactions in these models. Such parameterization requires insight in the diversity of diverging decision making models and the contextual conditions that may explain such diversity. To better include such social science information a promising direction is the re-analysis of existing case-studies and social science surveys to identify commonalities across locations as well as the role of context. The use of meta-analysis to achieve some of these objectives has revealed that information reported in case studies is often restricted and incomplete to make a full comparative analysis possible. Moreover, as narrative and econometric case studies are not conducted following a common structure or reporting protocol the necessary information to make a systematic review across case studies is often lacking in the scientific reports. Common reporting protocols to ensure that information is consistently documented have been successful in the individual-based and agent-based modeling communities. The ODD reporting protocol of individual-based and agent-based models (Grimm et al. 2006, 2010) is now common as supplementary material of all individual-based and agent-based model papers in peer-reviewed journals. A similar documentation protocol has been proposed by Seppelt and others (2012) for documenting ecosystem service assessments. If land change case-studies would apply similar documentation standards a wealth of information on land change processes and the underlying human-environment interactions worldwide would be disclosed.

Another constraint for parameterizing agent-based models is the limited information that standard land change case studies provide on the cognitive aspects of land change decisions (Meyfroidt 2012). Many studies describe the ways in which driving forces relate to land change decisions without considering the underlying cognitive processes and the way in which decision making adapts to changing conditions, including learning. While the investigation of such mechanisms is normally the field of environmental psychology, such insights are essential to understand transitions in decision making as are likely to take place under increasing influence of global markets, changing policy environments and climate change. Land change is happening in a dynamic socio-economic and environmental context, leading to dynamic decision making patterns in which we have yet insufficient insights.

Spatial simulation models are frequently used to reconstruct historic land changes (Klein Goldewijk et al. 2011) and explore future changes or evaluate the land change consequences of alternative policies. The comparison of simulation results with reality provides a measure of the extent to which we understand the human-environment interactions resulting in land change (Castella and Verburg 2007; Pontius et al. 2008). The wide diversity in modeling concepts and implementations serves the variation in research and policy questions as well as the different scales of analysis. Adequate land change models require the integration of social science perspectives and multi-agent models are an example of the possibility to do so. However, the challenges for better understanding and integrating human-environment interactions in land change models are still manifold. But, in the end, the development of land change models provides a platform for integrating the different disciplinary perspectives on the complex socio-ecological system governing land change. Advancing land change modeling, therefore, not only requires the efforts of individual disciplinary researchers, it especially takes the courage of all individual researchers to collaborate, contextualize findings and respond to the needs to translate findings across spatial scales.

Acknowledgments The research leading to these results has received funding from the European Research Council under the European Union's Seventh Framework Programme (FP/2007-2013)/ ERC Grant Agreement n. 311819 and the project VOLANTE Grant Agreement n. 265104.

References

Agarwal, C., Green, G. M., Grove, J. M., Evans, T. P., & Schweik,C. M. (2001). *A review and assessment of land use change models. Dynamics of space, time, and human choice.* Bloomington: Center for the Study of Institutions, Population, and Environmental Change, Indiana University; South Burlington: USDA Forest Service.

Boserup, E. (1965). *The conditions of agricultural growth: The economics of agrarian change under population pressure.* Chicago: Aldine.

Bousquet, F., & Le Page, C. (2004). Multi-agent simulations and ecosystem management: A review. *Ecological Modelling, 176*(3–4), 313–332.

Bousquet, F., Le Page, C., Bakam, I., & Takforyan, A. (2001). Multiagent simulations of hunting wild meat in a village in eastern Cameroon. *Ecological Modelling, 139*(1–3), 331–346.

Briassoulis, H. (2000). Analysis of land use change: Theoretical and modeling approaches. In S. Loveridge (Ed.), *The web book of regional science.* Morgantown: West Virginia University. http://rri.wvu.edu/resources/web-book-rs/

Bürgi, M., Hersperger, A. M., & Schneeberger, N. (2004). Driving forces of landscape change – Current and new directions. *Landscape Ecology, 19*(8), 857–868.

Castella, J. C., & Verburg, P. H. (2007). Combination of process-oriented and pattern-oriented models of land-use change in a mountain area of Vietnam. *Ecological Modelling, 202*(3–4), 410–420.

Chomitz, K. M., & Gray, D. A. (1996). Roads, land Use, and deforestation: A spatial model applied to Belize. *The World Bank Economic Review, 10*(3), 487–512.

Claessens, L., Schoorl, J. M., Verburg, P. H., Geraedts, L., & Veldkamp, A. (2009). Modelling interactions and feedback mechanisms between land use change and landscape processes. *Agriculture, Ecosystems and Environment, 129*(1–3), 157–170.

Coleman, J. S. (1990). *Foundations of social theory.* Cambridge: The Belknap Press of Harvard University Press.

Easterling, W. E. (1997). Why regional studies are needed in the development of full-scale integrated assessment modelling of global change processes. *Global Environmental Change Part A, 7*(4), 337–356.

Fujita, M., Krugman, P., & Mori, T. (1999a). On the evolution of hierarchical urban systems. *European Economic Review, 43*(2), 209–251.

Fujita, M., Krugman, P., & Venables, A. J. (1999b). *The spatial economy: Cities, regions and international trade.* Cambridge, MA: MIT Press.

Geist, H. J., & Lambin, E. F. (2002). Proximate causes and underlying driving forces of tropical deforestation. *Bioscience, 52*(2), 143–150.

Geist, H. J., & Lambin, E. F. (2004). Dynamic causal patterns of desertification. *Bioscience, 54*(9), 817–829.

Gibson, C. C., Ostrom, E., & Anh, T. K. (2000). The concept of scale and the human dimensions of global change: A survey. *Ecological Economics, 32*(217), 239.

Grimm, V., Berger, U., Bastiansen, F., Eliassen, S., Ginot, V., Giske, J., Goss-Custard, J., Grand, T., Heinz, S. K., Huse, G., Huth, A., Jepsen, J. U., Jörgensen, C., Mooij, W. M., Müller, B., Pe'er, G., Piou, C., Railsback, S. F., Robbins, A. M., Robbins, M. M., Rossmanith, E.,

Rüger, N., Strand, E., Souissi, S., Stillman, R. A., Vabø, R., Visser, U., & DeAngelis, D. L. (2006). A standard protocol for describing individual-based and agent-based models. *Ecological Modelling, 198*(1–2), 115–126.

Grimm, V., Berger, U., DeAngelis, D. L., Polhill, J. G., Giske, J., & Railsback, S. F. (2010). The ODD protocol: A review and first update. *Ecological Modelling, 221*(23), 2760–2768.

Happe, K., Kellermann, K., & Balmann, A, (2006). Agent-based analysis of agricultural policies: an illustration of the agricultural policy simulator AgriPoliS, its adaptation, and behavior. *Ecology and Society, 11*(1), 49. http://www.ecologyandsociety.org/vol11/iss1/art49/

Havlík, P., Schneider, U. A., Schmid, E., Böttcher, H., Fritz, S., Skalský, R., Aoki, K., Cara, S. D., Kindermann, G., Kraxner, F., Leduc, S., McCallum, I., Mosnier, A., Sauer, T., & Obersteiner, M. (2011). Global land-use implications of first and second generation biofuel targets. *Energy Policy, 39*(10), 5690–5702.

Hersperger, A. M., Gennaio, M.-P., Verburg, P. H., & Bürgi, M. (2010). Linking land change with driving forces and actors: Four conceptual models. *Ecology and Society, 15*(4), 1.

Hertel, T. W., Golub, A. A., Jones, A. D., O'Hare, M., Plevin, R. J., & Kammen, D. M. (2010). Effects of US maize ethanol on global land use and greenhouse gas emissions: Estimating market-mediated responses. *Bioscience, 60*(3), 223–231.

James, F. C., & McCulloch, C. E. (1990). Multivariate analysis in ecology and systematics: Panacea or Pandora's box? *Annual Review of Ecology and Systematics, 21*, 129–166.

Keys, E., & McConnell, W. J. (2005). Global change and the intensification of agriculture in the tropics. *Global Environmental Change Part A, 15*(4), 320–337.

Klein Goldewijk, K., Beusen, A., van Drecht, G., & de Vos, M. (2011). The HYDE 3.1 spatially explicit database of human-induced global land-use change over the past 12,000 years. *Global Ecology and Biogeography, 20*(1), 73–86.

Lambin, E. F., Turner, B. L., II, Geist, H. J., Agbola, S. B., Angelsen, A., Bruce, J. W., Coomes, O., Dirzo, R., Fischer, G., Folke, C., George, P. S., Homewood, K., Imbernon, J., Leemans, R., Li, X. B., Moran, E. F., Mortimore, M., Ramakrishnan, P. S., Richards, J. F., Skanes, H., Stone, G. D., Svedin, U., Veldkamp, A., Vogel, C., & Xu, J. C. (2001). The causes of land-use and land-cover change: Moving beyond the myths. *Global Environmental Change, 4*, 261–269.

Lambin, E. F., Geist, H. J., & Lepers, E. (2003). Dynamics of land-use and land-cover change in tropical regions. *Annual Review of Environment and Resources, 28*, 205–241.

Le, Q. B., Seidl, R., & Scholz, R. W. (2012). Feedback loops and types of adaptation in the modelling of land-use decisions in an agent-based simulation. *Environmental Modelling & Software, 27–28*, 83–96.

Liu, J., Dietz, T., Carpenter, S. R., Alberti, M., Folke, C., Moran, E., Pell, A. N., Deadman, P., Kratz, T., Lubchenco, J., Ostrom, E., Ouyang, Z., Provencher, W., Redman, C. L., Schneider, S. H., & Taylor, W. W. (2007). Complexity of coupled human and natural systems. *Science, 317*(5844), 1513–1516.

Liverman, D. M., & Cuesta, R. M. R. (2008). Human interactions with the earth system: People and pixels revisited. *Earth Surface Processes and Landforms, 33*(9), 1458–1471.

Marceau, D. J., & Hay, G. J. (1999). Remote sensing contributions to the scale issue. *Canadian Journal of Remote Sensing, 25*(4), 357–366.

Matthews, R., Gilbert, N., Roach, A., Polhill, J., & Gotts, N. (2007). Agent-based land-use models: a review of applications. *Landscape Ecology, 22*(10), 1447–1459.

Meyfroidt, P. (2012). Environmental cognitions, land change, and social–ecological feedbacks: An overview. *Journal of Land Use Science, 8*(3), 341–367. doi:10.1080/1747423X.2012.667452.

Nelson, G. C., & Hellerstein, D. (1997). Do roads cause deforestation? Using satellite images in econometric analysis of land use. *American Journal of Agricultural Economics, 79*, 80–88.

Neumann, K., Stehfest, E., Verburg, P. H., Siebert, S., Müller, C., & Veldkamp, A. (2011). Exploring global irrigation patterns: A multilevel modeling approach. *Agricultural Systems, 104*(9), 703–713.

Overmars, K. P., & Verburg, P. H. (2005). Analysis of land use drivers at the watershed and household level: Linking two paradigms at the Philippine forest fringe. *International Journal of Geographical Information Science, 19*(2), 125–152.

Overmars, K. P., & Verburg, P. H. (2006). Multilevel modelling of land use from field to village level in the Philippines. *Agricultural Systems, 89*(2–3), 435–456.

Overmars, K., de Groot, W., & Huigen, M. (2007). Comparing inductive and deductive modeling of land use decisions: Principles, a model and an illustration from the Philippines. *Human Ecology, 35*(4), 439–452.

Pan, W. K. Y., & Bilsborrow, R. E. (2005). The use of a multilevel statistical model to analyze factors influencing land use: A study of the Ecuadorian Amazon. *Global and Planetary Change, 47*(2–4), 232–252.

Parker, D. C., & Filatova, T. (2008). A conceptual design for a bilateral agent-based land market with heterogeneous economic agents. *Computers, Environment and Urban Systems, 32*(6), 454–463.

Parker, D. C., Manson, S. M., Janssen, M. A., Hoffman, M., & Deadman, P. (2003). Multi-agent systems for the simulation of land-use and land-cover change: A review. *Annals of the Association of American Geographers, 93*(2), 314–337.

Parker, D. C., Hessl, A., & Davis, S. C. (2008). Complexity, land-use modeling, and the human dimension: Fundamental challenges for mapping unknown outcome spaces. *Geoforum, 39*(2), 789–804.

Pfaff, A. S. P., & Sanchez-Azofeifa, G. A. (2004). Deforestation pressure and biological reserve planning: a conceptual approach and an illustrative application for Costa Rica. *Resource and Energy Economics, 26*(2), 237–254.

Piorr, A., Ungaro, F., Ciancaglini, A., Happe, K., Sahrbacher, A., Sattler, C., Uthes, S., & Zander, P. (2009). Integrated assessment of future CAP policies: Land use changes, spatial patterns and targeting. *Environmental Science & Policy, 12*(8), 1122–1136.

Pontius, R., Boersma, W., Castella, J.-C., Clarke, K., de Nijs, T., Dietzel, C., Duan, Z., Fotsing, E., Goldstein, N., Kok, K., Koomen, E., Lippitt, C., McConnell, W., Mohd Sood, A., Pijanowski, B., Pithadia, S., Sweeney, S., Trung, T., Veldkamp, A., & Verburg, P. (2008). Comparing the input, output, and validation maps for several models of land change. *The Annals of Regional Science, 42*, 11–37.

Priess, J. A., Schaldach, R (2008). Integrated models of the land system: A review of modelling approaches on the regional to global scale. *Living Reviews in Landscape Research, 2*(1). http://www.livingreviews.org/lrlr-2008-1

Rabin, M. (1998). Psychology and economics. *Journal of Economic Literature, 36*(1), 11–46.

Rindfuss, R. R., & Stern, P. C. (1998). Linking remote sensing and social science: The need and the challenges. In D. Liverman, E. F. Moran, R. R. Rindfuss, & P. C. Stern (Eds.), *People and pixels: Linking remote sensing and social science* (pp. 1–27). Washington, DC: National Academy Press.

Rindfuss, R. R., Walsh, S. J., Mishra, V., Fox, J., & Dolcemascolo, G. P. (2003). Linking household and remotely sensed data, methodological and practical problems. In J. Fox, R. R. Rindfuss, S. J. Walsh, & V. Mishra (Eds.), *PEOPLE AND THE ENVIRONMENT – Approaches for linking household and community surveys to remote sensing and GIS* (pp. 1–29). Boston: Kluwer Academic Publishers.

Rindfuss, R. R., Entwisle, B., Walsh, S. J., Mena, C. F., Erlien, C. M., & Gray, C. L. (2007). Frontier land use change: Synthesis, challenges, and next steps. *Annals of the Association of American Geographers, 97*(4), 739–754.

Robinson, D. T., Brown, D. G., Parker, D. C., Schreinemachers, P., Janssen, M. A., Huigen, M., Wittmer, H., Gotts, N., Promburom, P., Irwin, E., Berger, T., Gatzweiler, F., & Barnaud, C. (2007). Comparison of empirical methods for building agent-based models in land use science. *Journal of Land Use Science, 2*(1), 31–55.

Robinson, D. T., Murray-Rust, D., Rieser, V., Milicic, V., & Rounsevell, M. (2012). Modelling the impacts of land system dynamics on human well-being: Using an agent-based approach to cope with data limitations in Koper, Slovenia. *Computers, Environment and Urban Systems, 36*(2), 164–176.

Rounsevell, M. D. A., & Arneth, A. (2011). Representing human behaviour and decisional processes in land system models as an integral component of the earth system. *Global Environmental Change, 21*(3), 840–843.

Rounsevell, M. D. A., Robinson, D. T., & Murray-Rust, D. (2012). From actors to agents in socio-ecological systems models. *Philosophical Transactions of the Royal Society B: Biological Sciences, 367*(1586), 259–269.

Rudel, T. K. (2005). *Tropical forests. Regional paths of destruction and regeneration in the late twentieth century.* New York: Columbia University Press.

Rudel, T. K. (2008). Meta-analyses of case studies: A method for studying regional and global environmental change. *Global Environmental Change, 18*(1), 18–25.

Rudel, T. K., Schneider, L., Uriarte, M., Turner, B. L., DeFries, R., Lawrence, D., Geoghegan, J., Hecht, S., Ickowitz, A., Lambin, E. F., Birkenholtz, T., Baptista, S., & Grau, R. (2009). Agricultural intensification and changes in cultivated areas, 1970–2005. *Proceedings of the National Academy of Sciences, 106*(49), 20675.

Seppelt, R., Fath, B., Burkhard, B., Fisher, J. L., Grêt-Regamey, A., Lautenbach, S., Pert, P., Hotes, S., Spangenberg, J., Verburg, P. H., & Van Oudenhoven, A. P. E. (2012). Form follows function? Proposing a blueprint for ecosystem service assessments based on reviews and case studies. *Ecological Indicators, 21*, 145–154.

Smajgl, A., Brown, D. G., Valbuena, D., & Huigen, M. G. A. (2011). Empirical characterisation of agent behaviours in socio-ecological systems. *Environmental Modelling & Software, 26*(7), 837–844.

Souty, F., Brunelle, T., Dumas, P., Dorin, B., Ciais, P., Crassous, R., Müller, C., & Bondeau, A. (2012). The nexus land-use model version 1.0, an approach articulating biophysical potentials and economic dynamics to model competition for land-use. *Geoscientific Model Development, 5*(1), 571–638.

Thomson, A. M., Calvin, K. V., Chini, L. P., Hurtt, G., Edmonds, J. A., Bond-Lamberty, B., Frolking, S., Wise, M. A., & Janetos, A. C. (2010). Climate mitigation and the future of tropical landscapes. *Proceedings of the National Academy of Sciences, 107*(46), 19633–19638.

Turner, B. L., & Ali, A. M. (1996). Induced intensification: Agricultural change in Bangladesh with implications for Malthus and Boserup. *Proceedings of the National Academy of Sciences, 93*(25), 14984–14991.

Turner, B. L., & Fischer-Kowalski, M. (2010). Ester Boserup: An interdisciplinary visionary relevant for sustainability. *Proceedings of the National Academy of Sciences, 107*(51), 21963–21965.

Valbuena, D., Verburg, P. H., & Bregt, A. K. (2008). A method to define a typology for agent-based analysis in regional land-use research. *Agriculture, Ecosystems & Environment, 128*(1–2), 27–36.

Valbuena, D., Verburg, P. H., Bregt, A. K., & Ligtenberg, A. (2010a). An agent-based approach to model land-use change at a regional scale. *Landscape Ecology, 25*(2), 185–199.

Valbuena, D., Bregt, A. K., McAlpine, C., Verburg, P. H., & Seabrook, L. (2010b). An agent-based approach to explore the effect of voluntary mechanisms on land use change: A case in rural Queensland, Australia. *Journal of Environmental Management, 91*(12), 2615–2625.

Van Meijl, H., van Rheenen, T., Tabeau, A., & Eickhout, B. (2006). The impact of different policy environments on agricultural land use in Europe. *Agriculture, Ecosystems & Environment, 114*(1), 21–38.

van Noordwijk, M., Lusiana, B., Villamor, G., Purnomo, H., & Dewi, S. (2011). Feedback loops added to four conceptual models linking land change with driving forces and actors. *Ecology and Society, 16*(1), r1. URL: http://www.ecologyandsociety.org/vol16/iss1/resp1/

van Vliet, N., Mertz, O., Heinimann, A., Langanke, T., Pascual, U., Schmook, B., Adams, C., Schmidt-Vogt, D., Messerli, P., Leisz, S., Castella, J. C., Jörgensen, L., Birch-Thomsen, T., Hett, C., Bech-Bruun, T., Ickowitz, A., Vu, K. C., Yasuyuki, K., Fox, J., Padoch, C., Dressler, W., & Ziegler, A. D. (2012). Trends, drivers and impacts of changes in swidden cultivation in tropical forest-agriculture frontiers: A global assessment. *Global Environmental Change, 22*(2), 418–429.

Veldkamp, A., Verburg, P. H., Kok, K., De Koning, G. H. J., Priess, J., & Bergsma, A. R. (2001). The need for scale sensitive approaches in spatially explicit land use change modeling. *Environmental Modeling and Assessment, 6*(2), 111–121.

Verburg, P. H. (2006). Simulating feedbacks in land use and land cover change models. *Landscape Ecology, 21*(8), 1171–1183.

Verburg, P. H., Ritsema van Eck, J. R., de Nijs, T. C. M., Dijst, M. J., & Schot, P. (2004a). Determinants of land-use change patterns in the Netherlands. *Environment and Planning B: Planning and Design, 31*(1), 125–150.

Verburg, P. H., Schot, P. P., Dijst, M. J., & Veldkamp, A. (2004b). Land use change modelling: Current practice and research priorities. *GeoJournal, 61*(4), 309–324.

Walker, R. (2004). Theorizing land-cover and land-use change: The case of tropical deforestation. *International Regional Science Review, 27*(3), 247–270.

Walker, R., & Solecki, W. D. (2004). Theorizing land-cover and land-use change: The case of the Florida Everglades and its degradation. *Annals of the Association of American Geographers, 94*(2), 311–328.

Walsh, S. J., Evans, T. P., Welsh, W. F., Entwisle, B., & Rindfuss, R. R. (1999). Scale-dependent relationships between population and environment in Northeastern Thailand. *Photogrammetric Engineering & Remote Sensing, 65*(1), 97–105.

Chapter 9
Simulation as an Approach to Social-Ecological Integration, with an Emphasis on Agent-Based Modeling

Randall B. Boone and Kathleen A. Galvin

9.1 Introduction

In past decades ecological and social science research took pathways that intersected in meaningful ways only infrequently. In ecology, humans were viewed as causes of change external to the systems of interest, or as sources of variation controlled for in experiments so that human influences could be ignored. In anthropological research, ecological settings have been explored and debated as a means to understand human evolution, societal development and power over resources (e.g., Orlove 1980; Watts 1997; Boyd and Richerson 2005). Concepts were mutually borrowed by each discipline from the other (e.g., evolution, niche theory, commons theory), but active integration was uncommon and sometimes even discouraged. The roles that humans play as components of systems became a focus in the second half of the last century, and queries with humans considered as a component of ecosystems were more common (e.g., Rappaport 1967; Liverman et al. 1998; Little and Leslie 1999) (see Sect. 9.3). However this systems view of humans did not allow for agency or diversity and was highly criticized in the social sciences (e.g., Moran 2008). Today, questions regarding sustainability are so broad in scope and outcomes so important to societies that scientific fields are being invented to address new questions about linkages within systems (see Part I of this volume). An example of institutional

R.B. Boone (✉)
Natural Resource Ecology Laboratory, Colorado State University,
1499 Campus Delivery, B234 Natural and Environmental Sciences Building,
Fort Collins, CO 80523-1499, USA

K.A. Galvin
Department of Anthropology, Colorado State University,
1787 Campus Delivery, Fort Collins CO 80523-1787, USA

M.J. Manfredo et al. (eds.), *Understanding Society and Natural Resources*,
DOI 10.1007/978-94-017-8959-2_9, © The Author(s) 2014

recognition of the importance of understanding these linkages is in the *Dynamics of Coupled Natural and Human Systems* competition in the US National Science Foundation. Each year millions of dollars are put to increasing our understanding of linkages between humans and the ecosystems they inhabit.

A main tool to understand linkages of societies and their environments is through the use of computer simulation. Simulation is a broad term, describing "… a class of symbolic models, which are representations of particular facets of reality …" (Galvin et al. 2006). Here we confine our discussion to the kinds of simulation models often used to represent coupled natural and human systems. These are a class of models called discrete-event simulations, where analysis steps are simulated to represent the passage of time, and events are scheduled to occur at particular points in time. Simulations are often processed-based, where processes describing interactions between system elements are described mathematically in computer code, or rule-based, where thresholds and logical bifurcations are described in code and represent decision making or other system attributes. Simulations often include stochastic components. The simulations may be point-based, meaning that they represent a single element of a system such as a plant or a person, and the results from that plant or person are taken to hold for other plants or people in the area considered homogeneous by the model. Alternatively, simulations may be spatially explicit, meaning they represent real-world locations where questions of sustainability are at issue.

Simulation approaches have been used in ecological research for decades (e.g., Huston et al. 1988) and more recently in the social sciences (e.g., Brenner 1999; Kohler and Gumerman 1999). Simulation methods are transforming social sciences by adding experimentation to the toolbox of researchers. Hypotheses that may be impractical to assess in reality because of expense, complexity, or moral constraints may be assessed using computer simulations. In what follows, we describe the utility of simulation in general terms, then specific to integration of social and ecological sciences. A pathway we and others use to discovery called integrated modeling is described. Agent-based modeling (ABM) is defined and its role in scientific integration is described. Examples from our work and from the literature are then given to provide context, and we conclude.

9.2   Utilities of Simulations

Constructing computer simulations requires that researchers make the interactions and assumptions that are implicit in mental models explicit (Epstein 2008). Primary processes to be included (e.g., primary production) must be distinguished from processes judged appropriate to ignore (perhaps groundwater contributions to primary production), rules must be defined, and parameter values that help describe how elements interact are identified. In a collaborative effort team members of different disciplines come together to share ideas and data. Each team member uses implicit mental models to understand how the system functions, but many people have never made those models explicit. Making processes, rules, and parameters explicit can be an illuminating, rewarding, and challenging exercise. For example, for a group to work well together requires a understanding of required terms and some baseline desire and ability to communicate to scholars in other disciplines. Reaching common understanding on what the most salient components of a system may be and how those will be represented in a simulation promotes team building across disciplines (Axelrod 2006).

We agree with Epstein (2008) that the assumption many make of models is that their goal is to make predictions. Predictions can be made but often the assumptions of such models are so simplifying as to have little purchase in the real world. Myriad interactions and unforeseen changes make detailed predictions about future system states all but impossible in all but trivial circumstances. Prediction is rarely the goal of our work. Instead, we often seek to identify the magnitude and direction of change that may be expected in a system, for example, given the changes a particular policy or land management decision may make on the environment and for human wellbeing. Other work by ourselves and others uses hypothetical landscapes, and tests theory without being encumbered by specific circumstances (Griffin 2006).

More generally, simulation can explain relationships, which is distinct from prediction (Epstein 2008). Alternative core dynamics may be incorporated in simulations, and those dynamics treated as hypotheses to be tested in experiments (Peck 2004; Grimm and Railback 2006). For example, the influence of topography on animal behavior may be quantified by using the observed topography in simulations, then substituting a flat landscape. Simulation can guide data collection, with sensitivity analyses (i.e., varying a parameter across its reasonable range of values and exploring changes in output) identifying new questions and uncertainties and allowing data collection efforts to be prioritized. Gaps in understanding can be suggested if an application that incorporates current theory is unable to generate the expected responses. Complex patterns can be shown to have simple underpinnings (e.g., the classic graphic of the Mandlebrot set used to demonstrate the nature of fractals) and simple patterns may be shown to be produced by relationships more complex than assumed (Epstein 2008). Simulation is helpful where analytical, differential equation-based approaches may become mathematically complex and intractable. Lastly, simulation is helpful when manipulations to real systems would be too costly, disruptive, or unethical (Peck 2004).

9.3 Integrated Modeling

Ecologists have developed in-depth knowledge about many elements in systems, although much remains to be learned. Prior to the 1980s, a majority of experiments on species interactions were on plots of 1 m^2 or less (Kareiva and Andersen 1988). New pathways of exploration and enabling technologies fostered a new type of ecological research exploring spatial scale and macroecology (Gaston and Blackburn 2000; Schneider 2001). But the pace of integrating and synthesizing information has been slow (Carpenter et al. 2009). In the 1960s to 1980s social science borrowed ecological terms and analyses such as energy flow studies, adaptation studies and the ecosystem concept and several important studies emerged (e.g., Vayda and McCay 1975; Thomas 1976). But increasing complexity was brought into these studies of human-environment interactions including landscape history (Crumley 1994), policy and power (e.g., Brosius 1997; Escobar 1998) and cultural meanings (Peet and Watts 1996; Berkes 1999). These were aided by new tools such as geographic information systems (GIS) and participatory mapping. New conceptual models that included micro-cultural processes like perception and macro-societal processes such as globalization at various scales were recognized as important elements of research in human-environment interactions (e.g., Liverman et al. 1998).

Subfields in ecology and social science disciplines are once again rapidly developing in large part due to advances in tools such as remote sensing, GIS and modeling. Other current impetuses are a growing human population, increasing stressors on landscapes from local to global scales, and a demand by the public that science address real-world, practical problems likely to have societal impacts. Sustainability science has emerged to address complex problems at the intersection of ecological and social science, with contributions from engineering, atmospheric, and medical sciences. Sustainability science goes beyond traditional hypothesis testing, and instead addresses real-world problems that "blend[s] theory and analysis with political awareness and policy concerns" (Galvin et al. 2006:159). Transdisciplinary teams of ecologists, anthropologists, and others come together to address questions of resilience, adaptive capacities that includes issues of inequality, class, gender and justice, and the sustainability of social-ecological systems (e.g., Folke et al. 2002; Berkes et al. 2003; Leach et al. 2012).

At the core of sustainability science endeavors to understand coupled systems are often computer models that are linked together in an integrated way. In general, the goal of this integration is to have the services ecosystems provide (MEA 2005) influence the behavior and conditions of people and societies, and in turn, to have human decisions and behaviors influence ecosystem services. Different models simulate different components of a coupled system, and many blueprints are used. For example, a hydrology model may be used to represent river flows, an ecosystem model simulates forest growth and carbon sequestration, and an agent-based model (see Sect. 9.4) may represent timber harvesters (see Sect. 9.5 for examples from our work). Often these are well-established models that have been used in discovery for years. New is the effort to link these models together to create an integrated system that includes both humans and the environment. Team members think deeply about their own fields and the simulation tools that each uses, and consider the points of connection between fields. In the example, primary connections may include the harvest of timber, economic benefits from harvest, and increased water runoff from harvested hillsides. The team identifies secondary and tertiary connections as well, perhaps including temperature changes in streams or changes in microclimates (Beschta et al. 1987), and decides what is to be included in connections between models, and excluded. The models are then linked either loosely or tightly (Galvin et al. 2006), a continuum of connectedness depending on the models being used and the questions being asked. Methods used in linking models and other considerations are beyond our scope here [see An and López-Carr (2011) and other entries in that special issue for an introduction]. The goal is to create an integrated set of computer simulation models that support assessment of future conditions under different decision pathways. Generally, even with quite simple constituent models, coupled systems models provide sufficient so-called levers and other controls to address a variety of scenarios.

9.3.1   Ecological and Social Models

More than a decade of working with integrating models has given us the impression that ecosystem models are more advanced than social system models for understanding the system, though there have been decades of social research into economic behavior, evolutionary behavior and market behavior (e.g., Cangelosi and Parisi 2001; Camerer et al. 2003). Four reasons for this occur to us, although these reasons are inter-related. First, despite the great distance between ecological theory and physics, ecosystems as subjects of natural science include more components that can be modeled through processes rather than using the rule-based approach most often adopted in societal models. An example we use in teaching comes to mind – image tossing 100 chickens into the air along some compass bearing, and mapping their landing places. If those chickens were dressed and frozen, one could predict their landing places quite accurately. But if those same chickens were alive, prediction would be all but impossible. Instead we must be content with describing a mean landing place and some deviation around that. Adding the freewill of the animal makes all the difference. Nobel laureate Richard Feynman put it more succinctly when he said "Imagine how much harder physics would be if electrons had feelings!" In short, societal models are replete with behaviors influenced by the free will of agents. This leaves models of social dynamics more likely to include a rule-based approach.

Second, ecosystems are more self-similar than human communities. Technically, the spatial autocorrelation in ecosystems is higher than in human neighborhoods. Imagine two forest patches separated by perhaps 10 km. What we know about the first patch is likely to apply, to some large degree, to the second patch. Now image a neighborhood of a few city blocks. What we may know about one family is likely not applicable to a family a few blocks away. Indeed, families living as neighbors may be quite distinct from each other. Our third point is closely related to our second. Ecologists are more comfortable than anthropologists at treating their subjects as similar units. Ecologists consider a herd of animals and emphasize the similarities. Anthropologists consider a group of people and see variability. The variability that is inherent within human populations has implications for power, poverty, inequality, development and ultimately sustainability of the social-ecological system. This is not to say that generalizations are not sought after; they are, but differences have real implications for people and the environment.

The fourth reason that comes to mind may be an outcome of the reasons already cited. There has been a greater embrace of simulation modeling in ecosystem science than in social sciences, and for a much longer time. By the 1980s, manuscripts were published that summarized the strides made in simulation modeling in ecology (e.g., Huston et al. 1988), whereas international meetings on social simulation only began this century, and the community pursuing social simulation is much smaller.

9.3.2 Integrated Modeling with Stakeholders

A utility of simulation we will highlight because of its role in our work is its ability to facilitate incorporating stakeholder and local knowledge into a research and modeling effort. Residents of the systems of interest may be interviewed or join in focal groups or meetings to share information. That information may then be used to parameterize integrated models. A suite of methods are now used that formalize gathering of local knowledge, including participatory role games, participatory geographic information science and participatory simulation. When creating rules that describe decision making by people managing or competing for resources, model builders or facilitators gather stakeholders and have them play games that help inform researchers about the decision making process stakeholders use in an area (e.g., Janssen and Ostrom 2006). For example, participants may be asked to role-play as managers of a local fishery, and the decisions made by participants may be emulated in rules used in simulation. Participatory GIS entails meetings with residents to discuss the spatial location and timing of events, the location of entities, their spatial attributes, decision making in using spatial resources, etc. (e.g., Talen 2000). Landscape representations used in these efforts may be on computer screens, printed paper maps, topographic paper mâché models, or simple sketches in the sand. In participatory modeling (e.g., Becu et al. 2008), stakeholders are involved directly in developing computer simulations, either in the model used in discovery or in some simplified version. This work can be challenging, ensuring that concepts such as simulation and scenario analyses are conveyed to participants well, but can be effective. Reid et al. (2009) emphasizes that work intended to involve, educate, and empower local people should encourage frequent sharing of information.

An undervalued use of the products from integrated assessments that we have seen is that they can provide a common starting point in discussions. Meetings between stakeholders can be more focused by demonstrating simulation results. Results from "what if" questions they had proposed are presented and discussed. Participants then have a common starting point, or even a common antagonist, from which to build discussions. Models sometimes influence decision making – often not to the degree that modelers would prefer – but at least relationships that may be forgotten about or ignored in causal considerations are included in discussions that have as their starting point results from comprehensive computer simulations.

9.4 Agent-Based Modeling

Agent-based modeling is distinct from statistical analysis and interpretation of data, equation-based analytical approaches to discovery, and numerical modeling of system dynamics. In analytical approaches, formulas are used that represent changes in system states. An example is Mathusian population change through time: $P_t = P_0 e^{rt}$, where P_0 is an initial population size, r is a growth rate, t is time, and P_t is the population at that time (see Turchin 2001 for a more complete treatment). An evident feature of such an approach is that the entity being modeled is the entire population. The same is true for numerical modeling of systems (e.g., Stella is a popular example of software used in such work; IEEE systems, Lebanon, New Hampshire, US), although they adopt a simulation approach. Populations known as stocks are modified at rates set by model designers, and during simulations portions of populations flow into other stocks. Analytical approaches take a top-down approach to discovery, with the structure of the population represented by hypotheses the analyst has incorporated into equations (Grimm 1999).

Agent-based modeling uses a bottom-up approach to discovery (Grimm 1999) that recognizes that individuals are the basic units of decision making, and determine (at least in part) population responses when aggregated. Simulations are composed of autonomous agents that interact with other agents and the environment according to rules (e.g., Billari et al. 2006). Simulations are made, allowing agents to interact, and emergent responses are sought (i.e., formally, an aggregate response not obvious from the constituent parts; less formally, something that is unforeseen, unpredictable, and interesting). Rather than employing deductive or inductive approaches, an abductive approach is used to explore reasonable hypotheses that may explain a suite of observations (e.g., Griffin 2006; Lorenz 2009). These ideas give context to Epstein's quote describing generative social science and the usefulness of agent-based simulation, "if you didn't grow it, you didn't explain its emergence" (Epstein 1999:43). Being able to grow a pattern of interest and visualize the interactions of agents on a computer screen provides supporting evidence that the rules of interaction embedded in the model are good candidates to represent real interactions.

The rules describing interactions are hypotheses, and analysts use methods such as scenario analyses, a structured form of *in silico* experimentation (Peck 2004) in discovery. Different rules may be enabled or disabled in simulations representing different hypotheses of interaction, and the emergent patterns compared to observed patterns to judge the suitability of the competing hypotheses (Grimm and Railback 2006). Alternatively, scenarios may be used to address "what if" questions, where rules on assessed simulation models are varied to represent future conditions, different responses, or changes in policy or management (e.g., Boone et al. 2011).

The bottom-up approach of ABM that focuses on simulations of individuals has several benefits and costs when compared to top-down analytical approaches. (The following dichotomies refer to typical methods, and advanced methods can blur distinctions between the approaches.) Mathematical models such as the Mathusian population equation given yield precise solutions very quickly, and can be efficiently implemented, often with little data. In contrast, ABM models include stochastic components and the simulation of time passing that may slow the generation of results, and coding the models may be rapid or take significant time. Some ABMs use a great deal of data, but it is a misconception that the method requires more data than analytical approaches; some well-known and influential models require just one or a few parameters, such as Schelling's (1971) model of dynamic segregation, Axelrod's (1984) competitions using Prisoner's Dilemma, and Reynolds (1987) modeling of flocking using Boids, and all that it has spawned (see Macy and Willer 2002 for a review). Analytical models are highly stylized, to prevent problems from becoming mathematically intractable, whereas ABMs may be highly stylized or realistic. In contrast to the single scale of population models, results from ABM analyses are inherently multi-scale, in that analyses may be reported at the scale of agents, or any aggregate of interest. For example, some grand simulation of individuals across a broad region may report results for those individuals, summaries of households composed of those individuals, summarized by village, or for the entire population included in the simulation. The realistic approach used in many ABMs allows for a variety of scenarios to be addressed.

The top-down, population-based analytical approach implies that population members are identical and static. Treating each individual as identical can be a severe limitation in analytical approaches. Consider the simple biological example of forest stand growth from seedlings. Treating each seedling the same as the rest implies that through time the seedlings will mature at the same pace and a uniform forest will grow. In practice this is not the case. Variations between individuals cause some trees to shade others, and to grow more rapidly, yielding a more realistic result of a forest with diverse size classes. In ABM, variation between individuals is easily incorporated, and interactions between individuals with different initial attributes, such as body masses or livestock holdings, can yield more realistic results (Huston et al. 1988). This is why some social scientists have embraced the ABM approach, as some of the variability seen in the real world can be captured in these types of models. Examples where variation between individuals may be important are numerous in social settings, such as in economics, land use and tenure, altruism, and risk analyses. Here we have focused upon variation intrinsic to individuals as they are initialized, but in an ABM, agents that may be initialized identically each has its own experiences that yields differences in individuals, making results path dependent.

In contrast to static members of populations in analytical approaches, ABM can represent adaptation, learning, and evolution in agents. This makes ABM well suited to represent the complex and adaptive coupled systems that are a focus of current sustainability research. The rules used to control decision making may reflect adaptation. For example, in our work livestock owners may move their animals to more distant areas in drought conditions. Local interactions between agents make learning from neighbors straightforward, and imparting agents with different forms of memory is possible. Given a group of neighbors contacted by a given agent, that agent may ask if any member of the group is doing better in some objective way than the agent, and if so, adopt the practices of the neighbor. For example, a farmer in a valley may observe harvests by her neighbors, and if one of their harvests exceeds her own, adopt those cropping practices (see the example below of Lansing and Kremer 1993). Genetic algorithms, evolutionary programming, and other evolutionary computation techniques evolve adaptations using a framework adopted from biology. Agents with a given set of attributes perform better on some objective function than others in the group, and produce offspring that have related but mutated sets of attributes. Through selection, agents evolve to be well adapted to local conditions. Biological examples include Boone et al. (2006) and Boone (2010). Human evolutionary examples are given in Barton et al. (2011) and Barton and Riel-Salvatore (2012).

An important use of ABMs is in communication with audiences. Imagine a traffic modeler explaining to a lay audience about the number of vehicles that may be supported on a given road. The presenter may show a formula that describes maximum traffic flow, $q*$ (from Malone et al. 2001) : $q* = (\beta + 2\gamma^{1/2}L^{1/2})^{-1}$, where β is the reaction time of drivers, γ is the reciprocal of twice the maximum average deceleration of a following vehicle, and L is vehicle length, although the details are not relevant here. Alternatively, the presenter may show output from an ABM, where vehicles are moving along the road in reasonable ways, jams develop and clear, and vehicle densities may be reported directly (Fig. 9.1). Efforts have different purposes, of course, but in general, audiences identify with the visual nature of ABM output. People identify with agents in models and readily anthropomorphize; they themselves are individuals experienced in interactions with other individuals and environments.

Perhaps most important is the ability of agent-based modeling to integrate disciplines (Axelrod 2006). Almost by its nature, simulating complex system attributes involves interdisciplinary teams. Such a team may include a hydrologist, ecologists specializing in primary and secondary production, anthropologists and an economist, plus programmers and other specialists in technology. As the team creates rules that define interactions between the agents and the ways they interact with the environment, team members must break down their high level understanding of the causes of behaviors into something that may be represented logically. The rules must be conveyed to the technical team members with sufficient mutual understanding to allow them to be represented in computer code. The team must develop a common language and understanding. But beyond that, "[t]he creation of a model forces the articulation of any number of individual design decisions, and thoughtfully done, each can be a starting point for new understanding." (Johnston et al. 2007:82). Identifying commonalities between disciplines is particularly rewarding (Axelrod 2006). Some bodies of theory are applicable to diverse fields, such as theories that touch upon acquiring resources (e.g., with ties to economics, anthropology, animal foraging), altruism and cooperation (e.g., ethology, interpersonal relationships), issues of carrying capacity (e.g., grazers, hunters, members in markets), Tragedy of the Commons (Hardin 1968) or lack thereof (e.g., grazing dynamics, ocean fisheries management, group dynamics). Discussing these commonalities can strengthen a team.

In summary, agent-based modeling is a useful approach when: interactions between individuals and the environment are a focus; a model is complex with many interactions; non-linear relationships are important; variability between individuals

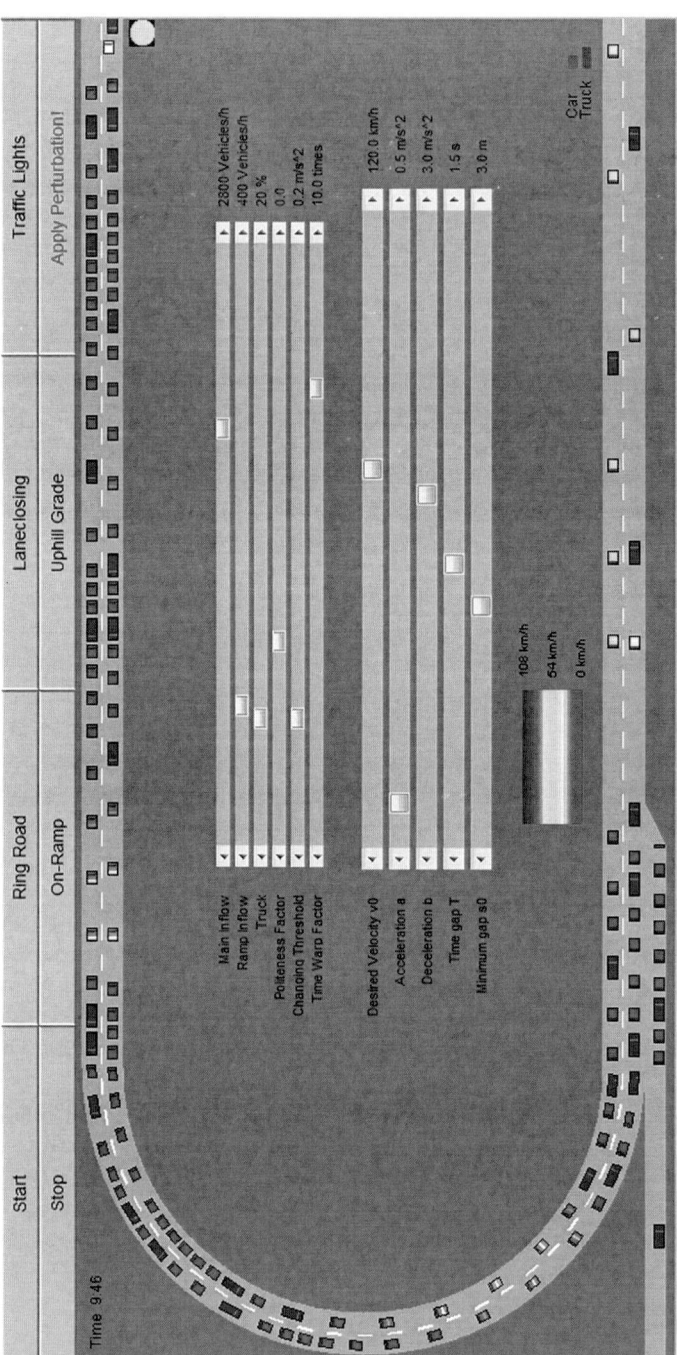

Fig. 9.1 Microsimulation of traffic flow, an instance of an online applet simulation tool by M. Treiber (2011). The work is introduced in Treiber and Kesting (2010)

is of interest; randomness and path dependency are relevant; restrictive assumptions are to be minimized; and visualization and understanding by stakeholders is an objective. We have provided a high-level overview of the utility of agent-based modeling. Some aspects of constructing agent-based models have been cited, but more detailed reviews of techniques (e.g., Kraus 1997; Gilbert and Terna 2000; Berry et al. 2002; Bonabeau 2002; Ramanath and Gilbert 2004; Goldstone and Janssen 2005; Grimm et al. 2005; Janssen and Ostrom 2006; Aumann 2007; Railsback and Grimm 2011), tools (Wilensky 2001; Gilbert and Bankes 2002; Railsback et al. 2006; Nikolai and Gregory 2009; Railsback and Grimm 2011), assessment pathways, a critically important aspect of simulation (Grimm et al. 2005; Wilensky and Rand 2007; Gilbert 2008), and comments (Bankes 2002; Richiardi et al. 2006) are available to those constructing models. Example models and other resources are available from the Network for Computational Modeling for SocioEcological Science (CoMSES Net, at http://www.openabm.org/).

9.5 Examples

Our examples are selected to highlight different aspects of integrating social and ecological information, using examples from our work where applicable and a classic example in agent-based modeling that demonstrates several aspects of interest here.

9.5.1 Integrated Assessments with SAVANNA and DECUMA

The ecosystem modeling tool we have used the longest in integrated assessments is SAVANNA. Indeed, M. Coughenour of Colorado State University began developing SAVANNA while working on one of the first large-scale projects to consider humans and their environment in an integrated way, the South Turkana Ecosystem Project of the 1980s (Ellis et al. 1993; Little and Leslie 1999). In the late 1990s, the SAVANNA model was the integrative tool we used to bring together data collection and analysis efforts that included anthropological surveys, ecological field sampling, literature review, and spatial data (Galvin et al. 2006). SAVANNA is a spatially-explicit, process-based model of ecosystem change. Landscapes are divided into a series of square cells, with geographic data layers informing the model of cell attributes. The model represents plant functional groups, such as palatable grasses, dwarf unpalatable shrubs, or acacia trees. Plant functional groups compete for water, nutrients, light, and space, based on cell attributes such as soil type, weather data that includes temperature and precipitation, plus a suite of parameter files that describe or control plant growth and competition. At each time-step, plants may produce seed, germinate, grow, outcompete other functional groups and gain ground cover, or be outcompeted by other plants and die. Wildlife and livestock are represented in the model as populations that feed on the plants to gain energy, and expend energy through basal metabolism, movement, thermal regulation, plus reproduction including gestation and lactation. Net energy increase goes to weight gain, and an energy deficit to weight loss. Body weight is compared to an expected standard to yield a condition index that affects birth, death, and other vital rates. The model uses a weekly time-step, where the state of the system is simulated once each week, and produces spatial and temporal output once per month.

The use of a comprehensive model such as SAVANNA is time intensive. That said, the trade-off is its great flexibility in addressing scenarios. We used SAVANNA in scenario analyses to address 15 management options available to the conservators of Ngorongoro Conservation Area (Boone et al. 2002). We explored effects of drought, changes in livestock access and stocking, the expansion of cultivation and its effect on the system, changes in veterinary care, changes in water supplies, and human population growth. In general these scenarios were represented by making relatively simple changes to the files used by SAVANNA. For example, to represent changes in water supply that affect livestock and wildlife distribution, water sources were added or removed in a GIS and distance-to-water surfaces used in the model were recalculated. To represent improved veterinary practices, the survival of livestock was increased slightly. We have used the model to help extend the utility of climate forecasts to South African livestock owners (Boone et al. 2004), by itself and as linked to a mathematical linear programming model that provided measures of economic benefit to livestock owners (Thornton et al. 2004). In recent years we have used the integrated system to explore effects of fragmentation on livestock, wildlife, and people (e.g., Boone et al. 2005; Boone 2007; Hobbs et al. 2008).

Early in our work, P. Thornton of the International Livestock Research Institute, Nairobi, led an effort to extend our integrated modeling of areas to include the livestock owners and their households. He created the PHEWS model (Pastoral Household Economic Welfare Simulator) as a population-based representation of Maasai households, which is tightly joined to the SAVANNA model (Thornton et al. 2003, 2006). That model represented decision making by household owners using a series of ordered rules, applied to a modest number (9–24) of groups of households, such as poor, medium, and rich business owners with livestock. The flows of food energy and currency were tracked in the model.

The population-based nature of PHEWS prevented us from simulating individual households who own their own livestock herds. It follows that we could not have local ecosystem services influence the decision making of household owners – local conditions cannot be defined for populations of hundreds of households. We converted the PHEWS model into an agent-based representation called DECUMA (DECision-making Under Conditions of Uncertainty for Modeled Agents, and also the name of a Roman fate that influences the length of life). In DECUMA, individual households are represented as occurring at specific locations on earth, and they own specific livestock herds. When linked to SAVANNA, that allows us to have local ecosystem services influence the decision making of pastoral people, and to have their decisions influence ecosystem services. Boone et al. (2011) describes the DECUMA model and linkage with SAVANNA in detail.

DECUMA has been applied in Kajiado District, southwest Kenya, and applications are ongoing in Samburu, Kenya, as well as Mali and Tibet. The Malian application provides an example of the usefulness of making tools used in integration portable. In that work, led by N. Hanan of South Dakota State University, we are exploring changes in the hydrology of lakes, the roles that pastoral people have had in those changes, and the benefits to them. A hydrological model (SWAT; Gassman et al. 2007) is being linked to the ecosystem model called ACE (African Carbon Exchange), which in turn is being linked to DECUMA. By programmatically isolating the materials DECUMA requires from an ecosystem model (see Boone et al. 2011 for details), we can relatively easily link the model with any ecosystem simulation tool that can provide the needed information (e.g., forage availability and forage acquired by animals).

Our ongoing analyses in Samburu, Kenya demonstrates this kind of integrated modeling. C. Lesorogol of Washington University, St. Louis, Missouri has gathered in-depth anthropological data for two study sites in southwest Samburu District, Mbaringon and Siambu. The sites differ in ecological settings, with Mbaringon at lower elevation and with less rainfall, for example. But the main difference of anthropological interest is that Siambu is subdivided, and Mbaringon remains communal lands (although somewhat fragmented). In the 1970s residents within some districts in Kenya began to subdivide into individually owned parcels. In Siambu, the land was divided into 240 small individually owned parcels. We are also investigating changes in Samburu norms, where the sense of reciprocity and sharing is less important in young peoples' lives.

Our integrated assessments are driven by both theoretical questions and by questions put forth by stakeholders (Reid et al. 2009). The eight scenarios (numbered below) we are addressing in Samburu reflect this, and highlight the flexibility of using comprehensive simulation tools such as SAVANNA and DECUMA. Central to our work are questions of subdivision and its effects. We are simulating sedentarizing people and their animals on individually owned parcels, and the effects of that on livelihoods (1). Another scenario asks about the influence of commercial cropping in Siambu and fence building in Mbaringon, and the effects of loss of access has on livestock (2). These types of scenarios are represented in the modeling system by altering spatial surfaces or agent behaviors so as to prevent animals from leaving home parcels or from using areas that are inaccessible. We describe a diversity of scenarios to demonstrate the utility of integrated modeling but discuss one (number 8) in more detail here.

Both Siambu and Mbaringon are grazing refuges for herders outside those areas. When drought conditions hold in other areas of Samburu, herders move their animals into these areas. In a scenario, we are adding additional livestock to each area, and summarizing effects on the resident animals (3). Plains zebras (*Equus quagga*) and occasionally Grevy's zebras (*Equus grevyi*) are joined by various antelopes in the Mbaringon study site. We will vary the numbers of wildlife by a factor of four in scenarios, with and without tourism benefits to local people, to judge effects on livestock numbers and household livelihoods (4). Livestock sales are increased in simulations, above the observed number of sales that is typical (5). This is an example of a scenario that Samburu residents asked us to include, given their interest in intensifying their livestock management. A program is in place now, led by C. Lesorogol, to introduce enhanced and highly productive goat breeds into Siambu. In a scenario, attributes of goats are modified to represent mixed herds of local and enhanced breeds, and effects on livelihoods judged (6). Residents asked us to explore the implications of improved crop yields on livelihoods (7). This scenario addresses a variety of management options, streamlined so as to be amenable to simulation with our tools. For example, land owners struggle to decide whether increased production from high-yield seed stocks outweigh their increased costs, whether to invest in chemical fertilizers (very few do in the region), investing in water projects for irrigation, and the usefulness versus costs of drought tolerant seeds.

The last scenario (8) compares the costs and benefits of increasing or decreasing veterinary care for livestock. Residents seek to balance the money they spend on veterinary care with the benefits they receive through improved livestock health and survival. We used an application of SAVANNA and DECUMA to Mbaringon to address this scenario. These preliminary results report outcomes from three simulations (baseline, increased livestock survival by 3 %, decreased by 3 %); in practice we do 20 or more simulations of each type to yield error estimates. Such a seemingly small change in survival for large herbivores can have dramatic impacts on population dynamics. In this example, Fig. 9.2 (top) shows about a 1 tropical livestock unit (TLUs) increase for each adult equivalent (AE). These metrics are methods of standardizing livestock of different species (e.g., a cattle is 1 TLU, and a sheep or goat is 0.1 TLU) and humans of different ages and sexes (e.g., an adult male is 1 AE, and a child 6–12 years old is 0.85 AE) (Boone et al. 2011). Figure 9.2 (bottom) demonstrates one of the many linkages within the SAVANNA-DECUMA integrated system. The average proportion of households' diets composed of supplemental food decreases when more funds are spent on veterinary care. This must be weighed against the costs of the improved care.

9.5.2   Balinese Water Temple Networks

We draw on a time-honored agent-based simulation study to demonstrate two aspects helpful to understanding how integrated modeling may address resilience, simulating adaptation and network modeling. Lansing and Kremer (1993) describe an agricultural system in Bali that is dependent upon irrigation fed by rivers carrying rainfall runoff. Blocks of terraces are planted in rice. A tension with three main dimensions exists within the system involving a balance between yield, water use, and pest damage. Individuals seek to maximize yields, but if everyone plants each year, there would likely be insufficient precipitation to irrigate. Also, if farmers planted each block, pest populations can expand and severely reduce yields. In the 1970s, crop management was disorganized, all blocks were planted, plants were likely water-stressed, and pests reduced yields by up to 50 %, far greater losses than seen in the 1990s. Farmers are organized into groups called *subaks*, which

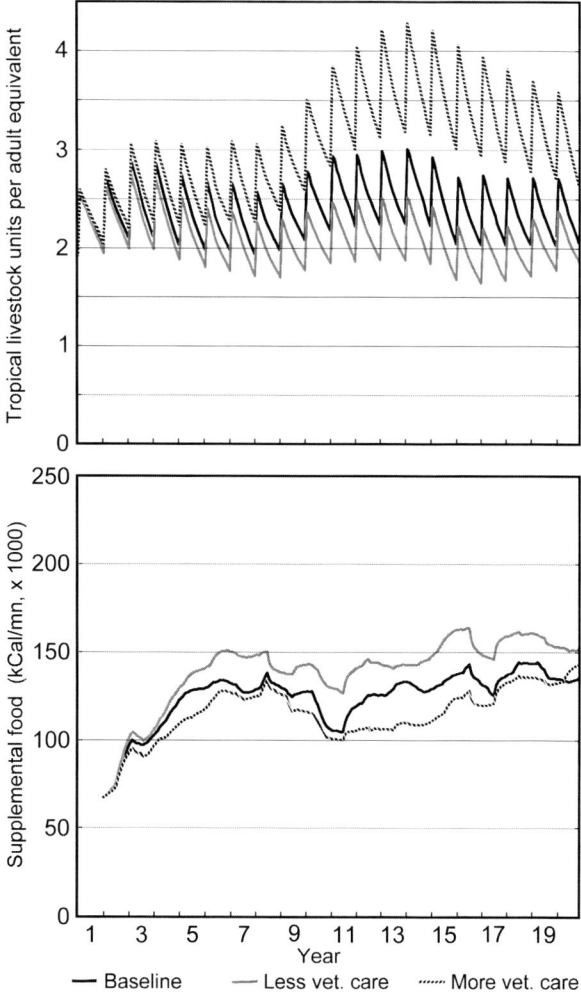

Fig. 9.2 Effects based on preliminary results of increasing or decreasing veterinary care on livestock in Mbaringon, Samburu District, Kenya. The scenario included increasing or decreasing livestock survival by 3 % reflecting changes in care. Increasing veterinary care increased numbers of livestock per person (*top*, *hashed line*) and reduced the need for supplemental foods (*bottom*, *hashed line*)

coordinate to balance whether blocks are planted or in fallow. An intricate network of temples and shrines is present that promotes coordination. Whether crop yield would be highest if coordinated at very local scales, at the scale of the entire region, or at a scale similar to that of the temples was of interest to Lansing and Kremer (1993). Lansing and Kremer (1993 and cites therein) built an agent-based model of 172 subaks, with each containing information about the basin in which it occurs, where water is drawn from, and other information. The spatial connectedness of subaks was represented as a network, such that neighbors were aware of the management and crop performance of their neighbors. The simulation included estimates of rainfall, irrigation demand, rice growth stage, and pest load. Harvests varied in response to water stress and pest load. In early simulations, the authors found that subak-scale coordination reduced yields due to pests. Coordination across the region caused plants to be water-stressed. The coordination that was best was at a scale on par with that of temples and shrines that are located within subaks. But the authors continued to question how temple networks would form, and how they may coordinate cropping. Lansing and Kremer (1993) incorporated a learning component into their network. Subaks residents looked to neighboring subaks, and if the management system used produced greater yields, it was adopted. After initializing crop management to randomly selected systems, within a decade of simulated learning crop yields almost doubled. This clear example of simulating adaptation, and specifically irrigation within the Bali Temples system, has spawned expanded analyses (e.g., Janssen 2007; Lansing et al. 2009).

9.5.3   Wet Season Versus Dry Season Livestock Dispersal

An example of the utility of a theoretical or stylized simulation relates to changes in livestock dispersal patterns by Maasai herders in Kajiado, Kenya. Forty years ago, herders moved their animals in a pattern echoing the movements of wildlife of the Amboseli Basin (Worden 2007). In the wet season, livestock were grazed broadly, using ephemeral water sources and eating forage distant from permanent water. As water became limiting, livestock herders moved their animals closer to permanent water sources. During the dry season, herds grazed areas around the permanent water sources. This pattern may be summarized as wet season dispersal of livestock.

It is reasonable to think that animals confined to a relatively small area at the height of resource shortages – the dry season – may reduce forage acquisition and livestock populations. In the 1980s, Maasai adopted a cultural and institutional system where elders imposed a wet season dispersal pattern for livestock (BurnSilver 2007). Herders graze their animals in the areas around permanent water sources (which tend to be near their permanent residences) during the wet season. Areas distant from permanent water are kept as grazing reserves. As the landscape dries and forage is depleted, elders open neighboring areas to grazing, in what is termed a "staged" approach (BurnSilver 2007). Herders graze animals there until forage is depleted, and another stage is opened. By the height of the dry season, higher elevation grazing reserves are being used, and animals are being grazed for 2–3 days, then walked back to permanent water to drink. They then return to the reserve to graze for 2–3 days, and the cycle repeats until new rains arrive.

We sought to assess the utility of wet season versus dry season dispersal for livestock, and chose a stylized ecosystem representation in NetLogo 5.0 (Northwestern University, Evanston, Illinois). Major aspects of the model are introduced here, with minor points omitted for brevity; the full model has been placed in the Community section of the NetLogo web site (http://ccl.northwestern.edu/netlogo/). A grassland was represented by a grid of patches 41×41, with each pixel approximating 1 km². The grid represented a torus to avoid edge effects, such that an animal moving off one edge of the grid appeared on the opposite edge. In each landscape cell, we implemented a forage growth model following Fryxell et al. (2005). That source provides formula that link stochastic precipitation with rates of growth of grasses. Rainfall included a 25 % inter-annual coefficient of variation, with precipitation distributed evenly throughout the growing season. A modified logistic growth curve represented forage production through time based on precipitation. Animals gained weight from the forage they ate, lost weight if there was insufficient forage, and reproduced at a rate related to their body condition.

A variable number of wells were distributed randomly throughout the grassland simulated. Animals were compelled to return to wells to drink every 3 days; if a measure of thirst for an animal exceeded a threshold (i.e., 7.5 days in this stylized simulations), the animal died. A switch on the simulation interface (Fig. 9.3) allowed the model to adopt a dry season or a wet season dispersal. We simulated dry season and wet season dispersal of livestock, and varied the number of wells on the landscape from 1 to 10. For each combination, we used 30 simulations to yield standard error estimates. Based on these stylized simulations, the utility of the newer dry season dispersal pattern to pastoralists may be questioned (Fig. 9.4). The usefulness of storing vegetation in areas distant from water sources for use in the dry season is outweighed by the ability of animals to graze more freely during the wet season. These results are not definitive, given the stylized application used, but they do suggest that more detailed follow-up analyses would be helpful and that field data be collected.

9.6 Summary and Conclusions

Models by their nature are incomplete representations of the realities they seek to describe. Some are caricatures of reality, some seek to emulate real-world higher-level patterns, and some seek quantitative agreement with patterns through space and time (Axtell and Epstein 1994). The correctness and utility of a model should be considered in the context of its purpose. A purpose becoming more common is to represent ecological and social systems, and the linkages in between. In general, ecological modeling is more advanced than modeling social systems – there are many opportunities for advancing knowledge and methods in social simulation. Agent-based modeling has been useful in representing human decision making. The method is highly flexible, and able to incorporate individual variation and path dependencies. If-then structures and the parameters used in them become hypotheses that may be tested, in direct analogy to field experiments. Linked ecosystem and agent-based models allow changes in ecosystem services to be reflected in the behaviors that people exhibit. In turn, the behaviors that people make can alter the services an ecosystem provides (Bonabeau 2002). For example, changes in forage availability may be simulated in an ecosystem model, which influences the ways in which people distribute their livestock, which in turn affects forage availability in later periods. In an example we demonstrate an integrated model of two areas in Samburu, Kenya. That is an example where we seek to be in quantitative

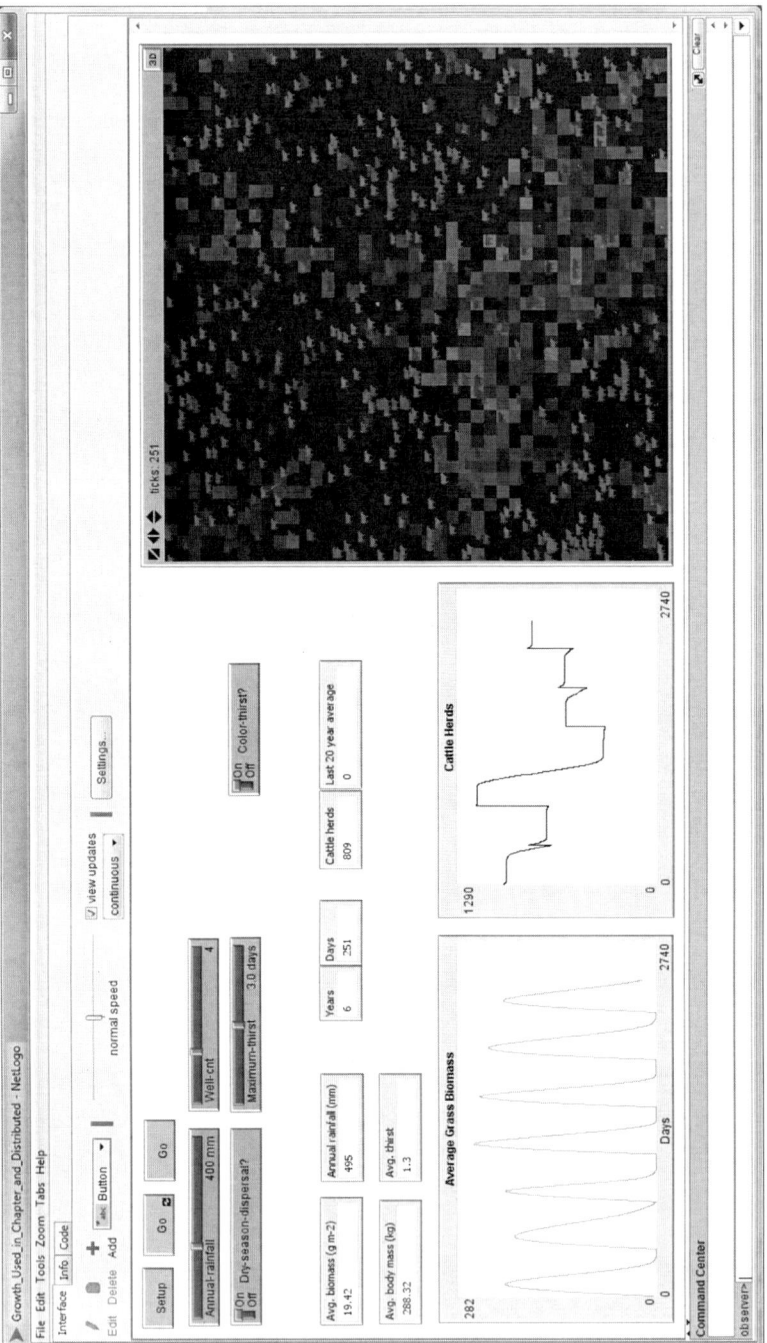

Fig. 9.3 The NetLogo interface for a simulation exploring the usefulness of dry season versus wet season dispersal for Maasai pastoralists and their animals

Fig. 9.4 The number of cattle herds that may be supported on a stylized landscape with different numbers of randomly distributed water wells, and either a pattern of wet season dispersal during grazing (*solid line*) or dry season dispersal (*dotted line*). Standard error bars are shown

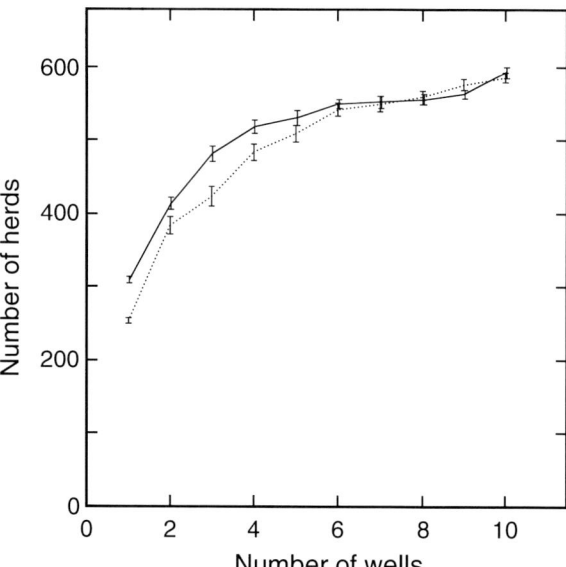

agreement with the observed system. It also demonstrates the flexibility (and challenge) that comes with using comprehensive tools. An example from the literature depicts agents learning and adapting, and the means in which agents may form networks where information is shared. Lastly, we demonstrate a stylized simulation of livestock dispersal patterns in grazing lands of Kenya.

Our emphasis on agent-based modeling should not suggest that its use is the only way to integrate across the natural and social sciences. As always, the questions to be addressed dictate the approach and tools to be used. Some other useful pathways have been cited here, such as participatory mapping and other participatory methods, which allow for the inclusion of indigenous knowledge in research. Other more mainstream means of modeling may be used in integration, such as empirical methods or systems modeling approaches. Spatial analyses using geographic information systems help bridge social and natural sciences, for example by providing geographic context to household survey results (e.g., Boone et al. 2000). Remote sensing allows effects of human activities to be placed in the context of broad spatial scales, with sampling through space and time and without undo expense. Land use change quantified using remotely sensed data is now a well-developed field. Planning and the use of scenario analyses allow interdisciplinary teams to integrate aspects of their work. For example, questions about changes in social systems may be framed by scenarios regarding changing climate or changes in the services that an ecosystem provides.

Though tools and pathways of integration are increasing, the ability to integrate across the sciences it is not without challenges. There remain issues of scale, including mismatched social-organizational scales, such as comparing administrative boundaries with landscape scales in which ecological flows such as water, wildlife and soil nutrients occur. There are scales of drivers and impacts that go between the local, regional, national, and to the global scale, such as climate change, land use and other policies, market influences and others that are difficult and sometimes intractable. There are also challenges associated with human populations that are important but sometimes difficult to include in integration efforts and include information on equity, gender, justice, class, ethnicity, power and history. These are important because they determine winner and losers of social ecological inquiry and they are central to realizing change in practice on the ground.

There are factors that help to integrate social sciences and natural resources. These include new theories and methods. Theories of political ecology (cf Robbins 2012), resilience and social-ecological systems (cf Folke 2006), and common property (cf Ostrom 2002) as examples can help us ask multi-scale questions, incorporate socially and culturally structured relations into the research (such as gendered decision-making roles or ethnicity/class and degraded or resource poor landscapes) and iteratively linked human decisions to environmental outcomes and vice versa. This inclusion of complexity calls for mixed methods; we are no longer tied to mainstream disciplinary methods but rather a set of mixed methods may be used to answer the problems at hand. These include Photovoice, videography, qualitative unstructured and semi-structured interviews, focus groups, workshops, participatory modeling, formal surveys and social network analysis. By coupling these methods with the ecological and geographical methods and tools mentioned above, including agent based modeling, we can continue to develop solutions to timely and important societal and environmental problems.

Acknowledgements Our thanks to our many colleagues who have participated in the research examples we have cited. Support for preparing this chapter was provided by US National Science Foundation grants DEB-1010465, BCS-0822752, and DEB-0919383.

References

An, L., & López-Carr, D. (2011). Understanding human decisions in coupled natural and human systems. *Ecological Modelling, 229*, 1–4.

Aumann, C. A. (2007). A methodology for developing simulation models of complex systems. *Ecological Modelling, 202*, 385–396.

Axelrod, R. (1984). *The evolution of cooperation*. New York: Basic Books.

Axelrod, R. (2006). Agent-based modeling as a bridge between disciplines. In L. Tesfatsion & K. L. Judd (Eds.), *Handbook of computational economics* (Agent-based computational economics, Vol. 2, pp. 1565–1584). Amsterdam: Elsevier.

Axtell, R. L., & Epstein, J. M. (1994). Agent-based modeling: Understanding our creations. *The Bulletin of the Santa Fe Institute, 9*(2), 28–32.

Bankes, S. C. (2002). Agent-based modeling: A revolution? *Proceedings of the National Academy of Science, 99*, 7199–7200.

Barton, C. M., & Riel-Salvatore, J. (2012). Agents of change: modeling biocultural evolution in upper Pleistocene western Eurasia. *Advances in Complex Systems, 15*(1&2), 1150003.

Barton, C. M., Riel-Salvatore, J., Anderies, J. M., & Popescu, G. (2011). Modeling human ecodynamics and biocultural interactions in the Late Pleistocene of western Eurasia. *Human Ecology, 39*, 705–725.

Becu, N., Neef, A., Schreinemachers, P., & Sangkapitux, C. (2008). Participatory computer simulation to support collective decision-making: Potential and limits of stakeholder involvement. *Land Use Policy, 25*, 498–509.

Berkes, F. (1999). *Sacred ecology*. New York: Taylor and Francis.

Berkes, F., Colding, J., & Folke, C. (2003). *Navigating social-ecological systems. Building resilience for complexity and change*. Cambridge, MA: Cambridge University Press.

Berry, B. J., Kiel, L. D., & Elliott, E. (2002). Adaptive agents, intelligence, and emergent human organization: capturing complexity through agent-based modeling. *Proceedings of the National Academy of Science, 99*, 7187–7188.

Beschta, R. L., Bilby, R. E., Brown, G. W., Holtby, L. B., & Hofstra, T. D. (1987). Stream temperature and aquatic habitat: Fisheries and forestry interactions. In E. O. Salo & T. W. Cundy (Eds.), *Streamside management: Forestry and fishery interactions* (pp. 191–232). Seattle: University of Washington, Institute of Forest Resources, Contribution No. 57.

Billari, F. C., Fent, T., Prskawetz, A., & Scheffran, J. (2006). Agent-based computation modeling: An introduction. In F. C. Billari, T. Fent, A. Prskawetz, & J. Scheffran (Eds.), *Agent-based computational modeling, contributions to economics* (pp. 1–16). Heidelberg: Physica-Verlag.

Bonabeau, E. (2002). Agent-based modeling: Methods and techniques for simulating human systems. *Proceedings of the National Academy of Science, 99*(10), 7280–7287.

Boone, R. B. (2007). Effects of fragmentation on cattle in African savannas under variable precipitation. *Landscape Ecology, 22*(9), 1355–1369.

Boone, R. B. (2010). Simulating species richness using agents with evolving niches, with an example of Galapagos plants. *International Journal of Ecology*, id150606. doi:10.1155/2010/150606

Boone, R. B., Galvin, K. A., Smith, N. M., & Lynn, S. J. (2000). Generalizing El Niño effects upon Maasai livestock using hierarchical clusters of vegetation patterns. *Photogrammetric Engineering and Remote Sensing, 66*, 737–744.

Boone, R. B., Coughenour, M. B., Galvin, K. A., & Ellis, J. E. (2002). Addressing management questions for Ngorongoro Conservation Area using the Savanna Modeling System. *African Journal of Ecology, 40*, 138–150.

Boone, R. B., Galvin, K. A., Coughenour, M. B., Hudson, J. W., Weisberg, P. J., Vogel, C. H., & Ellis, J. E. (2004). Ecosystem modeling adds value to a South African climate forecast. *Climatic Change, 64*(3), 317–340.

Boone, R. B., BurnSilver, S. B., Thornton, P. K., Worden, J. S., & Galvin, K. A. (2005). Quantifying declines in livestock due to subdivision. *Rangeland Ecology & Management, 58*, 523–532.

Boone, R. B., Thirgood, S. J., & Hopcraft, J. G. C. (2006). Serengeti wildebeest migratory patterns modeled from rainfall and new vegetation growth. *Ecology, 87*(8), 1987–1994.

Boone, R. B., Galvin, K. A., BurnSilver, S. B., Thornton, P. K., Ojima, D. S., & Jawson, J. R. (2011). Using coupled simulation models to link pastoral decision making and ecosystem services. *Ecology and Society, 16*(2), 6 [online]. URL: http://www.ecologyandsociety.org/vol16/iss2/art6/

Boyd, R., & Richerson, P. J. (2005). *The origin and evolution of cultures*. Oxford: Oxford University Press.

Brenner, T. (Ed.). (1999). *Computational techniques for modeling learning in economics*. Norwell: Kluwer Academic.

Brosius, J. P. (1997). Endangered forest, endangered people: Environmentalist representations of indigenous knowledge. *Human Ecology, 25*, 47–69.

BurnSilver, S. B. (2007). *Settlement pattern and fragmentation in Maasailand: implications for pastoral mobility, drought vulnerability, and wildlife conservation in an East African savanna*. Dissertation submitted to the Graduate Degree Program in Ecology, Colorado State University, Fort Collins.

Camerer, C. F., Loewenstein, G., & Rabin, M. (2003). *Advances in behavioral economics*. Princeton: Princeton University Press.

Cangelosi, A., & Parisi, D. (Eds.). (2001). *Simulating the evolution of language.* Dordrecht: Springer.

Carpenter, S. R., Armbrust, E. V., Arzberger, P. W., Chapin, F. S., III, Elser, J. J., Hackett, E. J., Ives, A. R., Kareiva, P. M., Leibold, M. A., Lundberg, P., Mangel, M., Merchant, N., Murdoch, W. W., Palmer, M. A., Peters, D. P. C., Pickett, S. T. A., Smith, K. K., Wall, D. H., & Zimmerman, A. S. (2009). Accelerate synthesis in ecology and environmental sciences. *BioScience, 59*(8), 699–701.

Crumley, C. L. (Ed.). (1994). *Historical ecology: Cultural knowledge and the changing landscapes.* Santa Fe: School of American Research Press.

Ellis, J. E., Coughenour, M. B., & Swift, D. M. (1993). Climate variability, ecosystem stability, and the implications for range and livestock development. In R. H. Behnke Jr., I. Scoones, & C. Kerven (Eds.), *Range ecology at disequilibrium: New models of natural variability and pastoral adaptation in African savannas* (pp. 31–41). London: Overseas Development Institute.

Epstein, J. M. (1999). Agent-based computational models and generative social science. *Complexity, 4,* 41–60.

Epstein, J. M. (2008). Why model? *Journal of Artificial Societies and Social Simulation, 11*(4), 12. http://jasss.soc.surrey.ac.uk/11/4/12.html

Escobar, A. (1998) Whose knowledge, whose nature? Biodiversity, conservation and the political ecology of social movements. *Journal of Political Ecology, 5*(1). http://jpe.library.arizona.edu/Volume5/Volume_5_1.html

Folke, C. (2006). The emergence of a perspective for social-ecological system analysis. *Global Environmental Change: Human and Policy Dimensions, 16*(3), 253–267.

Folke, C., Carpenter, S., Elmqvist, T., Gunderson, L., Holling, C. S., & Walker, B. (2002). Resilience and sustainable development: building adaptive capacity in a world of transformations. *AMBIO: A Journal of the Human Environment, 31*(5), 437–440.

Fryxell, J. M., Wilmhurst, J. F., Sinclair, A. R. E., Haydon, D. T., Holt, R. D., & Abrams, P. A. (2005). Landscape scale, heterogeneity, and the viability of Serengeti grazers. *Ecology Letters, 8,* 328–335.

Galvin, K. A., Thornton, P. K., Roque de Pinho, J., Sunderland, J., & Boone, R. B. (2006). Integrated modeling and its potential for resolving conflicts between conservation and people in the rangelands of East Africa. *Human Ecology, 34*(2), 155–183.

Gassman, P. W., Reyes, M. R., Green, C. H., & Arnold, J. G. (2007). The soil and water assessment tool: Historical development, applications, and future research directions. *Transactions of the American Society of Agricultural and Biological Engineers, 50*(4), 1211–1250.

Gaston, K. J., & Blackburn, T. M. (2000). *Pattern and process in macroecology.* Oxford: Blackwell.

Gilbert, N. (2008). *Agent-based models,* Quantitative applications in social sciences (Issue 7, Part 153). Thousand Oaks: SAGE University Papers.

Gilbert, N., & Bankes, S. (2002). Platforms and methods for agent-based modeling. *Proceedings of the National Academy of Science, 99,* 7197–7198.

Gilbert, N., & Terna, P. (2000). How to build and use agent-based models in social science. *Mind & Society, 1*(1), 57–72.

Goldstone, R. L., & Janssen, M. A. (2005). Computational models of collective behavior. *Trends in Cognitive Sciences, 9*(9), 424–430.

Griffin, W. A. (2006). Agent-based modeling for the theoretical biologist. *Biological Theory, 1*(4), 404–409.

Grimm, V. (1999). Ten years of individual-based modeling in ecology: What have we learned and what could we learn in the future? *Ecological Modelling, 115*(1999), 129–148.

Grimm, V., & Railback, S. F. (2006). Agent-based models in ecology: Patterns and alternative theories of adaptive behavior. In F. C. Billari, T. Fent, A. Prskawetz, & J. Scheffran (Eds.), *Agent-based computational modeling, contributions to economics* (pp. 139–152). Heidelberg: Physica-Verlag.

Grimm, V., Revilla, E., Berger, U., Jeltsch, F., Mooij, W. M., Railsback, S. F., Thulke, H.-H., Weiner, J., Wiegand, T., & DeAngelis, D. L. (2005). Pattern-oriented modeling of agent-based complex systems: Lessons from ecology. *Science, 310,* 987–991.

Hardin, G. (1968). The tragedy of the commons. *Science, 162*, 1243–1248.

Hobbs, N. T., Galvin, K. A., Stokes, C. J., Lackett, J. M., Ash, A. J., Boone, R. B., Reid, R. S., & Thornton, P. K. (2008). Fragmentation of rangelands: implications for humans, animals, and landscapes. *Global Environmental Change, 18*(4), 776–785.

Huston, M., DeAngelis, D., & Post, W. (1988). New computer models unify ecological theory. *BioScience, 38*(10), 682–691.

Janssen, M. A. (2007). Coordination in irrigation systems: An analysis of the Lansing-Kremer model of Bali. *Agricultural Systems, 93*, 170–190.

Janssen, M. A., & Ostrom, E. (2006). Empirically based, agent-based models. *Ecology and Society, 11*, 37 [online]. URL: http://www.ecologyandsociety.org/vol11/iss2/art37/

Johnston, E., Kim, Y., & Ayyanger, M. (2007). Intended and unintended: The act of building agent-based models as a regular source of knowledge generation. *Interdisciplinary Description of Complex Systems, 5*(2), 81–91.

Kareiva, P., & Andersen, M. (1988). Spatial aspects of species interactions: The wedding of models and experiments. In A. Hastings (Ed.), *Community ecology* (pp. 38–54). New York: Springer.

Kohler, T. A., & Gumerman, G. J. (Eds.). (1999). *Dynamics in human and primate societies: Agent-based modeling of social and spatial processes* (Santa Fe Institute studies in the sciences of complexity). Oxford: Oxford University Press.

Kraus, S. (1997). Negotiation and cooperation in multi-agent environments. *Artificial Intelligence, 94*, 79–97.

Lansing, J. S., & Kremer, J. N. (1993). Emergent properties of Balinese water temple networks: Coadaptation on a rugged fitness landscape. *American Anthropologist, 95*(1), 97–114.

Lansing, J. S., Cox, M. P., Downey, S. S., Janssen, M. A., & Schoenfelder, J. W. (2009). A robust budding model of Balinese water temple networks. *World Archaeology, 41*(1), 112–133.

Leach, M., Rockström, J., Raskin, P., Scoones, I., Stirling, A. C., Smith, A., Thompson, J., Millstone, E., Ely, A., Around, E., Folke, C., & Olsson, P. (2012). Transforming innovation for sustainability. *Ecology and Society, 17*(2), 11. http://dx.doi.org/10.5751/ES-04933-170211

Little, M. A., & Leslie, P. W. (Eds.). (1999). *Turkana herders of the dry savanna. Ecology and biobehavioral response of nomads to an uncertain environment.* Oxford: Oxford University Press.

Liverman, D., Moran, E. F., Rindfuss, R. R., & Stern, P. C. (1998). *People and pixels. Linking remote sensing and social sciences.* Washington, DC: National Academy Press.

Lorenz, T. (2009). Epistemological aspects of computer simulation in the social sciences. *Lecture Notes in Computer Science, 5466*, 141–152.

Macy, M. W., & Willer, R. (2002). From factors to actors: Computational sociology and agent-based modeling. *Annual Review of Sociology, 28*, 143–166.

Malone, S. W., Miller, C. A., & Neill, D. B. (2001). Traffic flow models and the evacuation problem. *UMAP Journal, 22*, 271–290.

MEA (Millennium Ecosystem Assessment). (2005). *Ecosystems and human well-being: Synthesis.* Washington, DC: Island Press.

Moran, E. F. (2008). *Human adaptability.* Boulder: Westview Press.

Nikolai, C., & Gregory, M. (2009). Tools of the trade: a survey of various agent based modeling platforms. *Journal of Artificial Societies and Social Simulation, 12*(2), 2. http://jasss.soc.surrey.ac.uk/12/2/2.html

Orlove, B. S. (1980). Ecological anthropology. *Annual Review of Anthropology, 9*, 235–273.

Ostrom, E. (2002). *The drama of the commons.* Washington, DC: National Academy Press.

Peck, S. L. (2004). Simulation as experiment: A philosophical reassessment for biological modeling. *Trends in Ecology and Evolution, 19*(10), 530–534.

Peet, R., & Watts, M. (1996). *Liberation ecologies: Environment, development, social movements.* London: Routledge.

Railsback, S. F., & Grimm, V. (2011). *Agent-based and individual-based modeling: A practical introduction.* Princeton: Princeton University Press.

Railsback, S. F., Lytinen, S. L., & Jackson, S. K. (2006). Agent-based simulation platforms: Review and development recommendations. *Simulation, 82*(9), 609–623.

Ramanath, A. M., & Gilbert, N. (2004). The design of participatory agent-based social simulations. *Journal of Artificial Societies and Social Simulation, 7*(4), 1 [online]. http://jass.soc.surrey.ac.uk/7/4/1.html

Rappaport, R. A. (1967). Ritual regulation of environmental relations among a New Guinea people. *Ethnology, 6*(1), 17–30.

Reid, R., Nkedianye, D., Said, M., Kaelo, D., Neselle, M., Makui, O., Onetu, L., Kiruswa, S., Ole Kamuaro, S., Kristjanson, P., BurnSilver, S., Goldman, M., Boone, R., Galvin, K. A., Dickson, N., & Clark, W. (2009). Evolving models to support communities and policy makers with science: Balancing pastoralism and wildlife conservation in East Africa. *Proceedings of the National Academy of Science.* doi:10.1073/pnas.0900313106.

Reynold, C. W. (1987). Flocks, herds, and schools: A distributed behavioral model. *Computer Graphics, 21*(4), 25–34.

Richiardi, M., Leombruni, R., Saam, N., & Sonnessa M. (2006). A common protocol for agent-based social simulation. *Journal of Artificial Societies and Social Simulation, 9*(1), 15. http://jasss.soc.surrey.ac.uk/9/1/15.html

Robbins, P. (2012). *Political ecology.* Malden: Wiley-Blackwell.

Schelling, T. C. (1971). Dynamic models of segregation. *Journal of Mathematical Sociology, 1*, 143–186.

Schneider, D. C. (2001). The rise of the concept of scale in ecology. *BioScience, 51*(7), 545–553.

Talen, E. (2000). Bottom-up GIS: A new tool for individual and group expression in participatory planning. *Journal of the American Planning Association, 66*(3), 279–294.

Thomas, R. B. (1976). Energy flow at high altitude. In P. T. Baker & M. Little (Eds.), *Man in the Andes* (pp. 379–404). Stroudsburg: Dowden, Hutchinson & Ross.

Thornton, P. K., Galvin, K. A., & Boone, R. B. (2003). An agro-pastoral household model for the rangelands of East Africa. *Agricultural Systems, 76*(2), 601–622.

Thornton, P. K., Fawcett, R. H., Galvin, K. A., Boone, R. B., Hudson, J. W., & Vogel, C. H. (2004). Evaluating management options that use climate forecasts: Modelling livestock production systems in the semi-arid zone of South Africa. *Climate Research, 26*(1), 33–42.

Thornton, P. K., BurnSilver, S. B., Boone, R. B., & Galvin, K. A. (2006). Modelling the impacts of group ranch subdivision on agro-pastoral households in Kajiado, Kenya. *Agricultural Systems, 87*(3), 331–356.

Treiber, M. (2011). *Microsimulation of road traffic flow.* Dynamic traffic simulation applet. http://www.traffic-simulation.de. Accessed 22 June 2012.

Treiber, M., & Kesting, A. (2010). An open-source microscopic traffic simulator. *IEEE Intelligent Transportation Systems Magazine, 2*(3), 6–13.

Turchin, P. (2001). Does population ecology have general laws? *Oikos, 94*(1), 17–26.

Vayda, A. P., & McCay, B. (1975). New directions in ecology and ecological anthropology. *Annual Review of Anthropology, 4*, 293–306.

Watts, M. J. (1997). Classics in human geography revisited: P.M. Blaikie: The political economy of soil erosion in developing countries. *Progress in Human Geography, 21*, 75–80.

Wilensky, U. (2001). *Modeling nature's emergent patterns with multi-agent languages.* Proceedings of EuroLogo 2001, Linz, Austria.

Wilensky, U., & Rand, W. (2007). Making models match: Replicating an agent-based model. *Journal of Artificial Societies and Social Simulation, 10*(4), 2. URL: http://jasss.soc.surrey.ac.uk/10/4/2.html

Worden, J. S. (2007). *Settlement pattern and fragmentation in Maasailand: implications for pastoral mobility, drought vulnerability, and wildlife conservation in an East African savanna.* Dissertation submitted to the Graduate Degree Program in Ecology. Colorado State University, Fort Collins.

Chapter 10
Inter-disciplinary Analysis of Climate Change and Society: A Network Approach

Jeffrey Broadbent and Philip Vaughter

10.1   Introduction

Networks matter. Whether a personal social network of contacts we use to navigate our daily lives, to globalized communication networks that connect governments, commerce, and social movements around the planet, networks are omnipresent. Social networking sites such as Facebook and Twitter have revolutionized how we network, as well as expanded the scale and sped up the time frame that we can now network on.

Social network analysis (SNA) examines the relationships among actors and ideas within social group. A relationship is any kind of transfer, from coercion and money to social approval and ideas, among people or groups of any size or level of organization. The vast and diverse network of transfers is central to reproduction of and change in social patterns and behavior, including how society relates to its ecological environment (Prell 2012). SNA is eminently applicable to studying the relations between society and the environment (Bodin and Prell 2011). SNA is inter-disciplinary in the sense that it can be used to trace the flow of scientific and other ideas into the realm of discourse within society. From there, this flow of ideas can be followed to its impact upon political action and its outcomes. At the same time, it can be applied to different levels of society, from the micro inter-personal dynamics to the macro-global scale flow of new norms and alliances. SNA can also be used to isolate different dimensions of society, such as the composition of the discourse

J. Broadbent (✉)
Department of Sociology, Institute for Global Studies, University of Minnesota,
909 Social Science Building, 267 19th Ave. S., Minneapolis, MN 55455, USA
e-mail: broad001@umn.edu

P. Vaughter
College of Education, University of Saskatchewan,
28 Campus Drive, Saskatoon, SK S7N 0X1, Canada

M.J. Manfredo et al. (eds.), *Understanding Society and Natural Resources*,
DOI 10.1007/978-94-017-8959-2_10, © The Author(s) 2014

field around a phenomenon, as well as the network of cooperation or information transfer among social actors. These flows and dimensions include the crucial feedback loops of grasping and framing a natural phenomenon wherein knowledge and belief, correct or not, are born and take life. Although providing the crucial substrate to social potential, molecular or genetic components cannot predict the formations and flow of society. SNA is inter-disciplinary, then, in the sense that it allows us to draw in the concepts and ideas about the natural and human worlds at all levels, and study how they work within the social and cultural arenas of collective human action.

Social scientists have taken note of the role that networks and networking play in social and political change, from improving safety conditions in nuclear power plants to negotiating new legislation on the supra-national level within the European Union. SNA has been used to analyze the discursive dimensions around political processes as well as coalition formation among organizations around environmental activism. The effectiveness of political or social movements is often determined by the nature of linkages between actors within a social network. Social network analysis can be used to examine political mobilization and the formation of advocacy coalitions as well as the spread of scientific knowledge and the dissemination of social or behavioral norms. By representing scientific knowledge as an information network along with other types of networks, a network approach can integrate the perspectives of different sciences (consilience) to study their conjoint and interactive effect upon the process of climate change production and solution.

And nowhere is the interaction between human society and the environment on a greater scale than in how our behaviors affect that greatest of global commons: the Earth's atmosphere. Increasing concentrations of greenhouse gases (GHGs) in the Earth's atmosphere from human industry have begun to change the planet's climate regime. And with the change in climate, have come changes to earth systems that humanity is dependent on. Sea levels are rising, threatening coastal communities; ice sheets have begun to melt, threatening fresh water supply; crops in some areas have begun to fail, threatening food supply. While human beings are adaptable, our capacity to do so will likely be overwhelmed as the scale of these impacts increase (IPCC 2007a). In order to reduce the emission of GHGs that spur climate change, norms of collective responsibility will need to be disseminated on a global scale (Broadbent 2010). But how will this happen? Indeed, *can* it happen?

When ozone depletion was operationalized as a problem within the 1980s, the driving force (production of chlorofluorocarbons or CFCs) was linked to a few specific activities (use of aerosol cans, use of specific refrigerants, use of certain packing materials) within a few economic sectors. The industries were able to substitute less harmful chemicals at low cost, and consumers did not have to radically change their behavior. Operationalizing climate change as a scientific certainty, let alone a problem, has been such a contentious debate because of the irreducible complexity of the issue. Greenhouse gases, which drive climate change, are not just produced by a few components of a few choice industries. Rather, they are omnipresent in virtually all economic activity and embedded within the production and maintenance of much of the globe's infrastructure. They are diverse in their source and type, creating debates over responsibility for their emission, as each greenhouse gas has its own global warming potential (GWP). Although carbon dioxide from industry and transport is the most abundantly produced anthropogenic greenhouse gas, methane emissions from agriculture cause more heat to be trapped within the atmosphere. To comprehensively tackle climate change as a problem, emissions from virtually every sector of the globalized economy must be addressed, not just a few choice "demon chemicals" from specialized sectors. Furthermore, because different societies around the globe have such varying sources of GHG emissions, national approaches to mitigation will have to be diverse rather than uniform. Mol (2001) has illustrated widespread norms and values in regards to the environment have diffused across the world, which seek to minimize the harm economic processes cause to the planet's ecosystems. However, this will take time to diffuse across different societies due to the scope and variety of climate change's drivers and impacts.

In the past, social scientists have studied how norms and values have been codified in international treaties on environmental issues (such as the Montreal Protocol) and how the design of these treaties have helped internalize environmental norms and values in societies around the globe (Schneider et al. 2002; Helm 2005; Speth and Haas 2006; Young 2002). However, constructs of environmental values or behavioral norms have been haphazard in regards to the threat of climate change, even with the drafting and implementation of the Kyoto Protocol. While some nations have made great progress in reducing their greenhouse gas emissions, others have not. Societies have varied greatly in their responses to climate change, and attention is now focused on what characteristics within societies are responsible for this variation. (Evans et al. 1993; Jacobson and Weiss 1998; Schreurs 2002, p. 261; Weidner and Janicke 2002, pp. 430–431) What factors have led to such varying norms and responses to climate change in societies around the world when global norms on policies around ozone depletion, ocean dumping, and pesticide use were embraced?

The authors of this chapter are part of a group of researchers, the Compon project, who propose that the next step in investigating these variables is to examine comparative policy networks in order to test hypotheses about social factors helping or hindering domestic responses to climate change. The project on Comparing Climate Change Policy Networks (Compon) project tests the effect of social organization, cultural meaning and political mobilization on a nation's response to climate change. The Compon project is a collaborative effort among teams of scholars using social network analysis to compare and contrast discourse and action around climate change and climate change policy within 19 societies around the globe. The societies currently within the study include Brazil, Canada, China, Germany, Greece, India, Indonesia, Ireland, Japan, Mexico, New Zealand, Portugal, South Korea, Sweden, Switzerland, Taiwan, the United Kingdom, Vietnam, and the United States. Using data collected by academic teams using the same instruments in these societies, social network analysis allows researchers to identify and compare patterns of belief, advocacy coalitions, mobilization and policy-formation as they shape the formation of mitigation policies and behaviors (Broadbent 2010; Broadbent 2013, #3577).

10.2   Structure, Function and Power in Social Networks

Network analysis concerns itself with the study of relationships among actors within a given network. New types of SNA approaches are incorporating not only actors, but also ideas or discourse, in the measured networks. One body of theory calls these ideas "actants," to distinguish them from willfully self-propelled "actors" (Latour 2005). The formal approach of Integrated Structurational Analysis (ISA) has been proposed to integrate the various dimensions in societal processes as network vectors among the units (actors and actants) (Broadbent 1998, 2003). The present section will discuss the approach and measures developed for social networks that can be applied to this kind of integrative synthesis and analysis. The essay will then distinguish social action and discourse networks and examine their interaction.

Depending on how the term 'actor' is operationalized, actors within social networks can be individuals (micro-level social networks), groups or organizations (meso-level social networks), or states in global relationships (macro-networks). To examine the networks that underlie and produce national-level policy formation, it is often appropriate to use organizations as the actor or social unit of analysis, as done here. The relationships between actors in a social network are described as *ties*, and represent a point of social contact between actors within the system. The social contact can consist of any type of interaction, be it the sharing of information or ideologies, the dissemination of a norm, or the exchange of support or resources. Much social scientific research, as is typical of survey research, has been conducted on samples of individualized actors. SNA differs because it also collects information on the relations or ties among the actors. If we want to study the ties as constituting a whole system, we have to study the group of actors that could potentially have direct ties among themselves. That rules out the random selection of actors from a large population (though one can study "ego-centric" networks that way). Rather, to study systems of relationships, we have to study the patterns of ties among a set of actors susceptible to relationships.

This kind of whole network study is applied to some kind of community, such as a classroom, a town, or in this case, a "policy domain." A policy domain refers to all the actors potentially influential upon a certain type of policy within a nation-state (or governmental area). The SNA approach takes into account both the qualities of the actors themselves, both their resources and their ideas, and the vital relationships that transfer those qualities as sanctions among the actors. The relational theory underlying the SNA approach argues that societal power is relational in that it involves the connection and mobilizations of numbers of actors and ideas. Hence, the relational ties are fundamental because they reveal the active flow of ideas and resources among actors that enable the power to affect policies and large scale societal changes.

One of the primary structural concerns within social network analysis (SNA) is identifying the "most important" or "most prominent" actors within a given social network (Wasserman and Faust 1997). The concept of actor *importance* is a measure of the property of actor *location* within a social network, with the most *important* actors being located in the most *strategic locations* within the network. Thus, an actor's role is characterized by its structural position within the network (Borgatti and Foster 2003). Actor centrality is the measure typically employed for quantifying an actor's role within a social network. A central actor is an actor with many ties to other actors within a network.

Bodin et al. (2006) note that a network actor with a high degree of centrality can effectively coordinate actors within the network during times of change. Burt (2003) characterizes these actors as "brokers" within social networks. In social networks the policy sphere, this means policies can be passed through a legislature more quickly, but Abrahamson and Rosenkopf (1997) contend this leads to centralized decision making within the network. Another implication is that actors within the network will have more limited access to other sources of information (Weimann 1982).

Another key structural feature of social networks is density within the network. Density within SNA is a quantifiable measure of connectivity between actors within the network and of the connectivity of the network as a whole. Density is not used as a measure of centrality per se; rather, it is a measure of the network's cohesion – the number of links between actors within the whole network, not on an individual basis as with centrality (Wasserman and Faust 1997). Density of a network is calculated by dividing the number of links by the number of nodes within the network. One of the structural characteristics of dense social networks is a buffering capacity referred to as redundancy (Bodin et al. 2006). In dense networks, if an actor is removed from the network, because of the many links between other actors within the network, the loss does not have as profound an effect on the overall network structure. For advocacy coalitions, this means even if a central actors is removed from the coalition, other actors can step into the position and assume the functions of the central actor (Folke et al. 2005).

Social networks with greater connectivity of knowing each other exhibit higher levels of trust among actors within the network (Granovetter 1985). Pretty and Ward (2001) theorize that greater network density increases the possibility of social control of the actors within the network, which facilitates top-down regulation of the environment by the state. Oh et al. (2004) caution that dense networks can streamline policy processes, but may also promote homogenization of both experience and knowledge. Moreover, Frank and Yasumoto (1998) caution that too many links between actors within a social network can lock certain actors into inflexible positions, making political change difficult.

A concluding example of a structural feature of social networks is modularity or betweenness. Betweenness can be measured within a social network by quantifying the distance between nodes within a network. In any given social network, groups with high internal density may be loosely connected to other groups with high internal density. This phenomenon is termed modularity, and describes groupings of actors within a social network (Bodin et al. 2006). Within a civil society, this can be characterized by businesses having dense ties to one another, but weak or peripheral ties to government ministries or environmental NGOs. The betweenness of actors within a social network is a measurement of diameter – it is the number of steps needed to reach from one node to another within then network.

A high degree of betweenness in social networks allows different blocks of actors to interpret knowledge and develop policy responses distinct from one another. This is often the case in social networks around ecological governance, with different blocks developing different interpretations of data about the environment (Ghimire et al. 2004). The more modular a social network is, the less trust is demonstrated between different blocks within the network (Borgatti and Foster 2003). Likewise, it is more difficult to transfer tacit and/or complex knowledge ("externalize" scientific knowledge) within social networks with a high degree of betweenness (Reagans and McEvily 2003). In turn, advocacy coalitions characterized by high modularity within the network are prone to fragment, as the removal of a single actor can disengage a block of actors from the rest of the network (Borgatti and Foster 2003).

SNA sometimes assumes that higher centrality gives an actor more power over the other actors, and hence over the behavior of the whole network. However, this assumption is greatly in need of empirical testing in actual policy systems. The policy network approach taken by the Compon project includes measures of actor power in the formation of policy. One measure is created by survey respondents checking off those actors in the list they think to be very powerful within the policy domain (in this case, climate change). This is a reputational measure of power. Another measure involves the actor scoring their degree of satisfaction with the outcome of a policy debate in which they were involved. The higher the satisfaction, the measure assumes, the greater the effective behavioral power of the actor. These power measures can be used to trace the relative influence of different actors, their coalitions, and their ideas and ideologies – in this case about climate change. These measures were developed in earlier policy network studies (Knoke et al. 1996).

10.3 Action Networks and Discourse Networks

How different societies around the world respond to the call for mitigation of their emissions associated with climate change is a complex process involving a number of different interacting factors. The relevant factors can be broadly modeled as two different types of networks – *discourse networks* and *action networks*. Both types of networks have their own systemic dynamics and properties. Since both types of network are social phenomena, they are therefore more than the sum of the individual ideas and individual actions between actors in the field or domain. Social network analysis can be used to analyze the whole topography or morphology of these multi-actor, multi-idea fields.

The Compon project uses both types of networks to examine the social and political dynamics of mitigation policy formation and outcome in a number of societies. Use of the Discourse Network Analyzer (DNA) software enables the examination of the discourse around climate change issues and policies within a society's media and/or legislative records. Pioneered by Philip Leifeld, DNA applies the methods of SNA to study the actors quoted in newspapers and the policy positions they advocate. When analyzed by network analysis techniques, this data reveals the ideational cleavage lines between actors or groups of actors within a media discourse field particular to a given society (Leifeld and Haunss 2012). As such, this quantitative technique provides empirical data to the theoretical position stressing the importance of collective representations of (ideas describing) phenomena developed in Actor-Network Theory through qualitative research methods (Latour 2005). The discourse field includes the ideas from scientific research that claim to accurately describe and predict natural phenomena as well as ideas welling up from less disciplined human processes that cloak such scientific claims in popular preferences and prejudices. This type of interaction has been deeply investigated in research on science and society (Jasanoff 2005). DNA allows the more precise identification of different types of discourse and their degree of support by political actors.

The other type of SNA, termed the policy network method, is used to investigate the relationships between actors active in the climate change policy sphere. The policy network method grew out of the quantitative network analysis developed in the 1960s and first applied to small groups or communities. The policy network analysis (PNA) approach turned that technique to study the policy formation process as influenced by organizations, including agencies within the state and associations with society as an interactive polity. Researchers first used PNA to examine American and German political processes during the 1970s (Laumann and Pappi 1976). They subsequently expanded the approach to compare Germany, the US and Japan (Knoke et al. 1996). As distinct from the newspaper discourse analyzed by DNA, the policy network survey gets responses about their ideas, resources and networks directly from representatives of the groups and organizations involved in the policy-influence process. The survey data therefore allows for a precise examination of the discourse (policy stances, beliefs, ideologies) held by organizations as well as their coalition formation, political pursuits and degrees of influence. The use of a standardized basic network survey in multiple cases (nation-states or areas) allows for rigorous cross-case comparison and the search for common causal factors leading to emissions trajectories (from 1990 to present, in sum increasing, reducing, or level) (Broadbent 2010).

10.3.1 Culture as Context in Social Network Analysis

When using SNA for comparative approaches (for instance in comparing different nations' carbon policy outcomes), it is important to remember that social networks around discourse and action emerge from and operate within a *context*. That is to say, neither actors, ideas or relationships are autonomous units. They differ in each situation, in this case in each nation-state policy network. In philosophical terms,

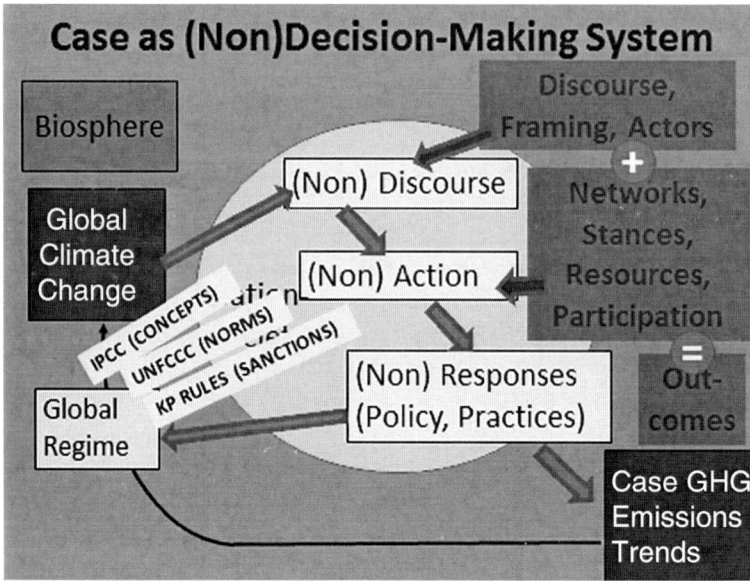

Fig. 10.1 Model of social response to climate change

those "units" take shape within and help constitute unique social/cultural ontologies (ways of being). The components are formed and operate under basic social conditions (variously referred by different schools of social science as institutions, structures or systems) that constrain, facilitate and channel the possibilities of both discourse and action within a given society. This can be more broadly defined as a nation's cultural framework. Existing cultural and social conditions constrain the emergence of new discourses and the possibilities of their application to create change. Depending on the nation in question, contextual factors can make the dynamics of either field more or less solid and enduring or fluid and volatile. The more fluid the system, the more that actions interact with discourses to produce new forms of power and in the current concern, change societal practices and political policies affecting climate change mitigation.

Figure 10.1 presents a hypothetical model of that process within the national arena and between national and international levels (Broadbent 2010). Climate change as a geophysical process driven by human-caused carbon emissions enters the society as conceptual information. These concepts are processed through societal discourse and action and eventually "constructed" or rejected as usable knowledge by different advocacy groups. This "construction" process is profoundly influenced by local factors of the society itself, such as culture, institutions, level of economic development and others. The mixed effect eventuates in decisions or non-decisions with effects upon the carbon emissions trajectories of the society. These emissions in turn feed back into the global geophysical situation and its propensity to produce climate change and disastrous effects on societies. As societies repeat these processes, they also build up through global negotiations a global climate regime – a set of ideas, norms and rules that may exert increasing influence upon the decision-making processes of member societies.

Many studies have tried to attribute attitudes on the environment and on the science that explains the state of it to different demographic characteristics of populations. Research has shown that public perceptions of risk are widely divergent within different national populations (Siegrist et al. 2005). Different social groups in different nations have different issues of contention around different scientific claims (Walls et al. 2004). Thus, demographic characteristics such as age, race, or sex may be poor predictors of attitudes towards science, risk, or environmental values within cross-cultural comparison for an issue such as climate change. Rather, a person's cultural framework serves as a better explanatory framework for how or why scientific knowledge (such as climate science) is valued or accepted (Jaeger et al. 1993). How successfully concepts of risk are understood (Slovic 1986) or scientific claims are communicated is largely a function of how science itself is framed within a given political and/or cultural environment (Jasanoff 1998). In cultures that employ a 'science-centered' paradigm, it is the duty of scientists to inform the ignorant state and to educate the irrational public as to what "real" risks are, and to provide advice on how to handle them (Tversky and Kahneman 1974). The robust environmental policies created around the globe between the 1960s and 1980s were facilitated by cultures that embraced this 'science-centered' paradigm; in these democracies the public considered science as both credible and relevant (Gustafsson and Lidskog 2012). In a political culture that values public opinion, but where the 'science-paradigm' is not accepted, environmental concern – and therefore the science that underpins it – may be marginalized by a public whose primary concern is the economy. In this 'economy-paradigm,' perceptions of risk to the economy may outweigh the perception of risks of damage to the environment.

Brown Weiss and Jacobson (1998) observe that environmental concerns tend to be brushed aside if they pose a risk to the economy within societies that value participatory democracy and employ the 'economy-paradigm.' Indeed, it has been argued that democracies are ill-equipped to deal with ecological concerns, as the public tends to vote in their short-term self-interest (Giddens 2009). Shearman and Wayne Smith (2007) speculate that cultures that value both democracy and economic growth will continue to ignore the implications of climate science, even if actors within democracies accept that science as valid, because transient issues will outweigh such a permanent and entrenched issue within the electorate. This tendency to ignore, if not all out reject, scientific information that is deemed inconvenient for a society is termed 'the triumph of short-termism' (Clayton et al. 2006). Giddens (2009) characterizes this tendency as 'loss aversion,' with the voting public more concerned about perceived losses than with future gains.

Cultures that tend to legitimize governance through majority rule, even when the majority may not be informed about an issue they are voting on, may have trouble cobbling together enough actors to form effective communication networks around climate change discourse or advocacy coalitions around climate change policy. This has opened up a transnational debate as to whether environmental regulation should be based on the advice of experts or whether regulation should be legitimized through democratic consensus (Collins and Evans 2007; Renn 2008). By the time climate change became a prominent environmental issue during the 1990s, a coalition of actors had arisen within the United States that was both critical and hostile towards scientific reports on climate change (Hamilton 2007). Among the criticisms launched at the climate science community were that the models of climate change should aspire to scientific certainty for the prediction of hazards (Baker 2007), that scientists were attempting to usurp the role of the state's authority (Jasanoff 1998), and scientific uncertainty around climate change had to be reduced before the climate science community could make policy recommendations (Brown 1992). All of these criticisms were criticisms that previous environmental issues had not had to contend with, or at least, not contend with to such an extreme. But because regulation based on science began to be framed as only legitimate through democratic consensus within American culture, a culture where scientific evidence could be contested because of democratic values became normative within the United States.

The issue of uncertainty has since emerged from American culture to become a rallying call to question all discourse around climate change and delay any mandatory regulation of greenhouse gas emissions within the United States (IPCC 2007b). The same tactics to create a culture of doubt around climate science have spread across the Pacific to democratic societies such as Australia and New Zealand (Hamilton 2007). The scientific community in many different democratic societies have been divided over how to respond to increased public scrutiny, with calls to reestablish a culture of communication between scientists and the public as authoritative rather than debatable (Collins and Evans 2007; Renn 2008), as well as a push to dialogue with the public over scientific knowledge (Jasanoff 2005; Lidskog 2008).

Thus, even if science is accepted within democracies, it may not be effective in setting policy objectives if the citizenry does not agree with the implications of science on regulation. This is not to suggest that cultures that value authoritative governance are uniquely equipped to deal with climate change; far from it. Indeed, authoritarian cultures have their own unique problems in addressing environmental concerns, including a lack of willingness to engage stakeholders and problems with adaptive management. Instead, the preceding examples are meant to illustrate to the reader the importance of context (both political and cultural) when examining both the discourse and action networks around climate change within a given society.

10.3.2   Discourse Networks Around Climate Change

The field of discourse represents the distribution of concepts (perceptions, beliefs, knowledge) and their meanings (interpretations, evaluations, frames, emotions) about prevalent in a society (Broadbent 2010). While resources and support are often traded between actors in a social network, so too are concepts and meanings, including an understanding of scientific principles. Engagement with science and an understanding of scientific inquiry have become embedded as norms within a number of societies around the globe (Beck 2002; Höijer et al. 2006). Individuals engage in discourse over science not only through formal education, but also increasingly though informal means, such as the news media and interpersonal communication (Van Dijk 2011). Organizations too must engage discourse over science, though the manner in which scientific knowledge and methods of scientific inquiry are institutionalized and normalized within organizations is not well understood. Watson (2002) contends that organizational learning is poorly conceptualized, meaning a systematic investigation of how organizations process and gain knowledge has been difficult to implement. Compounding this deficit in knowledge is the fact that the research that has begun to investigate how organizations learn has focused on the transmission of cultural, financial, or legal knowledge rather than beliefs institutionalized through gained scientific knowledge.

Communication networks are described by Hajer and Versteeg (2005) as discourse spaces where actors explain themselves in order to exert influence over other actors. Structural characteristics of discourse networks are important for understanding how effectively scientific knowledge is communicated between actors within a communication network. The density of a social network is the most widely used measure of group cohesion, with denser networks having more ties between actors (Blau 1977). Essentially, density quantifies network "knittedness" within SNA (Bott 1957). Dense networks facilitate the dissemination of scientific information in a communication network by increasing the accessibility of information in the network (Abrahamson and Rosenkopf 1997). In addition, networks with high density promote the development of universal norms in regards to natural resource management and environmental policy; they also promote compliance with these norms (Coleman 1990).

Because the geophysical and climatological processes involved in climate change are complex and require a specialized scientific background to understand them, non-specialists must rely heavily on scientists to frame and explain the problem. Scientists were relied upon to communicate risks on a number of environmental issues to a diverse network of actors during the latter half of the twentieth century. These issues included pesticide use, damage to the ozone layer and the impacts of radiation from nuclear weapons . However, faith in the scientific community has been heavily contested in regards to climate change. The framing of climate science as contentious rather than authoritative was facilitated by regulatory failures (Power 2007), an increasingly scientifically literate citizenry capable of questioning scientists (Nowotny et al. 2001), and a greater emphasis on individualization in a number of societies across the globe (Beck 1992). Scientific claims were no longer viewed as objective and scientists themselves were beginning to be viewed as untrustworthy. Climate scientists in particular began to be viewed as actors whose interests were in conflict with the interests of the public, the business community, and the state (Gouldson et al. 2007).

When information about climate change moves through the discourse field (contested or otherwise), evaluative norms come into play. Evaluative norms spread through discussion networks among organizations and individuals, as well as

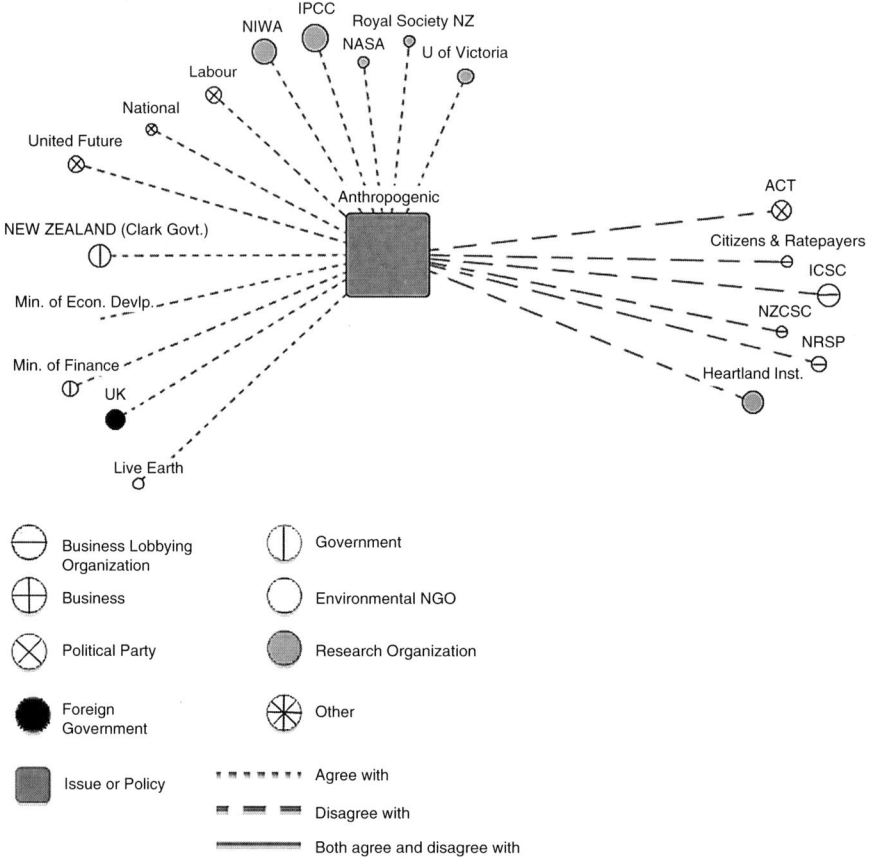

Fig. 10.2 Discourse network on legitimacy of climate science within New Zealand press, 2007–2008

through mass and specialized media (Broadbent 2010). The discourse network diagrams pictured below (Figs. 10.2 and 10.3) show the positions of organizations within New Zealand and the United States respectively on this issue (square) of climate science between 2007–2008. These diagrams were created using the software tool Discourse Network Analyzer (DNA) (Leifeld 2011) where the size of each actor node (circles) represents the number of statements made about the issue within a sample of the nation's news media.

Within the New Zealand discourse network (Fig. 10.2) around the validity of climate science, the media portrayed a wide consensus on the domestic stage around the validity of climate change science. Organizational actors who accepted the evaluative norm that climate change was real and anthropogenic include Clark's government ministries, most political parties (Labour, National, and United Future through statements to the press, Green, Māori, and New Zealand First through statements on television), as well as most research organizations (domestic and foreign)

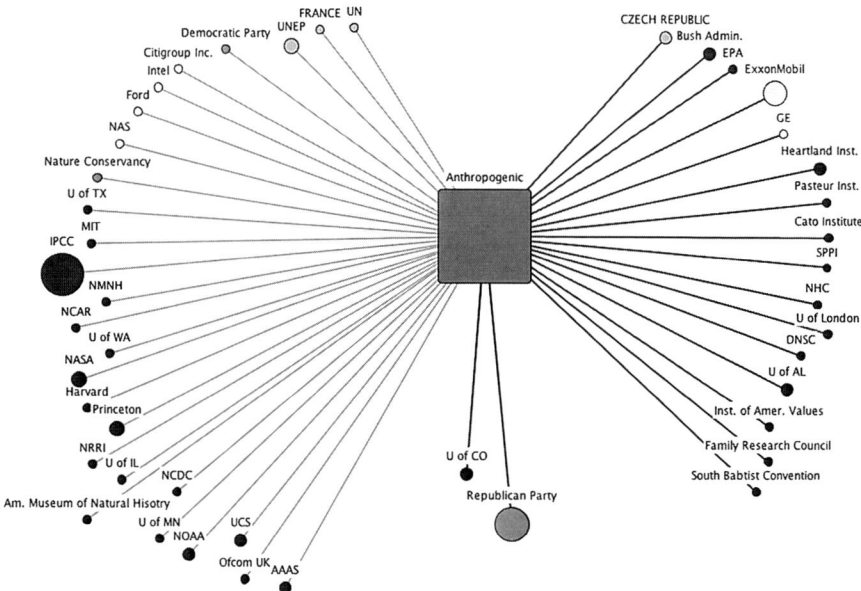

Fig. 10.3 Discourse network on legitimacy of climate science within United States Prestige Press, 2007–2008

quoted by the press (Vaughter 2013). Organizational actors who rejected or questioned the evaluative norm that climate change was real included a variety of business lobbying groups, though no businesses themselves. The ACT Party is the only political party that refuted the validity of climate science, and the only research institute (the Heartland Institute) to refute it within the New Zealand press was from the United States. Within the New Zealand press, the discourse over the legitimacy of climate science is unquestioned by the government, and the majority of all research organizations and political parties.

Because the science supporting climate change appears well accepted by both the scientific community and political actors within the New Zealand press, the majority of debate about climate change covered by the press is not about operationalizing climate change as a problem, but rather on how to implement a solution. Clark's Labour government and its constituent ministries, in coalition with the Green and New Zealand First Parties, comprise the majority of actors cited as pushing for a series of legislative solutions. This coalition of actors often stressed the 'science-paradigm' for legitimizing their proposed actions through citing both the IPCC and NIWA data in setting the time frame for implementing these policies. The opposition National Party in coalition with the Māori Party, accepted the scientific findings of the IPCC and NIWA as well and agreed climate change was an issue New Zealand needed to address. However, both parties viewed the Labour coalition's time frame for implementing an economy-wide ETS as reactionary, with negative repercussions for New Zealand's economy and the autonomy of indigenous groups. These actors often employed the 'economy-paradigm' by citing economic data from the New Zealand Institute to illustrate the costs of implementing an economy wide ETS as quickly as the Labour coalition wanted. Within this group of actors, the 'science-paradigm' was accepted but discounted in favor of the 'economy-paradigm.' Businesses and environmental NGOs stayed out any discussion of climate science, and instead advocated for or against specific pieces of climate legislation.

The discourse networks about anthropogenic climate change in the US (Fig. 10.3) contrast distinctly with those of New Zealand. In this discourse network the legitimacy of climate science is contested rather than accepted by government actors such as the Bush Administration. In addition, while the majority of research organizations within the New Zealand discourse network are quoted as supporting the legitimacy of climate science, there is a more even split between positions within the research organizations cited in the American discourse network. This spit frames climate science as more controversial, with reporting on both perspectives being more "balanced," despite little controversy within the climate science community itself (Boykoff and Boykoff 2004).

Actors within the American discourse network around climate change science appear much more divided in whether they accept or reject the 'science-paradigm' The cultural backdrop of the United States can also be glimpsed within this discourse network, with three fundamentalist religious groups weighing in on the issue, while religious groups are absent within the more secular New Zealand discourse network.

By portraying the climate debate as unsettled, the U.S. discourse network around climate change presents an inherent contradiction in its coverage of the issue. The debate over climate science occurs alongside discussions regarding the best ways to mitigate climate change. These two debates challenge each other's legitimacy, the former implying that the latter is premature, and the latter assuming that the former has already been settled (Burridge et al. 2013). The further disagreement and contradiction between cited research organizations over climate change further fuels this fire by failing to present a unified or even majority viewpoint on the issue. As Boykoff and Boykoff (2004) note, this tendency for the media to represent support for and skepticism of climate science in roughly equal proportions is not representative of the positions held by the science community, which overwhelmingly accepts the science behind climate change. The overall effect of this contradiction in media coverage produces a diffuse and convoluted definition of the issue that fails to identify what element or elements of the problem are actually in question. And without a clear definition of the problem, neither the public nor their elected officials will be able to begin operationalizing solutions with any success.

While discourse around climate change within the US is prominent and contains a number of diverse organizational actors, the stances of the predominant actors in the debate are inconsistent in regards to discourse around climate change policy (Fig. 10.4). Decision makers appear to bend to whichever way their constituency blows, with little agreement within political parties as to how climate change should

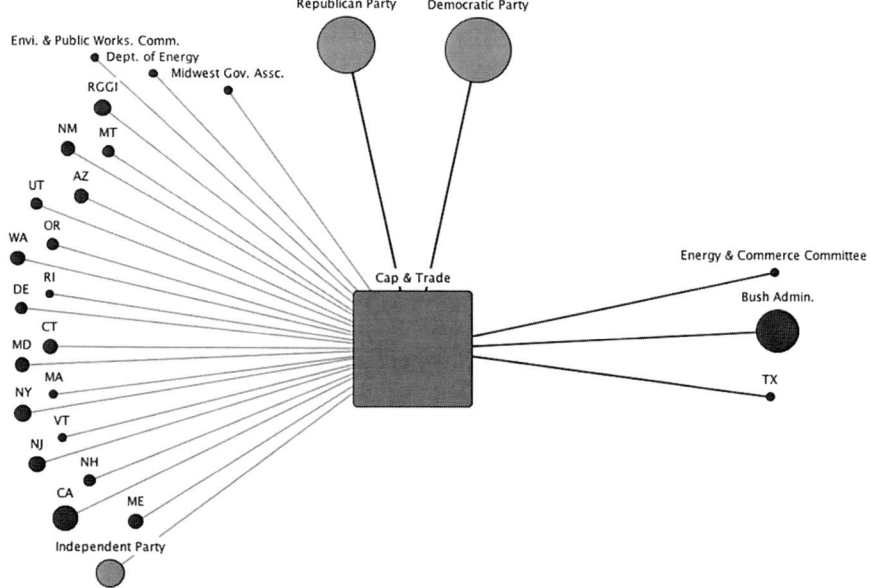

Fig. 10.4 Discourse network on federal cap and trade legislation in United States Prestige Press, 2007–2008

be conceptualized, let alone handled. The reluctance of either political party to operationalize solutions to the problems presented by climate change is unsurprising, given the lack of agreement among decision makers as to what the problem *is*. The Democratic and Republican parties both agree and disagree with a federal cap and trade mandate, at the same time the Republican Party is portrayed as both agreeing and disagreeing that climate change is a problem.

Discourse networks can be used not only to probe actors' normative stances, but can be employed to examine how information is disseminated within a society. Organizational actors learn through their networks as well as peer pressure about what evaluations (frames) to adopting regards to climate change. The dissemination of science, especially climate science, is never an easy task, because science itself is an iterative process, with an understanding of what is 'objective reality' changing and evolving over time. Some organizations can influence the flow of information to actors in a network, imposing frames of understanding upon them, or the diffusion can be interactive, through rational discussion among peers. In other instances, diffusion of knowledge may be blocked by certain actors within a network, or the implications of this knowledge may be discounted if they run counter to other concerns a given organization is facing. Actors using different normative standards will necessarily disagree about what to do. Only some actors will accept sufficient responsibility to seriously think about, evaluate and act upon the issue (Broadbent 2010).

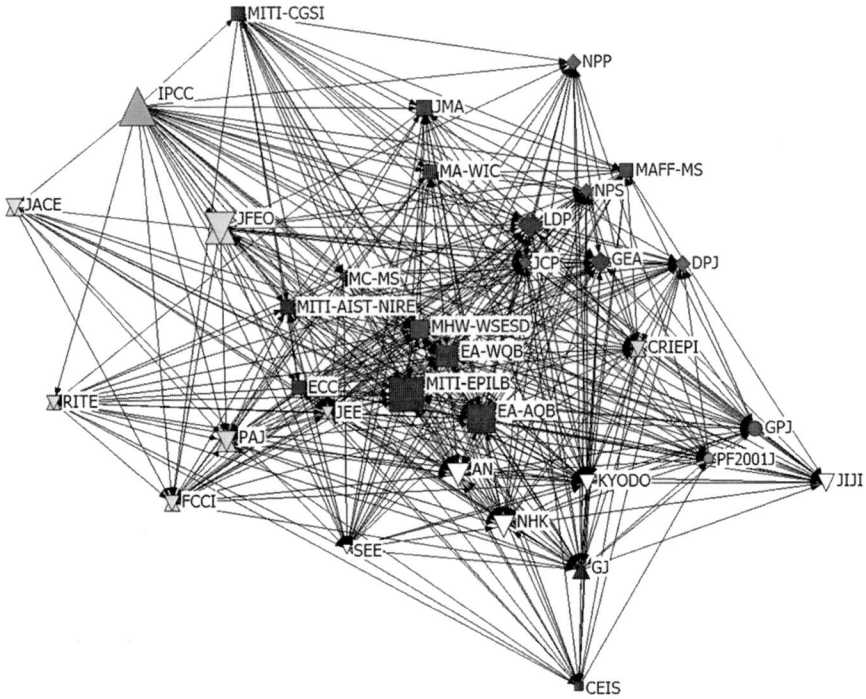

Fig. 10.5 Japanese organizational actors receiving information directly from the IPCC.

Tracing the flow of information and norms through networks will help indicate the function of organizations in a network. With optimal function, such forums may help the diffusion of scientific evidence and risk evaluation. For climate change, the Intergovernmental Panel on Climate Change (IPCC) has emerged as the predominant organization producing scientific information in relation to the topic. Thus, an examination of how information from the IPCC is disseminated from the organization into the civil society of a given nation can help inform how the IPCC functions as an actor within that nation's network of political actors. This in turn can illuminate how knowledge about climate change is conveyed within a society, and how actors evaluate the risks associated with climate change and construct responses to these risks. In the network data analysis, if a diversity of organizations have information networks to such a forum, and also hold scientific and action-oriented norms, it will indicate that the forums do indeed have the predicted function.

The network image in Fig. 10.5 shows how information about climate change was disseminated to a large and diverse set of organizations in Japan during 1997. The organizations include a large number of government agencies (blue squares), and large number of business organizations (yellow double triangles), some political parties (the brown and red diamonds), two environmental NGOs (green circles) and many media companies (white triangles).[1] The remaining brown triangle is Globe Japan, an international association of national politicians concerned about global environmental issues. The size of the icons reflects their perceived level of influence in Japan's domestic politics of global environmental issues (as determined by the number of respondents checking that organization as being "especially influential"). The communication network indicates that in the 1997 Japanese global environmental policy domain, the IPCC was among the big three influential organizations. Among the government ministries and agencies, the Air Quality Bureau of the Environmental Agency (AQ-EA) is second only to the Ministry of International Trade and Industry (MITI). The network image also reveals strong levels of perceived influence for the three news media clustered close to the government agencies. The Liberal Democratic Party is also assessed as highly influential, while the Japan Communist Party is diminutive. Business associations do not receive climate change information directly from the IPCC. Rather, businesses hand over this information-gathering task to a specialized business research institute, the Research Institute of the Electric Power Industry (CRIEPI), from which they probably get most of their information. Almost all of the domestic environmental NGOs do not receive information directly from the IPCC. Instead, the Japan branch of Greenpeace International serves as the primary information bridge-keeper to the domestic NGO community. This network figure indicates that in Japanese society the information bridge-keepers between outside and inside are relatively few, and those that perform this role have relatively high levels of political influence. This finding is in line with the network theory that being a bridge-keeper over a structural hole (a gap between clusters of organizations) gives power to the bridging actor (Burt 1992; Broadbent 2010).

Upon examination of the Japanese communication network about climate science, Japanese society appears receptive to the logic of scientific evidence – indeed the culture is enamored of technology and very successful in its innovation – and relatively free of powerful belief systems that would militate against accepting such logic. Compared to US media, Japanese news media are closely dependent upon government ministries for information and have rarely presented views questioning the validity of the IPCC findings and assertions. Japan's climate change science establishment is closely tied to and funded by the government. It seems that Japanese climate scientists rarely act as autonomous knowledge brokers among different sectors or in the policy making process, nor do they directly address the public contrary to current government policy (unlike, for instance, top climate scientist James Hansen in the US) (Broadbent 2010).

As an extension of this research to the global level, a publication based on the Compon media content analysis analyzed the comparative response to the 2001 and 2007 reports by the IPCC among five Asian societies (Broadbent et al. 2013). This study found different intensity of coverage in China, India, Japan, South Korea and Taiwan. However, the fact that Taiwan, not a UN member, always had the

---

[1] Note: the color version of this figure is available online in the Open Access version of this book, the print version includes grey scale images.

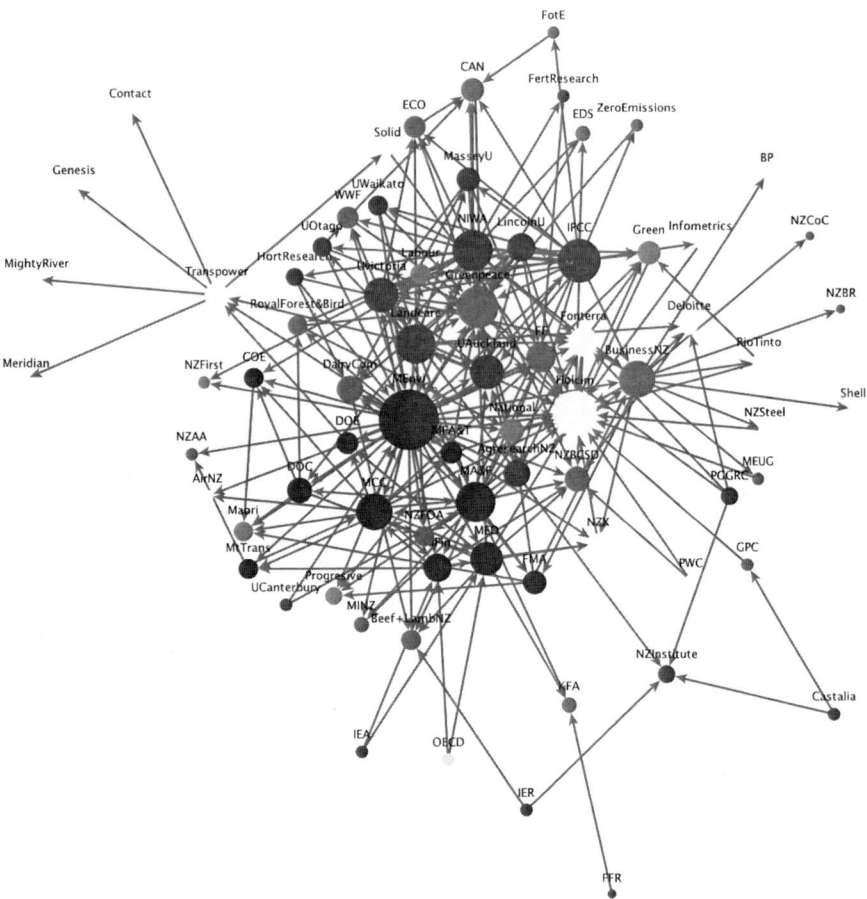

Fig. 10.6 New Zealand organizational actors receiving scientific information about climate change.

lowest coverage indicated the importance of belonging to the UN system for receptivity to UN-based ideas.

Figure 10.6 shows how information about climate change was disseminated within New Zealand in 2008, in the lead up to the implementation of its economy-wide emission trading scheme. Organizations featured in this network include government ministries (blue circles), business organizations (red circles), political parties (orange circles), environmental NGOs (green circles), and research organizations (purple circles).[2]

As with the Japanese network in Fig. 10.4, the size of the icon represents the level of perceived influence of the actor in disseminating knowledge about climate change.

---

[2] Note: the color version of this figure is available online in the Open Access version of this book, the print version includes grey scale images.

As within Japan, the IPCC is one of the three most central actors within the network. Another highly centralized actor within the network is the National Institute of Weather and Atmosphere (NIWA) a state owned and operated Crown Research Institute responsible for much of the climatological data submitted by New Zealand to the IPCC. The Ministry for the Environment (MEnvi) is the actor responsible for the majority of dissemination of scientific information on climate change to other actors, having the greatest centrality score of any of the actors within the network. The Ministry of Economic Development (MED) and the Ministry for Foreign Affairs and Trade (MFA&T) also figure prominently in the communication network, as well as several other Crown Research Institutes and domestic research universities. None of the political parties are presented as especially critical to information transfer, though the majority of political parties do appear within the network. Like in the Japanese case, businesses (yellow circles) tend to receive most of their information from business lobbying groups (red circles). While most of the domestic environmental NGOs do receive information from both the IPCC and NIWA, they do not in turn seem to give information about climate change to many actors within the network. This network figure indicates that within New Zealand the organizations conducting research on climate change themselves as well as the Ministry of the Environment act as bridge-keepers between actors within society, suggesting a high level of political influence.

In examining the New Zealand communication network around climate science, New Zealand also appears receptive to scientific evidence produced by the IPCC and its constituent organizations, many of which appear as Crown Research Institutes within this network. While the New Zealand media does present views questioning the validity of the IPCC's findings and assertions, the actors cited as doing so are not often research organizations as in the American media. While New Zealand climate change science organizations do not act exclusively as knowledge brokers among diverse clusters of actors, the do play an important role of being the sources of scientific information for those actors that do act as knowledge brokers.

10.3.3 Policy Networks Around Climate Change

The field of action represents the behavior of actors – individuals, organizations, states – as they interact to promote or oppose change, often though policy. National policies interact with and help or hinder the formation of global regimes (Broadbent 2010). Sabatier and Jenkins-Smith's (1993) concept of the Advocacy Coalition Framework (ACF) uses network structures in order to investigate action through policy processes. One of the original goals of the ACF was to investigate how actors mobilized within advocacy coalitions around scientific information to inform environmental policy (Weible et al. 2011). In order to do this, actors must identify allies with common objectives they are willing and able to enter into coalition with

(Weible and Sabatier 2005). The prevailing evidence indicates that actors with similar beliefs about the implications of scientific knowledge tend to coordinate with each other on actions (Zafonte and Sabatier 1998). However, recent research by Baldassarri and Diani (2007) has shown that support networks connect diverse clusters of organizations with common general beliefs but distinct organizational identities and priorities. Di Gregorio (2012) terms this process 'macro-integration' – where robust support networks are formed by organizations which do not necessarily have collective identity or set of values but their distinct identities and value sets are compatible enough to form a coalition. In this instance, action is most effective when actors with a common purpose but diverse identities network.

The study of advocacy coalitions has traditionally framed coalition building and function within the context of political contention. In this case, advocacy coalitions fit within the field of action. The theory of the Treadmill of Production (Schnaiberg et al. 2003) contends that measures to protect the environment will be met with severe opposition from industrial and exploitative actors within society. In this scheme, the only way to bring about change to the environmentally destructive status quo is through massive social mobilization. This social mobilization can manifest itself through demonstrations, boycotts, electoral victories, the passing of regulatory legislation, or some combination of all of the above. However, advocacy coalitions can also be framed as instruments of influence and instruction. In this case, advocacy coalitions fit within the field of discourse. The theory of Ecological Modernization (Janicke 2002; Mol and Sonnenfeld 2000) maintains that protections to the environment can be brought about through more passive means. In this instance, behaviors that protect ecosystems are brought about through the dissemination of norms, the diffusion of new ideas, and a non-politicized learning process (Broadbent 2010). Here, advocacy coalitions bring about change through consensus rather than through contention. In the following sections, we will be examining advocacy coalitions within the framework of the action field.

The implication of social network analysis (SNA) around environmental advocacy coalitions is an increased understanding of what features of social networks are necessary precursors for successful advocacy around environmental policy in general, and climate change policy in particular (Bodin et al. 2006; Crona and Bodin 2006). Tompkins and Adger (2004) put forward that social networks with more ties between stakeholders and regulatory actors builds resilience and adaptive capacity to environmental change in societies. However, there has been little research into the relation between network structure and specific policy outcomes. Additionally, while social movements are well studied within the social sciences, advocacy coalitions are less so. It can be difficult to tease apart the differences between an advocacy coalition and a social movement, but there is a growing need to within the literature, as advocacy coalitions continue to engage an increasingly diverse set of stakeholders around issues as complex as climate change.

In order to change behavior at a social level, the initial bearers of claims and norms must expand networks: persuade an increasing circle of adherents until their number and activity reaches a critical mass (Broadbent 2010). In this process, knowledge must be operationalized and social learning must turn into social mobilization. In order to effect change in response to the knowledge they are claiming as legitimate, advocacy coalitions need to garner enough political support to enable them to pass and enforce regulations and laws that demand and enforce certain environmental standards. These policies can become manifest through legislation such as an emissions trading scheme (cap and trade law) or some other form of emissions regulation (i.e., a carbon tax). In order to do this, advocacy coalitions must form connections with a larger and more diverse set of actors within a society to achieve this critical mass. When a mobilized advocacy coalition garners enough support to form a majority government, or gains enough support to push a government on a particular piece of policy, it begins to exert power within the state, through the legislative and policy-making process. From that vantage point, the new regime can establish the legal and policy conditions to bring about society-wide change in behavioral norms (by education, persuasion, inducement, regulation, new institutions and other means) (Broadbent 2010).

The political strength of advocacy coalitions in taking action to push for climate change legislation appears to vary depending on the cultural milieu of the society in question. In Sweden, where social corporatism is the norm within the political culture, there is a diverse representation of actors within the advocacy coalitions centered around climate change legislation. This included incorporation of a large number of environmental NGOs within the Swedish policy sphere. Within the US, the political culture is one of pluralism. Ironically, this leads actors to compete with one another for dominance within the policy sphere, with the wealthier business entities (often opposed to climate change legislation) exerting more influence. This can lead to more environmental NGOs being left outside the political process.

Organizational actors who work in tandem within an advocacy coalition on the political stage are central players within the study of social networks. Such networks often build upon longer existing relationships, such as the long-term exchange of mutual aid (reciprocity). These networks suffuse societies in different densities and patterns, helping give rise to different policy making processes. For instance, the reciprocity network penetrates the full Japanese field of labor politics very thoroughly, but in the US is only present among labor unions (Broadbent 2001, 2008). In the Japanese case, the presence of reciprocity networks increased the likelihood that the so-connected actors would transfer political support.

Broadbent (2005) notes that in Japan's action phase, advocacy coalitions have played a weak role in influencing national climate change policy. Frames concerning national prosperity and energy sufficiency formulated by the Ministry of Economy, Trade and Industry have dominated debates about climate change, rather than fears about the future disasters that climate change will bring such as presented by the Environment Ministry. The close alliance between the ruling Liberal Democratic Party and the corporatistic business sector led by the JFEO (Japan Federation of Economic Organizations or *Keidanren*) have further buttressed a weak political posture toward climate change insisting on voluntary action by business and no carbon tax on consumption rather than the imposition of regulations by government.

On the other hand, in New Zealand advocacy coalitions have played a vital role in passing domestic climate change legislation. In terms of scope, coverage, and speed of implementation, New Zealand's national emissions trading scheme (ETS) is arguably one of the world's most ambitious climate policies. Passed by the outgoing Labour government in late 2008, the ETS covers all six greenhouse gases (GHGs) listed within the Kyoto Protocol. The current incarnations of the European Union's ETS, as well as the Swiss and Norwegian ETS, cover only the emissions of carbon dioxide (CO_2). By regulating methane (CH_4) and nitrous oxide (N_2O) under the scheme, New Zealand opened up sectors of economic activity to emissions regulation, such as agriculture, forestry, and land use change, which had typically been ignored in Europe (Moyes 2008). What is remarkable about this is that New Zealand is by and large an agricultural export economy, unique within Annex I nations, with a large share of its greenhouse gas emissions coming from agriculture and forestry. The average proportion of GHG emissions from post-industrial nations are 83.2 % CO_2, 9.5 % CH_4, and 5.9 % N_2O (UNFCCC 2007). At the time the ETS was drafted, New Zealand's GHG emissions were 46.5 % CO_2, 35.2 % CH_4, and 17.2 % N_2O, with HFCs, PFCs, and SF_6 together accounting for the remaining 1.1 % of total GHG emissions (Ministry for the Environment 2007).

While New Zealand's policy instrument of choice for dealing with climate change generated substantial debate within both parliament and the press, a large proportion of organizational actors within New Zealand society accepted the frame that climate change was a long-term threat to the nation and that a policy approach was an appropriate response. A number of the ministries within the Labour government, three of the nation's major political parties, that nation's alternative energy companies, the airline industry, and a slew of businesses and environmental NGOs supported the creation of a comprehensive ETS. A coalition of agricultural and industrial business lobbying groups, and two of the nation's smaller political parties opposed the passage of the ETS. The opposition National Party conditionally supported the ETS, but later took up opposition against it as they objected to the speed at which it would be implemented across all sectors of the economy. Despite losing support from the National Party at the last minute, the Labour Government was able to cobble together support for climate change legislation that it was able to pass the ETS in 2008.

The social networks in focus here are advocacy coalitions which mobilized either for or against the ETS within New Zealand civil society in the lead up to the nation's Kyoto commitment period. The advocacy coalition around creating a domestic emissions trading scheme within New Zealand is characterized by a high degree of connectivity between actors within the network. Clark's Labour Government created the cabinet position of Minister of Climate Change and this actor holds a high degree of centrality within the network along with Ministry for the Environment and the Labour Party itself. While modularity is observable within the network (especially around the cluster of environmental NGOs), the degree of betweenness among all of the actors is relatively low, suggesting the network is relatively robust. Because the network is characterized by high measures of connectivity and low levels of betweenness, the centrally located Labour government actors were able to assert control within the network and direct the form carbon regulation would take.

10.4 Conclusion

This chapter has argued for the utility of Social Network Analysis for the inter-disciplinary investigation of climate change as a social issue and problem. SNA can integrate the effects of ideas from the social and natural sciences and the humanities upon the fields of discourse and action around climate change. Examining the structure and flow of different kinds of social networks around climate change reveals patterns of understanding and action that shape the social response to climate change and other problems. Examining both discourse networks and advocacy coalitions, the chapter has developed an initial comparison of differences in how three societies, Japan, New Zealand, and the United States, have been framing and responding to climate change, The chapter, drawing upon early results from the project Comparing Climate Change Policy Networks (Compon), illustrates the great potential of the network approach for the inter-disciplinary study of climate change and society cross-nationally.

References

Abrahamson, E., & Rosenkopf, L. (1997). Social network effects on the extent of innovation diffusion: A computer simulation. *Organization Science, 8*(3), 289–309.

Baker, V. R. (2007). Flood hazard science, policy, and values: A pragmatist stance. *Technology and Society, 29*, 161–168. (167)

Baldassarri, D., & Diani, M. (2007). The integrative power of civic networks. *American Journal of Sociology, 113*(3), 735–780.

Beck, U. (1992). *Risk society. Towards a new modernity*. London: Sage.

Beck, U. (2002). A life of one's own in a runaway world: Individualization, globalization and politics. In U. Beck & E. Beck-Gernsheim (Eds.), *Individualization institutionalized individualism and its social and political consequences* (pp. 22–29). London: Sage.

Blau, P. M. (1977). *Inequality and heterogeneity*. New York: Free Press.

Bodin, Ö., & Prell, C. (2011). *Social networks and natural resource management: Uncovering the social fabric of environmental governance*. Cambridge/New York: Cambridge University Press.

Bodin, Ö., Crona, B., & Ernstson, H. (2006). Social networks in natural resource management: What is there to learn from a structural perspective? *Ecology & Society, 11*(2), 395–402.

Borgatti, S. P., & Foster, P. C. (2003). The network paradigm in organizational research: A review and typology. *Journal of Management, 29*(6), 991–1013.

Bott, E. (1957). *Family and social network*. London: Tavistock.

Boykoff, M. T., & Boykoff, J. M. (2004). Balance as bias: Global warming and the US prestige press. *Global Environmental Change, 14*, 125–136.

Broadbent, J. (1998). *Environmental politics in Japan: Networks of power and protest*. Cambridge: Cambridge University Press.

Broadbent, J. (2001). Social capital and labor politics in Japan: Cooperation or cooptation? In J. Montgomery & A. Inkeles (Eds.), *Social capital as a policy resource in Asia and the Pacific Basin* (pp. 81–95). Boston: Kluwer Academic Publishers.

Broadbent, J. (2003). Movement in context: Thick networks and Japanese environmental protest. In M. Diani & D. McAdam (Eds.), *Social movements and networks: Relational approaches to collective action* (pp. 204–229). New York: Oxford University Press.

Broadbent, J. (2005). Japan's environmental politics: Recognition and response processes (chapter 5). In H. Imura & M. Schreurs (Eds.), *Environmental management in Japan.* Washington, D.C.:The World Bank and Northampton, MA: Edward Elgar.

Broadbent, J. (2008). *Japan's butterfly state: Reciprocity, social capital and embedded networks in U.S. and German comparison.* Unpublished.

Broadbent, J. (2010). Science and climate change policy making: A comparative network perspective. In A. Sumi, K. Fukushi, & A. Hiramatsu (Eds.), *Adaptation and mitigation strategies for climate change.* Tokyo: Springer.

Broadbent, J., Yun, S-J., Ku, D., Ikeda, K., Satoh, K., Pellissery, S., Swarnakar, P., Lin, T-L., Jin, J. (2013). Asian societies and climate change: The variable diffusion of global norms. *Globality Studies Journal* (#34). Retrieved from http://globality.cc.stonybrook.edu/wp-content/uploads/2013/07/032JBroadbent.pdf

Brown, G. E., Jr. (1992). Global change and the new definition of progress. *Geotimes, 18*(21), 19–21.

Brown Weiss, E., & Jacobson, H. K. (Eds.). (1998). *Engaging countries: Strengthening compliance with international environmental accords.* Cambridge, MA: MIT Press, 533pp.

Burridge, S., Vaughter, P., Fisher, D. R., & Weber, J. (2013). *A hot mess: Climate change issue definition (or a lack thereof) in the United States News Media.* Unpublished manuscript.

Burt, R. S. (1992). *Structural holes: The social structure of competition.* Cambridge: Harvard University Press.

Burt, R. (2003). The social capital of structural holes. In M. F. Guillen, R. Collins, P. England, & M. Meyer (Eds.), *The new economic sociology: Developments in an emerging field* (pp. 148–189). New York: Russell Sage.

Clayton, H. et al. (2006). *Is a cross-party consensus on climate change possible – or desirable?* (Report of the first inquiry of the All Parliamentary Climate Change Group). London: HMSO.

Coleman, J. S. (1990). *Foundations of social theory.* Cambridge: Belknap Press.

Collins, H., & Evans, R. (2007). *Rethinking expertise.* Chicago: University of Chicago Press.

Crona, B., & Bodin, Ö. (2006). What you know is who you know? Communication patterns among resource users as a prerequisite for comanagement. *Ecology & Society, 11*(2), 290–312.

Di Gregorio, M. (2012). Networking in environmental movement organization coalitions: Interest, values or discourse? *Environmental Politics, 21*(1), 1–25.

Evans, P. B., Jacobson, H. K. J., & Putnam, R. D. (1993). *Double-edged diplomacy: International bargaining and domestic politics.* Berkeley: University of California Press.

Folke, C., Hahn, T., Olsson, P., & Norberg, J. (2005). Adaptive governance of social–ecological systems. *Annual Review of Environment and Resources, 30,* 441–473.

Frank, K. A., & Yasumoto, J. Y. (1998). Linking action to social structure within a system: Social capital within and between subgroups 1. *American Journal of Sociology, 104*(3), 642–686.

Ghimire, S. K., McKey, D., & Aumeeruddy- Thomas, Y. (2004). Heterogeneity in ethnoecological knowledge and management of medicinal plants in the Himalayas of Nepal: Implications for conservation. *Ecology and Society, 9*(3), 6.

Giddens, A. (2009). *The politics of climate change* (pp. 59–73). Cambridge: Polity Press.

Gouldson, A., Lidskog, R., & Wester-Herber, M. (2007). The battle for hearts and minds. Evolutions in organisational approaches to environmental risk communication. *Environment and Planning C, 25*(1), 56–72.

Granovetter, M. (1985). Economic action and social structure: The problem of embeddedness. *American Journal of Sociology, 91,* 481–510.

Gustafsson, K., & Lidskog, R. (2012). Acknowledging risk, trusting expertise, and coping with uncertainty: Citizens deliberations on spraying an insect population. *Society and Natural Resources, 25,* 587–601.

Hajer, M., & Versteeg, W. (2005). Performing governance through networks. *European Political Science, 4*(3), 340–47.

Hamilton, C. (2007). *Scorcher: The dirty politics of climate change* (p. 147). Melbourne: Black Ink Agenda.

Helm, D. (2005). *Climate-change policy*. New York: Oxford University Press.

Höijer, B., Lidskog, R., & Uggla, Y. (2006). Facing dilemmas. Sense-making and decision-making in late modernity. *Futures, 38*, 350–366.

IPCC. (2007a). Climate Change 2007: The physical science basis: Intergovernmental Panel on Climate Change.

IPCC (Intergovernmental Panel on Climate Change). (2007b). Summary for policymakers. In: *Climate Change 2007: Impacts, adaptation and vulnerability. Contribution of Working Group II to the fourth assessment report of the Intergovernmental Panel on Climate Change.* Cambridge/New York: Cambridge University Press. Retrieved February 1, 2008, http://www.ipcc.ch/pdf/assessment-report/ar4/wg2/ar4-wg2-spm.pdf

Jacobson, H., & Weiss, E. B. (1998). A framework for analysis. In E. B. Weiss & H. Jacobson (Eds.), *Engaging countries: Strengthening compliance with international environmental accords* (pp. 1–18). Cambridge, MA: MIT Press.

Jaeger, C., Dürrenberger, G., Kastenholz, H., & Truffer, B. (1993). Determinants of environmental action with regard to climate change. *Climate Change, 23*, 193–211.

Janicke, M. (2002). The policy System's capacity for environmental policy: The framework for comparison. In H. Weidner & M. Janicke (Eds.), *Capacity building in national environmental policy: A comparative study of 17 countries* (pp. 1–18). Berlin: Springer.

Jasanoff, S. (1998). Contingent knowledge: Implications for implementation and compliance. In E. Brown Weiss & H. K. Jacobson (Eds.), *Engaging countries: Strengthening compliance with international environmental accords* (p. 65). Cambridge, MA: MIT Press.

Jasanoff, S. (2005). *Designs on nature. Science and democracy in Europe and the United States.* Princeton: Princeton University Press.

Knoke, D., Pappi, F., Broadbent, J., & Tsujinaka, Y. (1996). *Comparing policy networks: Labor politics in the U.S., Germany and Japan.* New York: Cambridge University Press.

Latour, B. (2005). *Reassembling the social: An introduction to actor-network-theory.* Oxford: Oxford University Press.

Laumann, E., & Pappi, F. U. (1976). *Networks of collective action: A perspective on community influence systems.* New York: Academic.

Leifeld, P., & Haunss, S. (2011). Political discourse networks and the conflict over software patents in Europe. *European Journal of Political Research, 50*(6).

Leifeld, P., & Haunss, S. (2012). Political discourse networks and the conflict over software patents in Europe. *European Journal of Political Research, 51*, 382–409.

Lidskog, R. (2008). Scientised citizens and democratised science. Re-assessing the expert-lay divide. *Journal of Risk Research, 11*(1–2), 69–86.

Ministry for the Environment. (2007). *New Zealand's greenhouse gas inventory 1990–2005* (The National Inventory Report and Common Reporting Format 17). Available at http://www.mfe.govt.nz/publications/climate/nir-jul07/nir-jul07.pdf

Mol, A. P. J. (2001). *Globalization and environmental reform: The ecological modernization or the global economy.* Cambridge, MA: MIT Press.

Mol, A., & Sonnenfeld, D. (2000). *Ecological modernization around the world: Perspectives and critical debates.* London: Frank Cass.

Moyes, T. E. (2008). Greenhouse gas emissions trading in New Zealand: Trailblazing comprehensive cap and trade. *Ecology Law Quarterly, 35*, 911–968.

Nowotny, H., Scott, P., & Gibbons, M. (2001). *Re-thinking science. Knowledge and the public in an age of uncertainty.* Cambridge: Polity Press.

Oh, H., Chung, M.-H., & Labianca, G. (2004). Group social capital and group effectiveness: The role of informal socializing ties. *Academy of Management Journal, 47*(6), 860–875.

Power, M. (2007). *Organized uncertainty. Designing a world of risk management.* Oxford: Oxford University Press.

Prell, C. (2012). *Social network analysis: History, theory and methodology.* Los Angeles/London: Sage.

Pretty, J., & Ward, H. (2001). Social capital and the environment. *World Development, 29*(2), 209–227.

Reagans, R., & McEvily, B. (2003). Network structure and knowledge transfer: The effects of cohesion and range. *Administrative Science Quarterly, 48*(2), 240–267.

Renn, O. (2008). *Risk governance. Coping with uncertainty in a complex world*. London: Earthscan.

Sabatier, P. A., & Jenkins-Smith, H. C. (1993). *Policy change and learning: An advocacy coalition approach*. Boulder: Westview Press.

Schnaiberg, A., Pellow, D., & Weinberg, A. (2003). The treadmill of production and the environmental state. In C. R. Humphrey, T. L. Lewis, & F. H. Buttel (Eds.), *Environment, energy, and society: Exemplary works* (pp. 412–423). Belmont: Wadsworth Thomson Learning.

Schneider, S., Rosencranz, A., & Niles, J. (2002). *Climate change policy: A survey*. Washington, DC: Island Press.

Schreurs, M. A. (2002). *Environmental politics in Japan, Germany, and the United States*. Cambridge/New York: Cambridge University Press.

Shearman, D., & Wayne Smith, J. (2007). *The climate change challenge and the failure of democracy* (p. 133). London: Praeger.

Siegrist, M., Keller, C., & Kiers, H. A. L. (2005). A new look at the psychometric paradigm of perception of hazards. *Risk Analysis, 25*(1), 211–222.

Slovic, P. (1986). Informing and educating the public about risk. *Risk Analysis, 6*(4), 403–415.

Speth, J., & Haas, P. (2006). *Global environmental governance*. Washington, DC: Island Press.

Tompkins, E. L., & Adger, W. N. (2004). Does adaptive management of natural resources enhance resilience to climatic change? *Ecology and Society, 9*(2), 10.

Tversky, A., & Kahneman, D. (1974). Judgment under uncertainty. Heuristics and biases. *Science, 185*(4157), 1124–1131.

UNFCCC. (2007, November 19). *Subsidiary body for implementation, compilation and synthesis of fourth national communications*. Executive Summary 5, UN Doc, FCCC/SBI/2007/INF.6.

Van Dijk, E. M. (2011). Portraying real science in science communication. *Science Education, 95*, 1086–1100.

Vaughter, P. (2013). *Can the Kiwi Ets Fly?: Climate Change Policy and Rhetoric within the New Zealand News Media*. Unpublished manuscript.

Walls, J., Pidgeon, N., Weyman, A., & Horlick-Jones, T. (2004). Critical trust: Understanding lay perceptions of health and safety risk regulation. *Health, Risk & Society, 6*(2), 133–150.

Wasserman, S., & Faust, K. (1997). *Social network analysis: Methods and applications*. New York: Cambridge University Press.

Watson, B. D. (2002). *Rethinking organizational learning*. Unpublished doctoral dissertation, University of Melbourne.

Weible, C. M., & Sabatier, P. A. (2005). Comparing policy networks: Marine protected areas in California. *Policy Studies Journal, 33*(2), 181–202.

Weible, C. M., Sabatier, P. A., Jenkins-Smith, H. C., Nohrstedt, D., Henry, A. D., & deLeon, P. (2011). A quarter century of the advocacy coalition framework: An introduction to the special issue. *Policy Studies Journal, 39*(3), 349–360.

Weidner, H., & Janicke, M. (2002). Summary: Environmental capacity building in a converging world. In H. Weidner & M. Janicke (Eds.), *Capacity building in national environmental policy: A comparative study of 17 countries* (pp. 409–443). Berlin/New York: Springer.

Weimann, G. (1982). On the importance of marginality: One more step into the two-step flow of communication. *American Sociological Review, 47*(6), 764–773.

Young, O. R. (2002). *The institutional dimensions of environmental change: Fit, interplay, and scale*. Cambridge, MA: MIT Press.

Zafonte, M., & Sabatier, P. A. (1998). Shared beliefs and imposed interdependencies as determinants of ally networks in overlapping systems. *Journal of Theoretical Politics, 10*(4), 473–505.

Chapter 11
Designing Social Learning Systems for Integrating Social Sciences into Policy Processes: Some Experiences of Water Managing

Kevin Collins

11.1 Introduction

Let me begin my introduction to this chapter with a short story. As a first year undergraduate, whilst browsing in my university library I came across a book on environmental politics. At the top of one of the chapters was a quote from a United Nations (UN) discussion some decades before about an ongoing policy conflict between two members of the UN. A delegate from another country, exasperated at the intractability of the situation, had stood up in the debate and shouted across the chamber to the two sparring representatives: '*Integrate damn you! Integrate!*'. Not appreciating then the importance of maintaining references, I have since been unable to find either where this quote originated or the book it was in – despite many hopeful attempts.

Apart from the lesson of keeping good references, this quote has always stuck in my mind as revealing some essential questions about integration: what is it; who does it; who determines what integrates with what; and under what conditions? Is integration even a choice – an invitation that can be refused? Perhaps most revealing, for the purposes of this chapter, is that the quote does not give much clue as to *how* the disputing representatives might integrate.

The desire for integration is not in question. Even a cursory review of current lifestyles, societies and policy agendas demonstrates that a desire for integration is now everywhere: especially in technology, society, business and governance. In environmental arenas, integration, at its most basic, and perhaps most challenging, is advanced as the connecting and harmonizing of environmental and human activity to achieve a socially desired state. Integration is used ubiquitously in environmental literatures, policy, and practice and is often the inescapable twin of the equally enigmatic *sustainability* in all its various forms.

K. Collins (✉)
Department of Engineering and Innovation, Open University,
Milton Keynes MK7 6AA, UK
e-mail: kevin.collins@open.ac.uk

M.J. Manfredo et al. (eds.), *Understanding Society and Natural Resources*,
DOI 10.1007/978-94-017-8959-2_11, © The Author(s) 2014

Integration is intuitively attractive and held up as the key to unlocking and progressing many policy situations in order to achieve sustainability. This is no more evident than the environmental policy 'event of the decade', the Rio + 20 conference in Brazil in 2012. Attended by heads of state and government, it is no exaggeration to say that hopes were high that agreements would be reached on a range of issues. With the global media present and watching every move of delegates, it was evident that 'something must be done' to address the lack of coherent policies and disjuncture with practice.

After 12 days of discussions and events, the main outcome was the non-binding document *The Future We Want* (UN 2012). Throughout, this document emphasized the need for integration of the economic, social and environmental dimensions of sustainable development in a holistic and cross-sectoral manner at all levels.

But the subsequent indications are that Rio + 20 was, at best, a political compromise that will achieve marginal gains in some areas, and, at worst, a failure. This rather bleak picture of the state of the political and global environment is all the more notable despite hundreds of international agreements and goals to tackle environmental issues and associated human welfare concerns. A recent report of the United Nations Environment Programme suggests that progress on a range of key issues relating to the Millenium Development Goals such as access to drinking water has been slow and in other areas, such as sanitation or wetland protection, non-existent (UNEP 2012). At the heart of many of these issues is the sense that there has been a failure to integrate social, economic and environmental concerns into policy and practice.

Efforts for integrating natural and social sciences are perhaps more advanced than integrating social science into policy, but, in either case, there is still much uncertainty about what integration actually entails and the methodologies that can be deployed.

This chapter explores some of these concerns in relation to integration and how a praxis (theory informed practice) based on systems and social learning approaches might be used to facilitate integration of the social sciences for policy and practice. In particular, the emphasis is on designing social learning systems as a methodological innovation for integration.

The discussion presented here draws on over a decade of designing social learning systems for engaging with and progressing complex environmental management situations relating to the governance of water resources in different contexts. At the core of this work is an emphasis on epistemological awareness as a means to enable integration of different disciplinary perspectives. The chapter does not discuss in detail the differences within the social sciences that give rise to a lack of integration – save to note that all disciplines have their own epistemologies. The approach to integration advocated in this chapter rests on surfacing epistemology and opening up opportunities for social learning. Thus, while the ideas and methodology discussed here have been developed in relation to integration of natural and social science integration, they apply as much to social science integration.

This chapter is divided into seven sections. Following this introduction, Sect. 11.2 explores the framing choices associated with many natural resource policy situations, before discussion of the links between integration and systems in Sect. 11.3. In Sect. 11.4, attention turns to social learning and designing social learning systems for integration. Some exemplars of social learning systems are explored in Sect. 11.5. A review of some constraints and opportunities relating to social learning systems is presented in Sect. 11.6. The chapter concludes with a short review of the implications for future research.

11.2   Framing Choices in Environmental Policy Situations

Efforts focused on integrating social sciences into policy require an appreciation of the kinds of situations which are encountered by policy-makers, practitioners and scientists of any discipline.

In the runup to Rio + 20, the *Planet under Pressure* conference in London 2012 was notable for its conference declaration that humans have entered the Anthropocene – the planetary era defined, for the first time, by human activity where 'many Earth-system processes and the living fabric of ecosystems are now dominated by human activities' (Brito and Stafford Smith 2012: 2). At the same time, the Global Environmental Outlook report published prior to the Rio + 20 conference also endorsed the message of human initiated climate change. In particular, it noted that 'As human pressures on the Earth System accelerate, several critical global, regional and local thresholds are close or have been exceeded. […] The impacts of complex, non-linear changes in the Earth System are already having serious consequences for human well-being. […] Because of the complexities of the Earth System, responses need to focus on the root causes, the underlying drivers of environmental changes, rather than only the pressures or symptoms' (UNEP 2012: 9).

Although debates will continue about the primacy or otherwise of anthropogenic climate change, these assessments make two things clear: (1) that our understandings of natural resource management are changing and (2) that understandings and practices that are 'more of the same' are no longer good enough (Schön 1995). At the core of prospects for integration and changing policy and practice is the need to change the way many environmental situations are conceptualised or framed (Schön and Rein 1994). The importance of being aware of how a situation is framed is as true for the social sciences as natural sciences. However, as Redman et al. (2004: 168) observe, while 'most scientists agree that interdisciplinary collaboration is essential […] our academic training and administrative barriers make that goal difficult to accomplish'. The extent to which the barriers can be reduced depends to some extent on how each discipline within the social sciences seeks to frame the situation through its particular disciplinary lens: whether political; psychological, geographical or economics for example.

Frances (1951) suggests that efforts towards the integration of the different branches in social science requires clarity of epistemological and ontological principles underpinning each of the disciplines (Frances 1951). Scrase and Sheate (2002), in their assessment of integration in environmental assessment, identify over 14 meanings of integration prevalent in practice, suggesting that value judgements as much as technical clarity are key to shaping understandings of integration.

Building on these concerns about existing framings, Dovers suggests any understanding of integration as a principle needs to acknowledge spatial and temporal dimensions, disciplinary boundaries, prevailing cultures and the dynamics of social systems, information and knowledge systems (Dovers 1997). In other words, the imperative for integration 'stems from recognition of the interdependence of human and natural systems, expressed in the research and policy agendas of sustainability' (Dovers 2005:3).

In recognising interdependence, the challenge of integration begin to centre on 'development of methods, processes, data streams, and so on to create *integrative capacity* [which, in turn,] demands a sophisticated understanding of the interactions between highly complex, non-linear, and often closely interdependent human and natural systems' (Dovers 2005: 3, emphasis in original).

Writing from a systems perspective, Ison also extends the boundary of our concern away from individual disciplines within social science, to the situation itself with which social scientists engage, arguing 'the nature of situations cannot be divorced from our own epistemological, theoretical and methodological commitments' (Ison 2008a: 244). In other words, the individual disciplines frame understandings of situations. Efforts for integrating social sciences must therefore pay attention to these framings.

Some possible framing choices include seeing situations either as *difficulties* or *messes* (Ackoff 1974); *tame or wicked* problems (Rittell and Webber 1973); or existing on the high ground of 'technical rationality' or part of the '*swamp*' of real life issues (Schön 1995).

The conventional environmental policy paradigm tends to focus on bio-physical systems and frames many natural resource situations as technical issues – ie more as 'problems' or difficulties rather than 'messy situations' (Collins and Ison 2009a). But some examples of re-framing are evident in policy. The Australian Public Service Commission noted that climate change is characterized as a 'wicked problem' because it is

> pressing . . . highly complex . . ., involving multiple causal factors and high levels of disagreement about the nature of the problem and the best way to tackle it (APSC 2007: 1).

Of particular interest from a methodological point of view, the ASC recognises the need to address 'wicked problems' with approaches that are, among other things, (i) holistic, (ii) innovative and flexible, (iii) work across agency boundaries, (iv) increase understanding and stimulate a debate, (v) engaging of stakeholders and citizens in understanding the problem and in identifying possible solutions; and (vi) tolerate uncertainty and accept the need for a long-term focus. This assessment by the ASC points to the need to understand more clearly how different elements of the situation inter-relate and give rise to the 'messiness'.

Drawing on a tradition of systems in a range of natural resource contexts, including agricultural extension, the EU funded Social Learning for Integrated Management of Water (SLIM) project identified a series of system-level characteristics of messy situations comprising: interdependency, complexity, uncertainty, controversy and multiple stakeholdings and thus perspectives (SLIM 2004a; Steyaert and Jiggins 2007). How these characteristics commonly frame a 'messy' situation, and thus what constitutes 'acceptable' responses, can be summarised as:

- multiple stakeholding – where diverse sets of actors actively construct their stake or interest in a situation
- interdependencies – existing when there is little agreement on the boundaries of an issue, or how it will be represented and communicated to others.
- complexity – arising from interdependencies and the diverse cause-effect relationships between local ecosystems, global climate systems and society. It is often linked to partial or complete lack of knowledge about a range of ecological and technical processes and risk (see Skidelsky 2008), social values and wants, and public policy-making imperatives.
- controversies – emerging from an interplay of the previous elements in particular contexts as seen through stakeholders' perspectives and value judgements, traditions of understanding (Russell and Ison 2007), and in the process of constructing their 'stakeholding' (after SLIM 2004b).

These system characteristics and elements combine and are expressed in different ways in different contexts: in some situations, complexity will be associated with data gaps and interpretation of cause and effect; in others complexity could be linked to scale issues or numbers and diversity of stakeholders involved. While it is impossible to pre-determine the exact mix, key to integration in these kinds of situations is epistemological awareness of how those involved in the situation are choosing to frame it. In turn, this can lead to an appreciation of the methodological choices that can be made for managing the natural resource situation in question and in particular what 'integration' among the diverse perspectives might entail.

Arising out of the SLIM project, Ison et al. (2007a) developed the following diagram to depict the importance of being aware of the different kinds of situations and the corresponding methodological approaches that can be chosen as part of the framing of situations.

On the left of Fig. 11.1, the problem is well defined and agreed and therefore a known set of responses, such as education or fiscal measures, can be readily deployed, accepted and used by those involved in this 'difficult' situation. On the right hand side, the situation is indeterminate (as shown by the incomplete boundary). The 'messy' situations extant on the right hand side of Fig. 11.1 can be seen in many natural resource policy contexts and are often evidenced by uncertainties about the nature of the situation itself; concerns about data gaps; and disagreements among stakeholders about what effort is required and how it should be focussed, to name but a few.

Where uncertainties, interdependencies and complexities are experienced, natural resource managing cannot be done by one or two actors in isolation. Instead, it requires a range of views and perspectives to be engaged in defining the situation and issues and determining an approach which is context relevant.

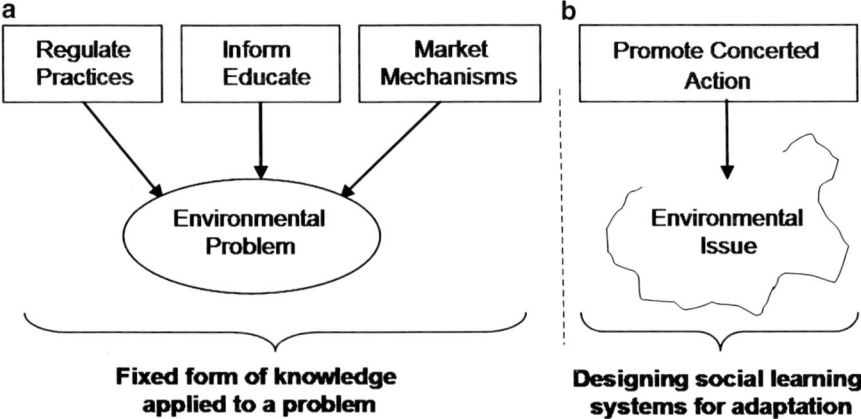

Fig. 11.1 The epistemological basis from which social learning for promoting concerted action can be developed as a purposeful choice. In (**a**), a known and agreed problem can be addressed by stakeholders using a known form of knowledge. In (**b**), social learning systems are required to determine responses to contested and incomplete understandings of the environmental issue (After SLIM 2004a; Ison et al. 2007a; Collins and Ison 2009b)

Appreciation of the characteristics of the situation through some form of learning can provide opportunities for a re-framing of the situation to enable integration of different disciplines.

The potential contribution of social science to the re-framing of research was explored by an EU expert group exploring research required for sustainable development. This group noted that:

> social scientists studying how research results get used or ignored in policy systematically come to the conclusion that a linear process does not work: there is not a clear domain of science, that produces knowledge, that feeds into or 'impacts' upon a separate system of policy. Rather, there is a set of multiple forms of knowledge, including a variety of research fields, which have to relate to a variety of policy areas and specific policies. (Anon 2007: 4)

Atkinson and Klausen's (2011) review of EU policies provides a similar assessment of integration. They note a 'high degree of comprehensiveness is necessary in order to accommodate concerns for the social and economic aspects of sustainability, not just the environmental aspects.' With this widening of concerns, however, they suggest consistency and aggregation are diminished leading to a well-known policy dilemma: 'the more aspects of an issue policy makers attempt to take into account, the more difficult it is to aggregate these aspects into a consistent policy' (246).

This perspective makes clear that while integration may be widely supported as an ideal principle in order to bring about improved policy and practice, a linear conceptualisation of the process of knowledge and practice, as typified by an emphasis on science – policy – action, will always be deficient as a means to enable integration of social science. Furthermore, an aggregation model of integration is conceptually and methodological flawed as policy-makers will struggle to cope with, literally, the added complexity.

These imperative of integration, constraints of framing and criticisms of linear policy suggest an alternative, more systemic conceptualisation of, and methodology for, integrating social science into policy is required. The potential for integrated natural resource managing from a *systems* perspective is explored next.

11.3   Integration and Systems

Integration is not a straightforward and linear task as might be implied by disarming phrases such as 'joined up thinking', 'adding' 'connecting' or indeed Wilson's 'jumping together' (Wilson 1998: 8).

Etymologically, integration comes from the Latin *integrare* – to make whole. Integration can be defined as the 'making up or composition of a whole by adding together or combining the separate parts or elements'; and 'combining of diverse parts into a complex whole' (OED 2012). In other words, integration is the process of making wholes.

The emphasis on making wholes is a fundamental aspect of systems theories and concepts. But what do we mean by system? Returning to etymology, the word system is derived from the Greek verb *synhistanai* meaning 'to place together'. According to the systems writer, Russell Ackoff, a system is

> a whole that cannot be divided into independent parts. […] The essential properties of a system taken as a whole derive from the interaction of its parts, not their actions taken separately. (Ackoff 1981: 64–65)

Peter Checkland, the systems writer, also emphasises the importance of the whole in understanding systems: 'the central concept 'system' embodies the idea of a set of elements connected together which form a whole, this showing properties which are properties of the whole, rather than properties of its component parts' (Checkland 1981: 3).

Thus a system can be described as 'a whole' comprising interdependent parts. With this simplified, but powerful understanding, the fundamental conceptual link between systems and integration becomes evident: ***integration is the making of wholes or systems***. In terms of methodology, the question then presents itself as: what methodological innovation can enable the making of wholes among diverse disciplines, including the social sciences, in policy contexts?

Before exploring some possible answers, it is important to be aware that integration does not automatically convey some positive quality to the system: not all systems and not all integration are positive or socially desirable. For example, an illegal waste system which ships and dumps waste computers from the EU into Africa, may be a highly integrated system, but one which, in this particular case, is considered socially undesirable and illegal.

Thus, from a systems point of view, integration, of itself, does not carry a value judgement in terms of ethics or outcomes, although of course evaluating the work done by the system may involve a wide range of measures of performance, including ethical considerations. In this sense, integration is a somewhat neutral concept – the definition and arrangement of the system for particular purposes by particular stakeholders determines whether integration achieves socially acceptable goals.

It is also important to be aware that any discussion on systems quickly reaches a point where an epistemological choice has to be made: do we see systems as 'out there', existing in the real world, or do we see systems as more observer dependent? This distinction is often linked to distinguishing between 'hard' systems which are claimed as an ontological reality and 'soft' systems (Checkland 1981) where emphasis is on the constructed nature of a system, dependent on the observer defining a system boundary. The different debates about these branches of systems and consequences for knowledge and knowing are not rehearsed here. Indeed, as Ison (2008b) suggests, the distinction between hard and soft systems has tended to create a dualism – a self-negating pair – rather than a duality – a connected and interdependent complementarity. It is, however, notable that the 'soft systems' tradition is more aligned to qualitative framings and methodologies which find resonance in the social sciences and for this reason the focus of this chapter remains on soft systems.

A central aspect of the soft systems perspective is raising awareness of the different boundary choices by different stakeholders in a situation. In this tradition of systems, it is accepted that stakeholders 'see' and value different elements and thus their boundary choices relating to their chosen system will vary. The choices of system boundary and system elements are fundamentally linked to the purpose ascribed to any system by the observer. Thus, within soft systems, a system can be described as 'a whole defined by someone as having a purpose'. The term 'system' is used as a shorthand for 'system of interest'– ie an observer-dependent formulation of what constitutes a whole made up of interdependent parts.

For example, within the same water catchment, a farmer's system of interest might comprise elements such as crop types, water supply, markets, land-use, financial concerns and family needs. A conservationist's system of interest in the same catchment might be local species, habitat and the river ecology. A business offering canoe trips in the catchment might be mostly concerned about river levels, water quality, habitat and access rights. Each stakeholders' system of interest may overlap in terms of having shared elements (eg the river), but this does not mean each system of interest is integrated with the others. Indeed, it is the sense in which these different systems of interests are experienced as competing rather than integrating, that gives rise to a sense of divergence of goals, controversy, disintegration of action and environmental loss.

The emphasis on wholes is an important conceptual framing in systems thinking, alongside the irreducibility of a system's characteristics to its component parts: ie a system cannot be reduced to its component parts. This has particular implications for the way we conceptualise integration. In short, integration is not a *thing* that can be applied to a situation. Rather, integration is a system-level property that ***emerges from*** the interaction and inter-dependencies of the different elements of the system.

This may seem self-evident, but its consequences are profound: integration cannot be *applied* to a situation of natural resource management by researchers (of any discipline), policy-makers or practitioners. Integration *arises* at the systems level – ie from the set of interactions between the different elements identified by someone as being part of the system. With this insight, the focus shifts away from trying to 'add' integration as some kind of ingredient into policy process which can be applied to a specific aspect, policy process or part of a situation. Instead, considerations centre about how to create the system-level conditions in which integration emerges.

Within this view of systems, natural resource management praxis is extended beyond the confines of engineering and biophysical science disciplines and into social sciences. Integration, understood as the making of wholes and an emergent property of systems, coupled with an emphasis on observer dependency, leads to the view that natural resource managing is fundamentally *a social process*. Methodologies are needed which recognise and engage with the social element and also the process element and which also recognise the complexity, controversy and multiple stakeholding in the way situations are framed. Drawing on exemplars from water resource managing, attention now turns to designing social learning systems as a key means to integrate social sciences into policy processes and outcomes.

11.4   Designing Social Learning Systems for Social Science Integration

Social learning is not a new idea or concept and its lineage can be traced in various literatures across social science branches including psychology, criminology, education and business studies. The concept is often linked to Bandura's work on social theory of learning (Bandura 1977) where individual learning takes place in a social context. Blackmore's (2007) review of social learning theories notes that social learning is likely to be interpreted and defined in accord with different theoretical traditions and interpretations. Of relevance to the discussion on natural resource managing, Blackmore also notes that social learning theory is part of the tradition of 'adaptive management' (Holling 1978) and is linked to Wenger's social theory of learning in communities of practice: defined as 'groups of people who share a concern, a set of problems, or a passion about a topic, and who deepen their knowledge and expertise in this area by interacting on an ongoing basis' (Wenger et al. 2002: 4–5; see also Wenger 1998).

With such origins and lineage it is perhaps not surprising that in the last decade there has been considerable attention from a range of authors in social learning for environmental managing (see Social Learning Group 2001) and in particular water resources (see, for example, Finger and Verlaan 1995; Daniels and Walker 1996; Woodhill and Röling 1998; Collins et al. 2007; Ison et al. 2007a, 2011; Pahl-Wostl et al. 2007; Pahl-Wostl et al. 2007, 2008; Mostert et al. 2007; Muro and Jeffrey 2008; see also Reed et al. 2010; Raadgever et al. 2012). The detailed distinctions and different interpretations of social learning by these authors are not rehearsed here.

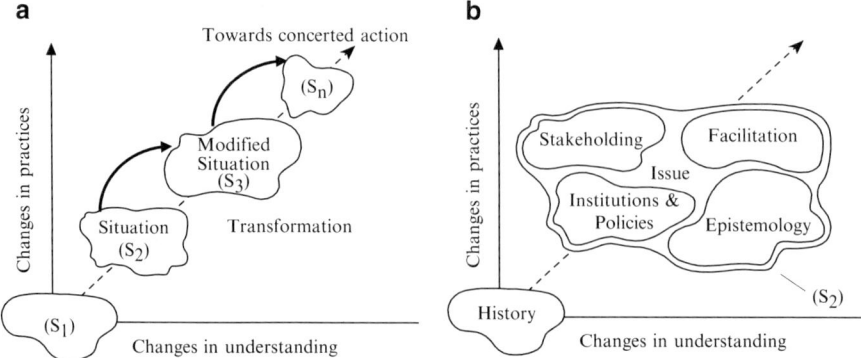

Fig. 11.2 A heuristic for social learning. (**a**) Social learning in complex, uncertain and contested situations over time enables transformation through changed understandings and practices leading to concerted action by stakeholders. (**b**) The SLIM social learning heuristic depicts how six key variables interact to shape issues and particular situations. These variables include history, stakeholding, facilitation, institutions and policies, and epistemology (After SLIM 2004a; Ison et al. 2007a)

Instead, the definitions of social learning developed by the previously mentioned SLIM project were used as the conceptual and methodological basis of the case studies reported below. In the SLIM project, social learning came to be understood as one or more of four potentially inter-related processes:

(i) the process of co-creation of knowledge, which provides insight into the history of, and the means required to transform, a situation;

(ii) the convergence of goals (more usefully expressed as agreement about purpose or purposes), criteria and knowledge leading to awareness of mutual expectations and the building of relational capital;

(iii) the change in behaviours that results from the understanding gained through doing ('knowing') that leads to concerted action; and

(iv) arising from these, social learning is thus an emergent property of the process of transforming a situation (SLIM 2004c, d; see also Ison et al. 2007a; Collins and Ison 2009a, b).

Social learning is thus based on the process of multiple stakeholders socially constructing an issue in which their understandings and practices change so as to transform the situation of concern. This interpretation of social learning refers to *collective* learning – ie learning at the system level – in a social context compared to Bandura's individual learning in a social context

A diagrammatic depiction of the process of social learning is shown in Fig. 11.2. The axes represent the relationship between changes in the situation (s1–>s2 etc.) arising from changes in understanding and action and the process of transformation over time. The key elements relating to the transformation process are shown: starting context; institutions; facilitation; stakeholding and epistemological constraints.

Appreciating the starting context is important so as to become aware of legacies, framings and previous experiences of those involved which may have led to previous actions, divisions, conflict or opportunities for new practices in the current situation. The institutions element is broadly interpreted and concerns those aspects of a situation which enable or constrain behaviours and practices such as laws, regulations, policies, organisations, traditions and customs. The element titled as facilitation is also broadly interpreted and refers to people, activities and/or things which enable stakeholders to engage in conversations and inquiry. This can be a professional facilitator but it can also involve some intermediary object (see Steyaert et al. 2007) around which new debates and practices are focussed. Instead of the more usual reference to stakeholder, the heuristic specifically refers to stakeholding to note that individuals actively construct their stake and that this can be changed as a result of engaging with others in a social learning process.

The final element refers to epistemological constraints. In the original SLIM work (see SLIM 2004e) this was titled as 'ecological constraints' as it was noted that there were often diverging and competing understandings and conceptual models of ecology in many natural resource management debates. The SLIM researchers later reworked this, expanding the title of this element to 'epistemological constraints' or 'epistemology' to denote that all conceptual models and ways of knowing from any disciplinary branch (including social sciences) or praxis could constrain the way the situation is experienced and understood.

The key aspect of this diagram is that it is intended as a heuristic rather than prescriptive model of social learning – what occurs within each element and the configuration and 'weighting' of each will vary according to specific context.

Within a social learning paradigm a priority in research practice is to know how to create the circumstances for social learning to occur – ie designing a social learning system. The heuristic depicted in Fig. 11.2 reveals that in social learning, the inquiry moves away from routine or first order learning, by questioning starting assumptions and making sense of context, thus revealing the second order framings used by stakeholders in the situation as they engage with epistemological differences. A social learning system should enable those involved, whether social scientists, scientists, policy makers and practitioners, or any combination thereof, to question framings, norms, policies and objectives in interactive processes involving multiple stakeholders.

In emphasising social learning, integration of social science in policy-making becomes centred on a concern about *designing* social learning systems (see Ison et al. 2007b) for natural resource managing, as a way of integrating different disciplines within social science to enable contributions to policy processes.

The work undertaken as part of the SLIM project found that to create the conditions for integration to emerge, a social learning system should have the following characteristics:

- Systemic features
    - Comprises elements or activities
    - Exhibits connectivity

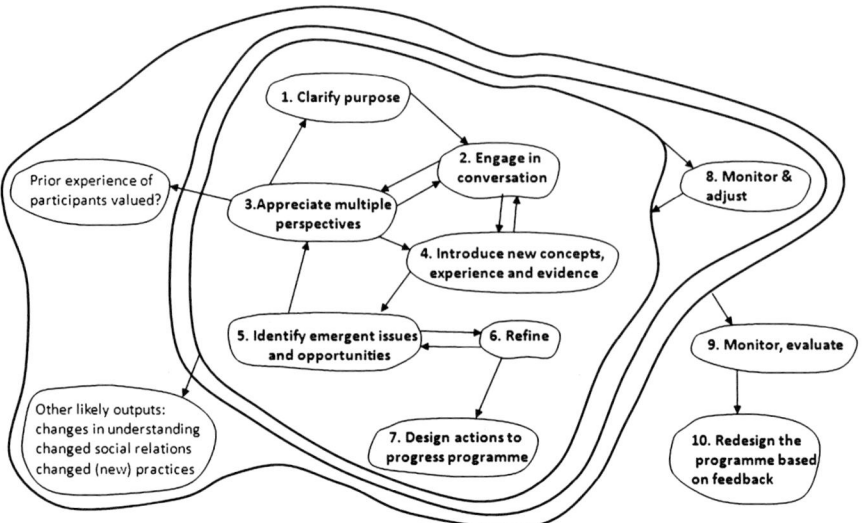

Fig. 11.3 Design for a social learning system in complex natural resource management situations (After Ison et al. 2009)

- – Results in transformation
- – Has emergent properties
- – Is bounded in some way

• Design/designer features

- – It is purposeful to those who participate
- – It is not deterministic
- – The 'designer' is aware that what is valid knowledge is contested.

This listing set out meta-level criteria for design that require more detailed interpretation in relation to specific contexts. One example of a how the meta criteria for design might be enacted as a social learning system is given in Fig. 11.3.

The design in Fig. 11.3 builds on the SLIM heuristic shown in Fig. 11.2 and sets out the key activities which can give rise to social learning. The first activity – clarifying purpose – emphasises the need for those involved in the design and process to be explicit about the purpose of engaging in the learning process. This is an iterative process which requires others to be engaged in the conversation (activity 2) and a (collective) appreciation of a range of stakeholders' views and perspectives (activity 3). As the process continues, new ideas, information and experiences can be introduced to help determine the situation more clearly and understand different framings. The exact means by which this is done varies according to context and need – in some cases a workshop bringing together social scientists and policy makers might be convened, in other instances, it might involve meeting with river managers on a river bank to understand their practices, concerns and framings of the situation. Methodologically, the aim is to enable conversations which enable participants to appreciate the multiple perspectives of those involved in the situation. The emphasis on appreciation is not just a requirement upon the researcher, but extends to all those in engaged in the learning system in that all participants 'sign up' or agree to this way of working.

Arising from the interplay and iteration of activities 1–4, issues and opportunities for progressing the situation begin to emerge and are refined over time (activities 5 and 6). This leads to developing agreement on a set of actions which can be enacted by individuals and/or organisations as appropriate. At this point, the performance of the learning system can be evaluated by recourse to a monitoring process (activity 8). The format of the monitoring process can again vary according to context, but to be consistent with the design of the learning systems, should have input from those involved in the process for determining performance criteria. This is particularly the case with the outputs from activity 3 which requires that the prior experiences of participants are valued in the learning process. Arising from activities 1–8, expected outputs of the learning system include changes in understanding, changes in relations between participants and new concerted practices as part of the transformation of the situation (see Fig. 11.2). It is in these outputs that integration of knowledges and integration of practices (as concerted action) emerges. Activities 9 and 10 relate to monitoring the performance of the learning system and its outputs and re-designing as appropriate.

The SLIM heuristic (Fig. 11.2) and the learning system show in Fig. 11.3 have been used in conjunction for researching and enabling social learning in various contested water resource situations by the author and colleagues from several institutions over the last decade. Aspects of some of this research in the UK, Australia and China provides some insight into the issues associated with designing social learning systems for progressing integration of social sciences into resource managing in different contexts.

11.5 Case Studies of Designing Social Learning Systems

The following account of the various case studies is not intended to be an extensive and detailed description of the individual case studies. Instead the aim is to draw out those aspects which shed light on the design of social learning systems. As noted above, not all of this work involves social science, but the general principles and experiences are of direct relevance to the inquiry about social science integration into policy.

11.5.1 Integrating Environment Agency Catchment Science into Policy

The first case study, which took place after the SLIM research mentioned above, centred on co-research beginning in 2006 with the Catchment Science group of the Environment Agency (EA) of England and Wales. The aim was to facilitate integration of a range of physical sciences with policy imperatives to progress implementation of the EU Water Framework Directive within the EA's science activities. Recognising the disparity between the different sciences, the aim of the research was to design a social learning system for the catchments scientists within the EA to enable them to integrate their different disciplinary perspectives and to contribute to the effective design of policy.

The work began by focussing upon clarification of purpose. This was done using a range of systems diagramming techniques to explore context and different conceptual models of catchment science (relating to epistemology in the SLIM heuristic). During the first workshop with the catchment scientists, it quickly became apparent that there was no shared conceptual framing of integrated catchment management or catchment science within the Science Group or across the EA as a whole or indeed the wider literature. It was also apparent that there were marked differences between scientists and policy makers as to defining their various roles and responsibilities: each thought the other should follow them.

A learning system was designed to work through the issues with the Science Group centred on a series of workshops over 18 months in which the nature of integrated science and the relationship between policy and science was explored in some detail. The research is described in more detail elsewhere (see Collins and Ison 2010).

Despite best intentions and commitment to learning, the scientists found it difficult to find ways to integrate their different sciences. Debates continued on which particular scientific discipline was the most important to understand catchments. Significant progress on integrating the sciences was only really achieved when it was realised by some members of the group that the science needed to be discussed in relation to policy. Policy-makers were then invited into the process and the social learning started anew, this time exploring how different framings of science and catchments were constraining or enabling integration of science and policy.

The research ended earlier than anticipated for a variety of institutional reasons, including organisational changes and personnel shifts. Even so, a key finding was that integration of the different scientific disciplines only really became possible at the level of, and in relation to, policy objectives rather than the level of scientific disciplines themselves. This is consistent with understanding integration as an emergent property of a system – in this case a system to develop integrated catchment science and policy. It also highlights the futility of expecting the sciences (whether physical or social) to be able to provide or pre-specify integration as a precursor or 'ingredient to' policy. Instead, for integration into policy to occur, sciences need to be reconceptualised not as the determinant of policy or receiver of policy, but as part of the system of managing for emergence of integration in natural resource managing.

11.5.2 Creating Water Sensitive Cities in Australia

The second case study vignette relates to research initiated to progress capacity building and shared understanding for Transitioning to Water Sensitive Cities in Australia – a notion that urban areas need to be more adaptive to a climate changing context and thus reconceptualise the relationship between a city and its hinterland. The research was the result of a partnership between the International Water Centre (IWC) and the National Urban Water Governance Program (NUWGP) and Monash University. A series of five national workshops, each of 2 days, were held in each of the state capitals during early 2009. The workshops involved over 500 participants from the water sector, ranging from policy-makers, practitioners and researchers from many different physical and social science disciplines. The workshops were designed and conducted as a 2 day social learning process, broadly based on the process depicted in Fig. 11.3. Again, based on systems concepts and diagramming techniques, the workshop design aimed to identify the following: issues and opportunities that enable or constrain transitioning to water sensitive cities; characteristics of water sensitive cities from participants' perspectives; priority actions required in each city; and personal enthusiasms for action.

To some extent the generic aims are perhaps common to many workshop processes. However, in this case, the design and the methodological techniques used ensured that participants engaged with each other at an epistemological level to reveal and learn about multiple framings, prior to discussions about the system of water governance in Australian cities, issues and actions to be taken. In other words, the design of the workshops aimed to surface second order as well as first order concerns and develop a water governance system based on the social learning emerging during the workshop.

The detailed findings are reported elsewhere (see Ison et al. 2009), but feedback from the workshop evaluations suggest that the learning design of the workshops enabled participants' multiple perspectives to be heard and contribute to an understanding of the issues and thus avoid a limited set of approaches and views from dominating. In summary the shifts reported by a majority of participants included:

(i) a substantial development in the conversations about water sensitive cities at interpersonal, inter- and intra-departmental and inter-organizational levels;
(ii) changes in conception about water sensitive cities;
(iii) embryonic changes in policies at, mainly, departmental levels;
(iv) increased advocacy by a wide range of stakeholders for policy and practices to move to water sensitive cities. (see Ison et al. 2009; Collins et al. 2009)

Some caveats are important to note. The longer terms effects of this research are still be judged as it was not possible to assess these within the research time frame. Also, the research involved many different disciplinary traditions in the engagement with policy. While it is not possible to comment on the social sciences per se, by extension, the potential for integration of the social sciences into policy is evident from the reported findings. With these caveats, it seems reasonable, on the early reporting from the workshops, that the learning design of the process represented a significant opportunity for participants to recognise different framings of the situation from multiple perspectives and thus engage in new conversations to help integrate research, policy and practice.

11.5.3 Social Learning for Ecosystem Services in Lake Baiyangdian, China

This research work in 2010–2011, funded by the UK research councils, was undertaken as part of wider research programme exploring the potential of managing ecosystems services as a means of poverty alleviation. Baiyangdian catchment lies in the middle of the North China Plain and is home to almost 700,000 people officially recognised as living in poverty. A large catchment area of 31,200 km^2, historically with nine rives, Lake Baiyangdian is the largest remaining semi-closed freshwater body in the northern part of China. As a result of industrial expansion and increased water use for agriculture to supply Beijing, rivers have dried up and the surface area of the lake greatly reduced with inevitable degrading of the catchment ecosystem and livelihood implications for the population.

Using a variation of the heuristic framework described earlier, a cross-disciplinary research group of ten scientists designed a process of inquiry involving researchers from science and social science backgrounds, working with managers and community groups in Lake Baiyangdian to engage in a social learning process about the history, context, and framings of issues in the catchment. A week long inquiry involving workshops, interviews and field visits was designed to be open to local contexts, allow for emergence of new insights from a range of actors with different geographical, organisational and political links; and to generate systemic understanding of the situation through social learning among those involved.

The detailed description of the work and the findings are reported in Wei et al. (2012). In terms of the present discussion, the findings reveal the limitations of current framings of Lake Baiyangdian water management, dominated by particular framings of nature, ecology and ecosystems that are considered manageable through engineering, understood as 'good practice' in Chinese water management. Insights from the social learning process were gained in several areas, including (i) understanding the diversity of participants' perspectives of the situation and how the disciplinary training and institutional framings of the researchers and policy-makers shaped their suggested focus for improving the catchment; (ii) opportunities and potential to progress water managing through the purposeful design and enactment of a catchment managing learning system; and (iii) that the lake could become a focus or 'mediating object' to enable new conversations about ways to transition to more ecologically sensitive decision making and greater social and institutional resilience.

Mindful of the particular cultural context, careful attention was paid to good process design, the presence and expression of stakeholders (including cultural outsiders), valuing multiple partial perspectives and effective facilitation – all key requirements if a more sustainable managing of the catchment is desired. As with the other cases reported above, social science integration was not a deliberate focus, but the design of the learning process was able to accommodate a wide range of physical and social science perspectives throughout the process, particularly when engaging with policy making in an attempt to facilitate new framings of the situation.

These examples are just some of the action research situations where designing a social learning system has been advanced as an appropriate conceptual and methodological innovation to bring about more integrated resource management which recognises the importance of both biophysical and social science. In the next section, we explore some limits and opportunities for designing social learning systems as means to enable integration of social science into policy.

11.6 Constraints and Opportunities for Social Learning for Integration

The opportunities and limitations of a learning approach to policy integration are still being discovered and appraised as more research in this area is undertaken. Within a social learning paradigm a priority in research practice is to know how to create the circumstances for social learning to occur as part of a learning system.

Perhaps surprisingly for readers of this volume, the constraints do not include difficulties associated with integration – even if this is an element perceived as a constraint by others engaged in natural resource management situations. The reason for this is because, from a systems perspective, integration is not the primary concern. Instead, in designing a social learning system, the focus is on how to create the conditions for enabling social learning for resource managing, from which integration of the different sciences with policy can emerge. However, there are several constraints which can undermine the potential for a learning system to function effectively and thus reduce the likelihood of integration. As noted above, these do not focus on social sciences explicitly since the constraints apply to the design of social learning systems.

The first constraint – and perhaps the meta constraint of social learning processes since it affects all others – is very simple to describe: trust. Understanding what leads to trust for multiple stakeholders and designing and creating the circumstances for trust is less simple. Both the social and the learning element of social learning systems require a level of trust to begin and continue the engagement with others which leads to action. Stakeholders' prior experiences and assessments of the usefulness, quality and likely outcomes of the current learning process (including any facilitation) will be significant factors in shaping trust. Hence, a key criterion for designing any social learning process can be summed up with the question: does this process and its constitutive elements contribute to or constrain the establishment and continuation of trust amongst participants and stakeholders? Answers to that question will vary widely according to context and stakeholders' varying perspectives. Developing the answers provides an opportunity for social sciences to contribute to the substantive content in understanding the situation in terms of trust as well as the design and evaluation of learning systems for NRM.

Linked to trust, and particularly important in terms of initiating social learning, there is often a significant time element in initiating and managing social learning processes. This extends to reporting on social learning processes where the changes arising from social learning might take several years to manifest, if at all. Time frames and lags can be especially problematic if the proposed initiative is seen as 'unusual' or 'threatening' or in some way counter to organisational remits, practices which are centred on wanting answers 'in the short-term'.

Allied to time constraints, the emphasis in social learning processes on participants being part of the co-research process brings a range of commitments and responsibilities which may not align easily with expectations.

This can be most evident in the difficulties of the researcher retaining epistemological awareness and avoiding being assigned the label 'consultant' and its ramifications, particularly the expectation of 'coming up with the answer' to policy-makers' questions. Designing a social learning system begins more with the participants 'coming up with a question'. This shift in thinking and practice required within a social learning process in the praxis of research and policy can be resisted by others who are keen (and under pressure from, for example, their organisations or funders) to find the solution, even though the nature of the situation has yet to be adequately defined. Added to this is that many organisations function with adherence to project management tools, such as PRINCE, which can fail to acknowledge or deal with messes and uncertainties and constrain efforts to move beyond projects as technical events.

Scale issues continue to be an issue for social learning processes. For example, how to move from localised, catchment level initiatives where trust can be built and actors have direct stakeholding, to wider, regional or national scales of policy-making. A systems approach can accommodate scale on the basis that any system of interest can also be a sub-system of a wider set of concerns. Thus a nested conceptualisation of systems for natural resource managing is possible. Even so, individual relationships, contact and commitments necessary for trust are difficult to establish and maintain across different levels of policy-making unless there is good awareness and understanding amongst those involved. Quite what this entails and how it can be enacted is context dependent, but design considerations should be led by earlier questions about trust.

The issue of scale also brings a compelling research question to the fore which positions policy-making as just one aspect or sub-system of a natural resource managing system. With such a view, the question becomes less 'what is the right scale?' and more 'what is the right knowledge and governance system for managing a particular resource'? This question address scale, but not as the determining factor, and frees up ways of connecting different scales as part of the social learning process.

This brief discussion on some constraints to social learning encountered also point to some opportunities for bringing methodological innovation into natural resource managing.

Perhaps the most important is skills development in systems thinking and practice such that researchers, policy-makers and practitioners are aware of how their histories, contexts and disciplinary training shape their understanding of a situation and the choice of management methods they deploy. Such epistemological and methodological awareness is key to moving from first to second order thinking.

An opportunity also presents itself in that the limits of participation as the means to achieve integration of multiple perspectives in situations of natural resource managing are increasingly recognised. As Collins and Ison (2009b) argue, participation is a necessary element of, but not sufficient for social learning to occur. This is because social learning is epistemologically different to participation.

Combining these two opportunities, perhaps a key opportunity for social science integration rests on the engagement with and reconceptualization of the social and the biophysical systems in natural resource managing. The notion of social-ecological system is explored widely in the literatures (see for Berkes et al. 1998; Folke et al. 2005; Armitage et al. 2009; Young 2012). Much of this literature explicitly calls for a reframing of the society-nature relationship, often associated with some element of learning. This reframing requires a new understanding of the relationship between biophysical sciences and social sciences where it is not a self-negating 'either/or' dualism, but a complimentary duality – ie each part contributing to a whole (see Collins and Ison 2009a). This brings us back to the beginning of the discussion on integration – defined earlier as the making of wholes. Contributing to the understanding of the duality (or the whole) in socio-ecological systems will be key to the ways in which social sciences can integrate into policy and become a central part of social learning systems for natural resource managing.

11.7 Concluding Comments and Implications for Future Integrated Policy-Making

This chapter has aimed to provide an account of a methodological innovation centred on designing social learning systems for integrating social science into policy.

The discussion on framing points to the possible framing choices available to scientists and policy-makers. The characteristics of messy natural resource management situations centred on interdependency, complexity, uncertainty, controversy and multiple stakeholdings suggest that no one individual or organisation is able to manage in isolation. Engaging with framing choices leads to epistemological awareness which is a key step towards integration.

Appreciating NRM situations as complex and 'messy' opens up possibilities for a complimentary methodological approach based on systems. Although much has been written on integration, it remains elusive as a concept and practice. However, the discussion on the link between integration and systems reveals their fundamental connection in that both are central to the process of 'creating wholes'. Within a soft systems tradition, the importance of boundary choices (a form of framing) is key to understanding different perspectives. A systems view also gives rise to the idea of integration as an emergent property of a system, rather than a 'thing' that can be added. The central concern then for integration of social science into policy is creating and thus designing system level conditions for integration as an emergent property of social learning systems. Integration of social science per se becomes less important than a concern with designing social learning systems from which integration emerges.

Integration, understood as the making of wholes and an emergent property of systems, coupled with an emphasis on observer dependency, leads to the view that natural resource managing is fundamentally *a social process*. In this framing, the imperative for social learning – understood as the process of stakeholders socially constructing an issue in which their understandings and practices change so as to transform the situation of concern through concerted action – becomes clear.

The key elements relating to the transformation process include an appreciation of the starting context; institutions; facilitation; stakeholding and epistemological constraints.

The vignettes of the case study research point to ways in which social learning systems can be designed and enacted in a variety of NRM situations and contexts, although the exact detail is beyond the scope of this chapter. Perhaps the key design criterion relates to ways in which trust is established and developed over time.

The discussion of some constraints and opportunities associated with a social learning approach as an innovative methodology for integrating social sciences into policy is only the beginning of our understanding as more research on social learning becomes available. But it would seem that social sciences has much to offer in terms of substantive content to natural resource managing as well as research on the design, process and evaluation of social learning processes.

A note of caution is necessary, however, as designing and enacting social learning systems requires epistemological, temporal and financial investment from policy-makers and scientists from all disciplinary backgrounds. It also requires developing skills and competency in systems concepts and ideas – not least integration as an emergent property of a social learning system and all that this entails for policy and science processes.

Looking forward, integrating social science into policy through a social learning systems is not a given. It requires commitment and willingness to engage in second order concerns about the nature of knowledge and understanding of complex, messy situations. The increasing use and acceptance of the concept of socio-ecological system as a coupled, co-evolving system would seem to be a central arena for future research. It is here that designing social learning systems as a methodology for enabling integration of social science with policy and biophysical sciences is likely to be of most import.

References

Ackoff, R. L. (1974). *Redesigning the future*. New York: Wiley.
Ackoff, R. L. (1981). *Creating the corporate future*. New York: Wiley.
Anon. (2007). *Research for sustainable development: how to enhance connectivity?* Report of an EC Workshop, Brussels, 7–8 June 2007. Background paper by expert group for Research and

Development for Sustainability (RD4SD). Available at: http://ec.europa.eu/research/sd/pdf/background_info/report_halfman.pdf. Accessed 18 Nov 2012.

APSC (Australian Public Services Commission). (2007). *Tackling wicked problems. A public policy perspective*. Canberra: Australian Public Service Commission.

Armitage, D. R., Plummer, R., Berkes, F., Arthur, R. I., Charles, A. T., Davidson-Hunt, I. J., Diduck, A. P., Doubleday, N. C., Johnson, D. S., Marschke, M., McConney, P., Pinkerton, E. W., & Wollenberg, E. K. (2009). Adaptive co-management for social–ecological complexity. *Frontiers in Ecology and the Environment, 7*, 95–102.

Atkinson, R., & Klausen, J. E. (2011). Understanding sustainability policy: Governance, knowledge and the search for integration. *Journal of Environmental Policy & Planning, 13*(3), 231–251.

Bandura, A. (1977). *Social learning theory*. Englewood Cliffs: Prentice Hall.

Berkes, F., Folke, C., & Colding, J. (1998). *Linking social and ecological systems: Management practices and social mechanisms for building resilience*. Cambridge: Cambridge University Press.

Blackmore, C. (2007). What kinds of knowledge, knowing and learning are required for addressing resource dilemmas? A theoretical overview. *Environmental Science and Policy, 10*(6), 512–525.

Brito, L., Stafford Smith, M. (2012, March 26–29). *State of the planet declaration*. In Planet under pressure: New knowledge towards solutions conference, London. Available at www.planetunderpressure2012.net/pdf/state_of_planet_declaration.pdf. Accessed 10 Apr 2013.

Checkland, P. (1981). *Systems thinking, systems practice*. New York: Wiley.

Collins, K. B., Ison, R. L. (2009a). Living with environmental change: adaptation as social learning. *Environmental Policy & Governance, 19*(6), 351–357. (Editorial, special edition)

Collins, K. B., & Ison, R. L. (2009b). Jumping off Arnstein's ladder: Social learning as a new policy paradigm for climate change adaptation. *Environmental Policy & Governance, 19*(6), 358–373.

Collins, K. B., & Ison, R. L. (2010). Trusting emergence: Some experiences of learning about integrated catchment science with the environment agency of England and Wales. *Water Resources Management, 24*(4), 669–688.

Collins, K., Blackmore, C., Morris, D., & Watson, D. (2007). A systemic approach to managing multiple perspectives and stakeholding in water catchments: Some findings from three UK case studies. *Environmental Science and Policy, 10*(6), 564–574.

Collins, K. B., Colvin, J., & Ison, R. L. (2009). Building 'learning catchments' for integrated catchment managing: Designing learning systems and networks based on experiences in the UK, South Africa and Australia. *Water Science & Technology, 59*(4), 687–693.

Daniels, S., & Walker, G. (1996). Collaborative learning: Improving public deliberation in ecosystem-based management. *Environmental Impact Assessment Review, 16*, 71–102.

Dovers, S. (1997). Sustainability: Demands on policy. *Journal of Public Policy, 16*, 303–318.

Dovers, S. (2005). Clarifying the imperative of integration research for sustainable environmental management. *Journal of Research Practice, 1*(2), Article M1. Available from http://jrp.icaap.org/index.php/jrp/article/view/11/30. Accessed 15 Oct 2012.

Finger, M., & Verlaan, P. (1995). Learning our way out: A conceptual framework for social–environmental learning. *World Development, 23*, 505–513.

Folke, C., Hahn, T., Olsson, P., & Norberg, J. (2005). Adaptive governance of social-ecological systems. *Annual Review of Environmental Resources., 30*, 441–473.

Frances, E. (1951). History and the social sciences: some reflections on the re-integration of social science. *Review of Politics, 13*(3), 354–374.

Holling, C. S. (Ed.). (1978). *Adaptive environmental assessment and management*. Chichester: Wiley.

Ison, R. L. (2008a, Oct 29–31). *Understandings and practices for a complex, coevolutionary systems approach*. In: Proceedings of the international symposium: Selected topics on complex systems engineering applied to sustainable animal production, Instituto Tecnolo del Valle de Morelia, Morelia, Michoac, Mexico.

Ison, R. L. (2008b). Methodological challenges of trans-disciplinary research: Some systemic reflections. *Natures Sciences Societes, 16*(3), 241–251.

Ison, R. L., Röling, N., & Watson, D. (2007a). Challenges to science and society in the sustainable management and use of water: Investigating the role of social learning. *Environmental Science & Policy, 10*(6), 499–511.

Ison, R. L., Blackmore, C. P., Collins, K. B., & Furniss, P. (2007b). Systemic environmental decision making: Designing learning systems. *Kybernetes, 36*(9/10), 1340–1361.

Ison, R. L., Collins, K., Bos, J., Iaquinto, B. (2009). *Transitioning to water sensitive cities in Australia: A summary of the key findings, issues and actions arising from five national capacity building and leadership workshops.* Melbourne: NUWGP/IWC, Monash University. Available from http://www.watercentre.org/resources/publications/attachments/Creating%20Water%20Sensitive%20Cities.pdf. Accessed 11 Apr 2013.

Ison, R., Collins, K., Colvin, J., Jiggins, J., Roggero, P. P., Seddaiu, G., Steyaert, P., Toderi, M., & Zanolla, C. (2011). Sustainable catchment managing in a climate changing world: new integrative modalities for connecting policy makers, scientists and other stakeholders. *Water Resources Management, 25*(15), 3977–3992.

Mostert, E., Pahl-Wostl, C., Rees, Y., Searle, B., Tàbara, D., & Tippett, J. (2007). Social learning in European river-basin management: barriers and fostering mechanisms from 10 river basins. *Ecology and Society, 12*(1), 19 [online]. Available from http://www.ecologyandsociety.org/vol12/iss1/art19/. Accessed 10 Oct 2012.

Muro, M., & Jeffrey, P. (2008). A critical review of the theory and application of social learning in participatory natural resource management processes. *Journal of Environmental Planning and Management, 51*(3), 325–344.

OED (*Oxford English Dictionary*). (2012). Oxford University Press. Oxford.

Pahl-Wostl, C., Craps, M., Dewulf, A., Mostert, E., Tabara, D., & Taillieu, T. (2007). Social learning and water resources management. *Ecology and Society, 12*(2), 5 [online]. Available at http://www.ecologyandsociety.org/vol12/iss2/art5/. Accessed 20 Oct 2012.

Pahl-Wostl, C., Mostert, E., & Tàbara, D. (2008). The growing importance of social learning in water resources management and sustainability science. *Ecology and Society, 13*(1), 24. [online] Available at http://www.ecologyandsociety.org/vol13/iss1/art24/. Accessed 20 Oct 2012.

Raadgever, G. T., Mostert, E., & van de Giesen, N. C. (2012). Learning from collaborative research in water management practice. *Water Resources Management, 26*(11), 3251–3266.

Redman, C. L., Morgan Grove, J., & Kuby, L. H. (2004). Integrating social science into the long-term ecological research (LTER) network: Social dimensions of ecological change and ecological dimensions of social change. *Ecosystems, 7*, 161–171.

Reed, M. S., Evely, A. C., Cundill, G., Fazey, I., Glass, J., Laing, A., Newig, J., Parrish, B., Prell, C., Raymond, C., & Stringer, L. C. (2010). What is social learning? *Ecology and Society, 15*(4), r1 [online]. Available from: http://www.ecologyandsociety.org/vol15/iss4/resp1/. Accessed 20 Oct 2012.

Rittell, H. W. J., & Webber, M. M. (1973). Dilemmas in a general theory of planning. *Policy Science, 4*, 155–169.

Russell, D. B., & Ison, R. L. (2007). The research-development relationship in rural communities: An opportunity for contextual science. In R. L. Ison & D. B. Russell (Eds.), *Agricultural extension and rural development: Breaking out of knowledge transfer traditions* (pp. 10–31). Cambridge: Cambridge University Press.

Schön, D. A. (1995). Knowing in action: The new scholarship requires a new epistemology. *Change,* November/December, 27–34.

Schön, D. A., & Rein, M. (1994). *Frame reflection: Toward the resolution of intractable policy controversies.* New York: Basic Books.

Scrase, J. I., & Sheate, W. R. (2002). Integration and integrated approaches to assessment: What do they mean for the environment? *Journal of Environmental Policy & Planning, 4*(4), 275–294.

Skidelsky, R. (2008). *Morals and the meltdown* [online]. Available from http://www.skidelskyr.com/site/article/morals-and-the-meltdown/. Accessed 14 June 2012.

SLIM. (2004a). *SLIM framework: Social learning as a policy approach for sustainable use of water*. SLIM, Open University, Milton Keynes. Available at http://slim.open.ac.uk. Accessed 8 Sept 2012.

SLIM. (2004b). *Stakeholders and stakeholding in integrated catchment management and sustainable use of water* (SLIM Policy Brief No. 2). SLIM, Open University, Milton Keynes. Available from http://slim.open.ac.uk. Accessed 10 Sept 2012.

SLIM. (2004c). *Introduction to SLIM publications for policy makers and practitioners*. Introductory Policy Briefing. SLIM, Open University, Milton Keynes. Available at http://slim.open.ac.uk

SLIM (2004d). *The role of learning processes in integrated catchment management and sustainable use of water* (SLIM Policy Brief No. 6). SLIM, Open University, Milton Keynes. Available from http://slim.open.ac.uk. Accessed 11 Sept 2012.

SLIM (2004e). *Ecological constraints in sustainable management of natural resources* (SLIM Policy Briefing No. 1). SLIM, Open University, Milton Keynes. Available from http://slim.open.ac.uk. Accessed 11 Sept 2012.

Social Learning Group. (2001). *Learning to manage global environmental risk*. Cambridge, MA: MIT Press.

Steyaert, P., & Jiggins, J. (2007). Governance of complex environmental situations through social learning: A synthesis of SLIM's lessons for research, policy and practice. *Environmental Science and Policy, 10*(6), 575–586.

Steyaert, P., Barzman, M. S., Brives, H., Ollivier, G., Billaud, J. P., & Hubert, B. (2007). The role of knowledge and research in facilitating social learning among stakeholders in natural resources management in the French Atlantic coastal wetlands. *Environmental Science and Policy, 10*(6), 537–550.

UN. (2012). *The future we want. Outcome of the UN Rio + 20 conference*. UN: Rio de Janeiro. Available at: https://rio20.un.org/sites/rio20.un.org/files/a-conf.216l-1_english.pdf.pdf. Accessed 1 Feb 2013.

UNEP. (2012). *GEO5 global environmental outlook: Summary for policy-makers*. UNEP: Nairobi. Available at http://www.unep.org/geo/pdfs/GEO5_SPM_English.pdf. Accessed 10 Jan 2013.

Wei, Y., Ison, R. L., Colvin, J., & Collins, K. (2012). Reframing water governance in China: A multi-perspective study of an over-engineered catchment. *Journal of Environmental Planning & Management, 55*(3), 297–318.

Wenger, E. (1998). *Communities of practice: Learning, meaning and identity*. Cambridge: Cambridge University Press.

Wenger, E., McDermott, R., & Snyder, W. (2002). *Cultivating communities of practice: A guide to managing knowledge*. Boston: Harvard Business School Press.

Wilson, E. O. (1998). *Consilience: The unity of science*. New York: Knopf.

Woodhill, J., & Röling, N. (1998). The second wing of the eagle: The human dimension in learning our way to more sustainable futures. In N. G. Röling & M. A. E. Wagemakers (Eds.), *Facilitating sustainable agriculture: Participatory learning and adaptive management in times of environmental uncertainty*. Cambridge: Cambridge University Press.

Young, O. (2012). Navigating the sustainability transition: Governing complex and dynamic socio-ecological systems. In E. Brousseau, T. Dedeurwaerdere, P.-A. Jouvet, & M. Willinger (Eds.), *Global environmental commons: Analytical and political challenges in building governance mechanisms* (pp. 80–104). Oxford: Oxford University Press.

Author Bios

Leslie Acton is currently a Ph.D. student and joint member of the Campbell lab and the Basurto lab at Duke University, studying Marine Science and Conservation. She received her M.E.M. from Duke University and B.S. from the University of North Carolina at Chapel Hill.

Abigail Bennett is a Ph.D. Candidate at the Duke University Marine Lab. She received a B.S. from the University of South Florida in Environmental Science and Policy.

Randall B. Boone is a Research Scientist at the Natural Resource Ecology Laboratory and Associate Professor in the Department of Ecosystem Science and Sustainability at Colorado State University. He is also a faculty member of the Graduate Degree Program in Ecology. He received his B.S. from Oregon State University and both his M.S. and Ph.D. from the University of Maine.

Jeffrey C. Bridger is Senior Scientist in the Department of Agricultural Economics, Sociology, and Education at Penn State University. He currently teaches in the Community Environment and Development Program and in the Masters of Professional Studies Program in Community and Economic Development. He received his B.S. from George Mason University and his M.S. and Ph.D. in rural sociology from Penn State University.

Jeffrey Broadbent is a Professor in the Department of Sociology and at the Institute of Global Studies at the University of Minnesota. Dr. Broadbent received his B.A. in Religious Studies with an emphasis on Buddhism from the University of California Berkeley and he completed his M.A. in Regional Studies with a focus on Japan and a Ph.D. in Sociology, both at Harvard University.

Tom R. Burns is Professor Emeritus at the Department of Sociology at the University of Uppsala in Uppsala, Sweden and a Visiting Scholar at Stanford University's Woods Institute. He received a B.S. in Physics and an M.A. and Ph.D. in Sociology, all from Stanford University.

Kevin Collins is a Lecturer on Environment and Systems in the Department of Engineering and Innovation at The Open University, UK. He received his Ph.D.

M.J. Manfredo et al. (eds.), *Understanding Society and Natural Resources*,
DOI 10.1007/978-94-017-8959-2, © The Author(s) 2014

from University College London and teaches a range of courses on systems approaches to environmental managing.

Robert Costanza is Professor and Chair in Public Policy at Crawford School of Public Policy at Australian National University. Dr. Costanza is also currently a Senior Fellow at the National Council on Science and the Environment in Washington, DC; Affiliate Fellow at the Gund Institute of Ecological Economics, Rubenstein School of Environment and Natural Resources, The University of Vermont; and a Senior Fellow at the Stockholm Resilience Center in Stockholm, Sweden. Dr. Costanza received B.A. and M.A. degrees in architecture and a Ph.D. in environmental engineering sciences from the University of Florida.

Joan M. Diamond is Deputy Director and Senior Scenarist at the Nautilus Institute for Security and Sustainability. She is the Executive Director of the Millennium Alliance for Humanity and Biosphere, a Stanford University and a visiting scholar at the Center for the Advanced Studies in the Behavioral Sciences at Stanford. She also oversees the Institute of Foresight Intelligence.

Esther A. Duke is the Director of Special Projects and Programs in the Human Dimensions of Natural Resources Department at Colorado State University. She holds a B.A. in International Studies/Latin American Studies and English/Creative Writing from Illinois Wesleyan University and an M.S. in Human Dimensions of Natural Resources from Colorado State University.

Paul Ehrlich is the Bing Professor of Population Studies in the department of Biological Sciences and Senior Fellow at the Woods Institute for the Environment at Stanford University. He is a fellow of the American Association for the Advancement of Science, the American Academy of Arts and Sciences, the American Philosophical Society, the Beijer Institute of Ecological Economics, and a member of the National Academy of Sciences. Dr. Ehrlich received his Ph.D. from the University of Kansas.

Graham Epstein is a student in the joint Ph.D. program in Public Policy at the School of Public & Environmental Affairs and the Department of Political Science at Indiana University Bloomington. He holds a B.S. in Ecology from the University of Waterloo and an M.S. in International Rural Planning and Development from the University of Guelph.

Melissa L. Finucane is a Senior Behavioral and Social Scientist at the RAND Corporation in Pittsburgh Pennsylvania and a Senior Fellow at the East-West Center in Honolulu, Hawai'i. She received an M.Psych. and a Ph.D. in psychology from the University of Western Australia.

Jefferson Fox is a Senior Fellow at the East-West Center in Honolulu, Hawai'i. He is also a member of the affiliate graduate faculty in geography and anthropology, University of Hawai'i. Dr. Fox received his Ph.D. in Development Studies from the University of Wisconsin-Madison.

Jörg Friedrichs is Associate Professor in Politics at the Department of International Development (Queen Elizabeth House) and Fellow of St Cross College, University of Oxford. He received his Dr. Phil. from the University of Munich.

David Fulton is a Research Ecologist with the U.S. Geological Survey and Assistant Unit Leader at the Minnesota Cooperative Fish and Wildlife Research Unit. He is also Adjunct Professor in the Department of Fisheries, Wildlife and Conservation Biology at the University of Minnesota. Dr. Fulton holds a B.S. from Texas A&M University, M.S. from Washington State University, and Ph.D. from Colorado State University.

Kathleen A. Galvin is Professor in the Department of Anthropology, and Senior Research Scientist at the Natural Resource Ecology Laboratory at Colorado State University. Professor Galvin is also Associate Director for Education in the School of Global Environmental Sustainability helping to develop curricula in sustainability. She received her B.A. and M.A. at Colorado State University and her Ph.D. at SUNY University, Binghamton.

Michael C. Gavin is an Assistant Professor and Co-director of the Conservation Leadership Masters Program in the Human Dimensions of Natural Resources Department at Colorado State University. Dr. Gavin received a B.A. in Environmental Studies and Biology from Bowdoin College and a Ph.D. in Ecology from University of Connecticut.

Rebecca Gruby is an Assistant Professor in the Department of Human Dimensions of Natural Resources at Colorado State University. She holds a B.S. in natural resource conservation from the University of Florida and a Ph.D. in environmental sciences and policy from Duke University.

Ilan Kelman is a Reader in Risk, Resilience and Global Health at University College London (UCL) and a Senior Research Fellow at the Norwegian Institute of International Affairs (NUPI). His B.A.Sc. and M.A.Sc. are in engineering from the University of Toronto in Canada while his Ph.D. was completed in the Department of Architecture at the University of Cambridge, England.

Donald Kennedy is president emeritus of Stanford University, the Bing Professor of Environmental Science and Policy, emeritus, and a senior fellow by courtesy of the Freeman Spogli Institute for International Studies at Stanford University. He currently serves as a director of the Carnegie Endowment for International Peace, and as co-chair of the National Academies' Project on Science, Technology and Law. He received A.B. and Ph.D. degrees in biology from Harvard University.

Kristin Ludwig is a Staff Scientist in the Natural Hazards Mission Area at the US Geological Survey and Adjunct Faculty at George Mason University. Dr. Ludwig earned a B.S. in Earth Systems from Stanford University and a Ph.D. in Oceanography from the University of Washington.

A.E. Luloff is Professor of Rural Sociology and Human Dimensions of Natural Resources and the Environment in the Department of Agricultural Economics,

Sociology, and Education at Penn State University. He is one of two co-organizers (along with James C. Finley) of the Human Dimensions of Natural Resources and the Environment dual-title, inter-college graduate degree program there. He is cofounder and past Executive Director of the International Association for Society and Natural Resources. He received his B.S. in rural sociology at Cornell University, M.S. in sociology at North Carolina State University, and Ph.D. in rural sociology at Penn State University.

Nora Machado is an Associate Researcher at the Center for Research and Studies in Sociology at the Lisbon University Institute, and Lecturer at the Department of Sociology at the University of Gothenburg in Sweden. She earned her Ph.D. in Sociology and obtained degree of Docent (Reader) in Sociology at the University of Uppsala in Sweden.

Gary E. Machlis is Science Advisor to the Director of the U.S. National Park Service, and Professor of Environmental Sustainability at Clemson University. He is also co-Leader of the Department of the Interior's Strategic Sciences Group, which conducts science-based assessments during major environmental crises. Dr. Machlis received his bachelor's and master's degrees from the University of Washington in Seattle, and his Ph.D. in human ecology from Yale.

Michael J. Manfredo is a professor and Head of the Human Dimensions of Natural Resources Department at Colorado State University. He currently serves on the board of the North American Section of Society for Conservation Biology and is an executive committee member of the Black Caucus Institute's 21st Century Council. He is also co-organizer of an interdisciplinary joint masters of development practice program with Ecosur University, Mexico. He holds a B.A. in Anthropology and an M.S. in Recreation and Parks from Pennsylvania State University as well as a Ph.D. in Recreation Behavior and Social Psychology from Colorado State University.

Mateja Nenadovic is currently a Ph.D. student in the Basurto Lab at Duke University, studying Marine Science and Conservation. He received a M.S. from the University of Maine in Marine Biology and Marine Policy and a B.S. from Suffolk University in Marine Science.

Lennart Olsson is Professor of Geography at Lund University in Sweden and the founding Director of LUCSUS – Lund University Centre for Sustainability Studies. Lennart is also the Coordinator for a Linnaeus programme, LUCID, sponsored by The Swedish Research Council for the period 2008–2018. He received his Ph.D. in Physical Geography from Lund University.

Rajendra Kumar Pachauri has served as the chairman of the Intergovernmental Panel on Climate Change since 2002. He is the director general of TERI, a research and policy organization in India, and chancellor of TERI University. He has been appointed as Senior Adviser to Yale Climate and Energy Institute (YCEI) from July 2012 prior to which he was the Founding Director of YCEI (July 2009 – June 2012). Dr. Pachauri received his M.S. in Industrial Engineering and his Ph.D. in Industrial Engineering and Economics from North Carolina State University.

Andreas Rechkemmer is Professor and American Humane Endowed Chair at the Graduate School of Social Work, University of Denver. He holds a Masters Degree (M.A.) in Philosophy and Political Science from Ludwig Maximilian University of Munich, Germany, and a Ph.D. (Dr. rer. pol.) in International Relations from Free University of Berlin, Germany.

Eugene A. Rosa was the Boeing Distinguished Professor of Environmental Sociology, the Edward R. Meyer Distinguished Professor of Natural Resource and Environmental Policy, Professor of Sociology, and Affiliated Professor of Fine Arts at Washington State University, and a Visiting Scholar at Stanford University. He held a B.S. from the Rochester Institute of Technology, an MA and Ph.D. from Syracuse University, and postdoctoral credentials from Stanford University. Sadly, Gene passed away in February 2013 as this chapter was being finalized. He is sorely missed by his friends, family, and by MAHB.

Sumeet Saksena is a Fellow at the East-West Center in Honolulu, Hawai'i. He is also a member of the affiliate graduate faculty in urban and regional planning, University of Hawai'i. He received his Ph.D. in environmental sciences and engineering from the Indian Institute of Technology.

James H. Spencer is the Chair of the Department of Planning, Development and Preservation, and a Professor of City & Regional Planning at Clemson University. He is also an Adjunct Senior Fellow at the East-West Center in Honolulu, Hawai'i. Dr. Spencer holds a B.A. in social anthropology from Amherst College, M.E.M. from the Yale University School of Forestry and Environmental Studies, and a Ph.D. in urban planning from UCLA.

Tara L. Teel is an Assistant Professor in the Department of Human Dimensions of Natural Resources at Colorado State University. She received her Ph.D. in Human Dimensions from Colorado State University and M.S. and B.S. degrees in Fisheries and Wildlife Management from Utah State University.

Gene L. Theodori is a Professor and Chair of the Department of Sociology at Sam Houston State University. He is also an Adjunct Graduate Professor in the Department of Ecosystem Science and Management at Texas A&M University. He received his B.A. in sociology from California University of Pennsylvania, his M.S. in sociology from Texas A&M University, and his Ph.D. in rural sociology from Penn State University.

Jerry J. Vaske is a Professor in the Department of Human Dimensions of Natural Resources at Colorado State University. He is the founding co-editor of the journal *Human Dimensions of Wildlife*. He received his B.A. and M.A. degree from the University of Wisconsin and his Ph.D. from the University of Maryland.

Philip Vaughter is a Postdoctoral Fellow at the University of Saskatchewan. He received his Ph.D. in Conservation Biology from the University of Minnesota.

Peter H. Verburg is a Professor of Environmental Spatial Analysis and Head of the Department of Spatial Analysis and Decision Support at the Institute for

Environmental Studies/Amsterdam Global Change Institute at VU University Amsterdam. Dr. Verburg is chair of the SSC of the Global Land Project (IHDP/IGBP), and a member of the scientific steering committee of the Global Land Project. He completed his M.S. and his Ph.D. in Agricultural and Environmental Sciences at Wageningen University.

Index

Printed by Books on Demand, Germany